Science and Technology in Post-Mao China

Harvard Contemporary China Series: 5

edited by
DENIS FRED SIMON
and MERLE GOLDMAN

Published by
THE COUNCIL ON EAST ASIAN STUDIES / HARVARD UNIVERSITY
Distributed by the Harvard University Press
Cambridge (Massachusetts) and London 1989

Science and Technology in Post-Mao China

Printed in the United States of America

Index by Katherine Frost Bruner

The Council on East Asian Studies at Harvard University publishes a monograph series and, through the Fairbank Center for East Asian Research and the Reischauer Institute of Japanese Studies, administers research projects designed to further scholarly understanding of China, Japan, Korea, Vietnam, Inner Asia, and adjacent areas. Publication of this volume has been assisted by a grant from the Shell Companies Foundation.

Index by Katherine Frost Bruner

Library of Congress Cataloging in Publication Data

Science and technology in post-Mao China / edited by Denis Fred Simon and Merle Goldman.
p. cm. – (Harvard contemporary China series ; 5)
Bibliography: p.
Includes index.
ISBN 0-674-79475-3 : $14.00
1. Science–China–History–20th century. 2. Technology–China–History–20th century. 3. China–History–1976– I. Simon, Denis Fred. II. Goldman, Merle. III. Series.
Q127.C5S333 1988
338.95106–dc19 *88-28456*
CIP

CONTRIBUTORS

RICHARD BAUM is Professor of Political Science at the University of California, Los Angeles. He has written extensively on China's post-Mao scientific, technological, and political-legal reforms. Currently, he is working on a book dealing with the social physics of complex systems.

WILLIAM FISCHER, Professor of Operations Management at the School of Business Administration, University of North Carolina, Chapel Hill, served, from 1980–1984, as a member of the American Faculty at the National Institute for Industrial Science and Technology Management Development at Dalian, in the People's Republic of China. Since then he has visited and conducted research on Chinese managerial issues in several other contexts. Dr. Fischer has worked in technology-management positions in U.S. industry and in the U.S. Army and has consulted variously on technology management in developing countries.

WENDY FRIEMAN is Director of the Asia Technology Program at Science Applications International Corporation, where she is responsible for research on science and technology in China and Japan. Her areas of specific interest include Chinese military research and development, Japanese industrial research policy, and technology diffusion in China and Japan. She is co-author of *Gaining Ground: Japan's Strides in Science and Technology* (Ballinger, forthcoming).

MERLE GOLDMAN is Professor of Chinese History at Boston University. Her most recent books are: *China's Intellectuals: Advise and Dissent* (Harvard University Press, 1981) and, with Timothy Cheek and Carol Lee Harmin, *China's Intellectuals and the State: In Search of a New Relationship* (Harvard Council on East Asian Studies, 1987). Currently, she is at work on a book about the Post-Mao democratic movement.

ROY F. GROW, Chair of the Political Science Department at Carleton College, has, in the last two years, written twenty articles about technology transfer and foreign commercial activity in China. He is the author of a forthcoming book called *Competing in China: Japanese and American Businessmen in a New Market* (Harper and Row/Ballinger, forthcoming).

NINA HALPERN is Assistant Professor of Political Science at Stanford University. She is the author of several articles on Chinese economic policymaking and the role of experts. Her current research concerns methods of policy coordination in the Chinese bureaucracy and the relationship between political and economic change in the PRC and other Communist nations.

GAIL HENDERSON, Assistant Professor in the Department of Social Medicine, University of North Carolina School of Medicine, combines interests in social science and medicine in her research on the Chinese health-care system. She has written, with Myron S. Cohen, *The Chinese Hospital: A Socialist Work Unit* (Yale University Press, 1984) and several articles on the rise of high-technology medicine in China. Her ongoing studies in China include the evolution of rural health care in a county in Shandong; family change in Shanghai; and a national-level longitudinal study of nutrition, health status, and health-care utilization.

ATHAR HUSSAIN is Reader at the University of Keele and Senior Research Fellow at the London School of Economics, engaged in a research project on price and market structure in the Chinese economy. He is co-editor of *The Chinese Economic Reforms* and also of the forthcoming two-volume *Paths of Development in China.*

WILLIAM C. KIRBY is Associate Professor of History and Director of the Asian Studies Program at Washington University, St. Louis. He is the author of *Germany and Republican China* (Stanford University Press, 1984) and various articles on Chinese economic development during the Republican era. His present research on the rise of modern China's economic bureaucracy will be published as *The International Development of China's Economy: Nationalist Industrial Policy and Its Heirs* (Stanford University Press, forthcoming).

LEO A. ORLEANS is an independent researcher. He was for many years a China specialist at the Library of Congress and is the author of numerous monographs and articles on Chinese population, science and technology, education, and other development issues. His latest book, *Chinese Students in America: Policies, Issues, and Numbers,* was supported by the Committee on Scholarly Communications with the PRC and published by the National Academy Press.

DETLEF REHN, a senior researcher in the Ostasien-Institut, Bonn, is especially interested in the development of China's high-tech and electronics industries. His recent publications include *Die Reform der chinesischen Industrie-wirtschaft Das Beispiel der Electronikindustrie* (Cologne, 1987) and, with Denis Fred Simon, "Innovation in China's Semiconductor Industry: The Case of Shanghai," *Research Policy,* October 1987. Rehn and Simon are co-authors of *Technological Innovation in China: The Case of the Shanghai Semiconductor Industry* (Ballinger, 1988).

TONY SAICH is Associate Professor at the Sinologisch Instituut, Leiden, and a Research Fellow of the International Institute for Social History, Amsterdam. His most recent book is *China's Science Policy in the 1980s* (Manchester University Press, 1988). He has edited the Sneevliet (Maring) papers on China for publication and currently is participating in a project at the John King Fairbank Center, Harvard University, to produce a new documentary history of Chinese Communism.

LAURENCE SCHNEIDER, Professor of Chinese History and Associate Dean of the Faculty of Social Sciences, State University of New York at Buffalo, has written widely about continuity and change in twentieth-century Chinese cultural history. He is currently studying the transmission of modern natural sciences from the West to China and is completing a case study of the development of modern genetics in China. He is the author of *A Madman of Ch'u: The Chinese Myth of Loyalty and Dissent* (University of of California Press, 1980) and the editor and author of the introductory chapter of *Lysenkoism in China: Proceedings of the 1956 Qingdao Genetics Symposium* (M. E. Sharpe, 1986).

OTTO SCHNEPP, Professor of Chemistry, University of Southern California, served from 1980 to 1982 as Science Attaché and Science Counselor at the U.S. Embassy in Beijing. His research in China's science and technology includes manpower development, evaluation, and technology transfer. Two recent publications, now in press, are, with A. Bhambri and M. A. Von Glinow, "Procedural Guidelines for Technology Transfer Projects," in *Technology Transfer to the People's Republic of China* (Woodrow Wilson Center of the Smithsonian Institution) and "Science and Technology Personnel in the PRC," in *China's Science and Technology*, ed. Denis Simon (Professors World Peace Academy).

DENIS FRED SIMON, Associate Professor of International Business and Technology, Fletcher School of Law and Diplomacy, Tufts University, specializes in the study of international technology transfer and comparative business-government relations, with an emphasis on East Asia, where he is a frequent visitor. He is currently undertaking a study of China's computer development. His most recent book, with Detlef Rehn, is *Technological Innovation in China: The Case of the Shanghai Semiconductor Industry* (Ballinger, 1988).

RICHARD P. SUTTMEIER, Henry P. Bristol Professor of International Affairs at Hamilton College, has written extensively on Chinese science and technology affairs. During 1987, he directed the Beijing office of the National Academy of Sciences/Committee on Scholarly Communications with the PRC. He has also worked as a senior analyst for the Office of Technology Assessment on a project dealing with technology transfer to China.

ABBREVIATIONS LIST

AH	Academia Historica (Taiwan)
AS	Academia Sinica (Taiwan)
ATEC	agro-technical center(s) (PRC)
AWPM	American War Production Mission (to China in WWII)
BNDB	*Beijing Nongye Daxue bao* (Beijing Agricultural University journal)
BR	*Beijing Review*
CAAS	Chinese Academy of Agricultural Sciences (PRC)
CAD	computer-assisted design
CAS	Chinese Academy of Sciences (PRC)
CASS	Chinese Academy of Social Sciences (PRC)
CAST	Chinese Association for Science and Technology (PRC)
CCP	Chinese Communist Party
CEIEC	Chinese Electronics Import and Export Corporation (PRC)

CEROF	China's National Import/Export Corporation for Cereals, Oils, and Foodstuffs
CIA	Central Intelligence Agency (Washington)
CIMMYT	International Maize and Wheat Improvement Center (Mexico)
CITIC	China International Trust and Investment Corporation (PRC)
CMC	Central Military Commission (PRC)
COCOM	The Coordinating Committee for Multilateral Export Controls (NATO)
COSTIND	Commission for Science, Technology, and Industry for National Defense (PRC)
CPSU	Communist Party of the Soviet Union
CR	Cultural Revolution
CRT	cathode ray tube
CUTC	China United Trade Corporation (PRC)
ERC	Economic Research Center (PRC)
ETSRC	Economic, Technological, and Social Research Center (PRC)
FAO	Food and Agriculture Organization (UN)
FBIS	*Foreign Broadcast Information Service*
FDR	Franklin D. Roosevelt Library, Hyde Park
FRUS	*Foreign Relations of the United States* (Washington, USGPO)
FYB	*Fertilizer Yearbook* (Printing Office, FAO)
FYP	Five-Year Plan
GLD	General Logistics Department (PRC)
GMD	Guomindang
GMRB	*Guangming ribao* (Guangming daily, Beijing)
GPO	Government Printing Office (Washington)
GSD	General Staff Department (PRC)
GYTL	*Zhongguo jindai gongyeshi ziliao* (Materials on the modern history of Chinese industry), ed. Chen Zhen (Beijing, 1961)
ha	hectare
HST	Harry S. Truman Library, Independence, Missouri
HYV	high-yield plant varieties
IC	integrated circuit

ICBM	intercontinental ballistic missile
IMH	Institute of Modern History
IPR	Institute of Pacific Relations
IRBM	intermediate-range ballistic missile
IRRI	International Rice Research Institute (Philippines)
JPRS	*Joint Publication Review Service*
KB	kilobit
KXTB	*Kexue tongbao* (Science journal, CAS)
LAN	local area networking
LSI(C)	large-scale integrated circuits
MB	megabyte(s)
MEI	Ministry of Electronics Industry (PRC)
MIPS	million instructions per second
MMBs	Machine-Building Ministries (PRC)
MNC(s)	multi-national corporation(s)
MOA	Ministry of Agriculture (PRC)
MOE	Ministry of Education (PRC)
MOFERT	Ministry of Foreign Economic Relations and Trade (PRC)
MSI	Ministry of Space Industry (PRC)
MTBF	mean time between failures
NA	National Archives (Washington)
NAS	National Academy of Sciences (Washington)
NDIO	National Defense Industries Office (PRC)
NDSTC	National Defense Science and Technology Commission (PRC)
NIC(s)	newly industrialized countries
NPC	National People's Congress (PRC)
NRC	National Resources Commission (Ziyuan Weiyuanhui, GMD)
NRCSTD	National Research Center for Science and Technology Development (PRC)
NSC	National Seed Corporation (PRC)
NSF	National Science Foundation (Washington)
PAC	Party Archives Commission of the Guomindang
PC	personal computer
PLA	People's Liberation Army (PRC)

PRC	People's Republic of China
PYB	*Production Yearbook* (of FAO)
R&D	research and development
RAM	random access memory
SEZ	Special Economic Zone
SHA	Second Historical Archives of China, Nanjing (PRC)
SMBC	State Machine-Building Commission (PRC)
SPC	State Planning Commission (PRC)
SSTC	State Science and Technology Commission (PRC)
S&T	science and technology
STLG	Science and Technology Leading Group (PRC)
S-T-P	science, technology, political systems
SWXT	*Shiwuxue tongbao* (Biology journal, Biology Department, CAS)
SYB	*Statistical Yearbook of China*
TDC	technology development center
TERC	Technical Economics Research Center
UNDP	United Nations Development Program
UNRRA	United Nations Relief and Rehabilitation Administration
USDA	US Department of Agriculture
USIA	US Information Agency
USTC	University of Science and Technology of China
WB	World Bank (Washington)
WDR	*World Development Report*
WWH	Weng Wenhao papers
ZBFT	*Ziran bianzhengfa tongxun* (Dialects of nature journal, CAS)
ZYGB	*Ziyuan weiyuanhui gongbao* (Official gazette of the NRC), 1941–1948
ZYJK	*Ziyuan weiyuanhui jikan* (NRC quarterly), 1941–1946
ZYYK	*Ziyuan weiyuanhui yuekan* (NRC monthly), 1939–1941

CONTENTS

Introduction

INTRODUCTION

The Onset of China's New Technological Revolution

MERLE GOLDMAN *and* DENIS FRED SIMON

When the Chinese Communist Party came to power in 1949, one of its major goals was to modernize China's science and technology. Like the late-nineteenth-century self-strengtheners, they regarded the development of science and technology as a means to achieve wealth, power, and status in the international community. With the exception of the Cultural Revolution years, 1966–1976, and despite the twists and turns of the Party's policies, this goal has remained consistent. After the Cultural Revolution, when China had fallen further behind the West and even the other developing economies of East Asia, this goal became something of an obsession. As stated by the scientist Qian Xuesen, President of the Chinese Association of Science and Technology, "In the twenty-first century, if a particular country fails to be in the lead in science and technology, it will be difficult for it to maintain its economic activities and international standing." Because of the advances in technology, Qian predicts that the rivalries among nations will be primarily in the scientific and technological realm: "This is an intellectual war and the twenty-first century will be an era of this kind of 'warfare.'"[1]

The beginnings of a "new technological revolution" are taking place in post-Mao China. Recognizing the intimate relationship between economic and military advance *and* scientific and technological progress, the leadership has made a strong commitment to close the prevailing technological gap between China and the industrialized world by the early twenty-first century. Accordingly, Chinese leaders have introduced a series of reforms to enable China to become a major participant in the emerging scientific "warfare" described by Qian. But China is also burdened by a number of constraints, primarily constraints in the traditional culture and in the Marxist-Leninist one-party state, which may limit its ability to compete.

THE CONSTRAINTS OF THE TRADITIONAL CULTURE

Traditional China created a great civilization, which among other things also produced extraordinary technological achievements—in hydrology, astronomy, metallurgy, ceramics, printing, agriculture, and oceanography. Yet China did not experience the "scientific revolution" and the accompanying "industrial revolution" that transformed the West in the eighteenth and nineteenth centuries. The scholar-bureaucrats, the literati, who governed China were not interested in developing science and technology. They were wholly engaged in governance, humanistic pursuits, and moral questions. Moreover, as the historian of science Nathan Sivin points out, they looked for patterns they perceived in the human world when viewing the non-human world.[2] The study of nature, therefore, was not a separate entity, but was incorporated into the Confucian framework which explained and proscribed behavior in human relationships. Thus, science and technology were inextricably enmeshed in the political-social system.

Not only the human focus of the literati's philosophy, but also their mental processes limited the development of science and technology. The creators of China's most significant scientific and technological achievements were not literati; they were primarily artisans and craftsmen who had a master-apprentice relationship, which meant that few of their discoveries were written down. Moreover, the literati, whose educational attainments might have made possible more substantial forms of theoretical conceptualization, did not build models, establish hypotheses, or conduct experiments in a systematic fashion. They did not develop the mental ethos associated with scientific investigation—such as skepticism, innovation, and inquiry into the unknown, processes associated with scientific and technological development. As Richard Baum has pointed out elsewhere, "China's major achievements were in the areas of observation rather than conceptualization, concrete thinking rather than theoretical speculation, deduction rather than induction, arranging ideas in patterns rather than developing theories to explain patterns."[3]

At about the same time that the scientific revolution was getting under way in the West, Jesuit missionaries introduced Western mathematics and astronomy to China in the late seventeenth century. For the first time, China's astronomers used mathematical models to predict as well as to describe phenomenon. But this brief encounter with Western science did not move China in new directions. In fact, it revived traditional ideas of astronomy and led to the rediscovery of older methods rather than the introduction of new ones. The literati did not reconsider or even question their own understanding of nature. From the seventeenth to the early nineteenth century, the Evidential Research movement used comparative and quantitative methods to recover the classics in their original form. This movement even included the beginnings of peer review. But their "scientific" methods were used on their sacred texts, not on the nonhuman world. In the nineteenth century, the self-strengtheners sought to combine Western techniques with Chinese civilization, but again the orthodox Confucian framework overwhelmed the Western technology; China was unable to meet the Western challenge.

Given the character of the cultural and sociopolitical milieu that dominated traditional Chinese society and educational institutions, it is not surprising that China did not develop modern science and had difficulty in absorbing Western science and technology. But, except for the West, no other great civilization has developed modern science either. What were the unique institutions in the West that made possible the emergence of modern science and technology? As explained by Sivin, when Galileo was condemned by the Church in 1633, he and his colleagues established a new intellectual community outside the old establishment and the orthodox framework. With a degree of independence, they experimented with and formulated theories on nature and created a body of knowledge that only those trained and accepted by their peers understood. In traditional China, there were different interpretations of Confucianism, but no literati institutions ever challenged the overall Confucian framework, and few literati developed norms and practices distinct from other literati.

In the early decades of the twentieth century, the Westernizers in the May Fourth era trumpeted the slogan of "science" along with "democracy" to discredit the traditional culture. Their humiliation by the Western nations and disillusionment with Confucianism created a spiritual vacuum which was filled with the idealization of Western science and technology as a solution to all China's problems, an idealization that has many parallels in the post-Mao era. Their faith in science became what Baum has called "the functional equivalent to their previous faith in Confucianism."[4] But their uncritical embrace of Western science and technology reflected little understanding of what it was; they did not absorb even the spirit of skepticism and inquiry into the unknown necessary for its development.

Nevertheless, during both the warlord period following the 1911 Revolution and the weak, disorganized Nationalist Government established by the Guomindang (GMD) from 1927–1945, there was the start of contacts and interchange with Western scientists and the emergence of a small number of intellectuals educated in the universities who were trained in Western science. As William Kirby points out, this period witnessed the maturing of the first and second generation of self-conscious professional scientists and engineers, which bequeathed to the Chinese Communists institutions and approaches that could have developed modern science and technology. Although the universities were not independent of political control, and the GMD tried to reimpose a Neo-Confucian view on the intellectuals, the government's incompetence and weakness made it possible for a small number of scientists to engage in professional activities with a degree of autonomy and consistency. Their achievements, though modest, were a beginning.

THE CONSTRAINTS OF THE SOVIET AND MAOIST RESEARCH MODEL

With the establishment of the People's Republic of China in 1949 and the emulation of the Soviet Stalinist model in the 1950s, China for the first time had a government that took direct and explicit responsibility for developing science and technology. Along with receiving considerable Soviet assistance in the form of visiting experts, training, and equipment, the Chinese also received the Soviet system of scientific and technological organization. This system tended to deemphasize the role of research in the universities and stress the centralization of research in institutes under an Academy of Sciences, with high priority given to research related to military and heavy-industry needs. As described by Richard P. Suttmeier elsewhere, it was expected that the scientific and technological innovations developed in the research institutes would flow spontaneously into industry and the economy.[5] But, in China as in the Soviet Union, that flow did not occur. Research activities were isolated from production activities. Moreover, the various institutes, concentrated primarily in the Chinese Academy of Sciences (CAS) and the industrial ministries, were guided by a central plan and funded by an annual budget from the central government. Consequently, power over research was concentrated in the hands of political leaders who had little understanding of science or how to get innovations from the research laboratories onto the factory floors. Also, because the research system was vertically organized with little horizontal interchange, CAS created rigid, narrow structures which were not conducive to developing theoretical science or its practical application into readily usable production technologies.

As Suttmeier further points out, the Soviet model also imposed strong ideological constraints on the process of innovation. Since Marxism-Leninism regards scientific and technological knowledge as belonging to the whole people, it considers such knowledge a free good. Therefore, it undervalues such knowledge and removes practical "material" incentives for innovation. Those individuals who do become "innovators" are usually characterized as heroic figures because of their great ability to overcome the barriers to innovation but receive little remuneration. An equally important constraint was that Chinese science in the 1950s was engulfed by the same ideological torrents engulfing Soviet science. Lawrence Schneider's study of the damage of Lysenkoism, the Soviet politicization of genetics on China's budding professional geneticists, is a case in point. The effect of these ideological disputes was to politicize science in China in the same way that Confucianism had done. Thus, the few sprouts of S&T development that had peeped through in the middle decades of the twentieth century began to wither.

Periodically, this encrusted Soviet model was shaken up by Mao's own vision of scientific and technological development, specifically during the Great Leap Forward of 1958–1960 and the Cultural Revolution period. In these episodes, Mao imposed his, rather than the Soviet Union's, social and political criteria on technological and economic activities. He emphasized mass participation, egalitarianism, self-reliance, and indigenous scientific and technological development. Party cadres and representatives of the masses were to take over technological and scientific tasks from the professionals, who were criticized for their "ivory-tower" mentality and were sent to the countryside or to factories to learn from the masses. Campaigns involving political mobilization and ideological exhortation rather than scientific study and laboratory experiments were supposed to achieve desired technological breakthroughs. And, in fact, as Genevieve Dean has noted elsewhere, whether or not due to the inspiration of the thought of Mao Zedong, a few scientific and technological achievements were made during these periods of political turbulence.[6]

These Maoist episodes certainly loosened the rigidity and narrowness of the Soviet system. And they also led some intellectuals to explore alternative models for China. Nonetheless, they also obstructed rather than advanced China's scientific and technological development. The central planning apparatus responsible for implementing China's S&T development was disrupted, and power devolved to institutes and ministries, which made for even less interaction than before. Even more damaging were the political campaigns which demoralized and decimated several generations of scientists who had been trained in the West and the Soviet Union. In the Cultural Revolution, in particular, even though Mao initially had wanted to exclude scientists from the attacks on other experts and intellectuals, as the movement gained momentum it swept up virtually all intellectuals, including scientists. The exceptions were a small number working on missile development, though recent evidence indicates that even those individuals were not immune from attack. Most academic, professional, and scientific research was stopped as scientists and engineers, along with writers and historians, were reeducated and sent for "thought reform" to the countryside and factories. When a small number of universities and institutes were reopened in the early 1970s, they still used political rather than scientific criteria to carry on their work. The damage that had been done to the country's modest S&T infrastructure and education system would take years to repair.

THE POST-MAO SCIENTIFIC AND TECHNOLOGICAL REFORMS

Much of the rhetoric of the Deng Xiaoping regime, which came to power in late 1978, resonates with the May Fourth Movement's celebration of science and technology as the key to solving all China's problems. The disillusionment with the prevailing ideology—this time, Marxism-Leninism-Maoism—once more created an ideological vacuum that the leadership and intellectuals again sought to fill with the idealization of advanced science and technology. As in the early decades of the twentieth century, thousands of students and scholars were sent abroad to study in the West and Japan, and numerous foreign experts were invited to China. After decades of relative isolation, ties were reestablished with the international scientific and engineering community. Although China's leaders were looking for the same quick fix as earlier, this time there was some acknowledgment that the process might be more complicated than merely transferring technology. While, at first, the main aim was to resurrect the Soviet-style system that had been shattered by the Cultural Revolution, gradually the reform leadership moved away from the Soviet and Maoist models, which it regarded as obstacles to the development of advanced science and technology. Tony Saich describes the efforts to reduce the stifling effects of Soviet-style centralized control by devolving resources and authority to lower levels. Relatedly, Leo Orleans describes the reemergence of universities as centers not only for education but also research and development, as they had been before 1949.

As a result of Chinese investigation of the experience of such places as Stanford, Harvard, and M.I.T, university-industry relations are being strengthened. Chinese research institutes and universities can now compete for projects and can contract their services to economic enterprises. Presumably, as this practice takes hold, institutes will become more self-supporting and the central government will gradually diminish its funding. Over time, as the dependence on central funding decreases, the research sector may become increasingly, albeit gradually, independent of political control. As the chapter by Detlef Rehn reveals, some research institutes in key cities such as Shanghai have even formed joint ventures with economic enterprises and have begun the process of merging with them to form so-called "R&D-production" alliances (*lianheti*). Another means for bringing science and technology closer to the needs of production is the use of market, as described by William Fischer. The dynamics of the market, including increased consumer demand and concern with quality and style, Fischer points out, may finally bring technological innovation into the factory, a process the Soviet model discouraged. Otto Schnepp points out that the students who study abroad have some voice in the jobs to which they are assigned when they return, though many are still placed in positions irrelevant to their training. These reforms plus contract research, private consulting, and moonlighting may help break down the existing vertical rigidities and allow some movement of scientific and technical personnel. They may even introduce some flexibility into what has been an inflexible labor market. Without such freer movement, an important mechanism for the diffusion of technology, which has proven so important in places such as South Korea and Taiwan, will be missing.

Revision of ideology also buttresses the use of the market in science and technology. As Suttmeier's concluding chapter explains, in the post-Mao period, science and technology are no longer treated as a free good, but as a commodity that can be bought and sold. This revision is institutionalized in a patent system established in 1985, which legally provides material incentives for innovation. For those engaged in basic research, as well as applied research, a Chinese Science Foundation, patterned on the United States National Science Foundation, was established in February 1986 and allocates funds through peer review rather than by administrative decrees. Other reforms that may insulate science from political interference are the election of directors of institutes by their colleagues, which began in 1981, when the president and vice- president of CAS were elected by their peers. Although the practice has been carried out somewhat erratically, as professionals assume greater responsibility in directing science and technology, more technically qualified and competent persons have become directly involved in advising the political leadership. Nina Halpern's discussion of the input of experts into economic policymaking describes this process.

Another example of the move away from the Soviet model can be seen in the military-industrial complex, as described by Wendy Frieman, which, in the post-Mao era, has received a seemingly smaller share of investment and foreign-exchange resources in relation to the modernization of industry, agriculture, and science. Faced with a declining budget and new opportunity for profit in the civilian economy, China's military industries, also motivated by material incentives, have gradually moved from producing obsolete tanks and weapons to producing refrigerators and television sets for a hungry consumer market. It is estimated that over 50 percent of what the military industry produces is consumer goods. If this "civilianization" of the military-industrial complex continues and the military is forced to compete in civilian markets, the military as well as civilian industries may become more efficient and may inject new technological expertise into the military, thus enhancing rather than detracting from the effectiveness of that sector.

The "open-door" policy adopted by Deng Xiaoping and his cohorts is directly linked to the reform program in science and technology. A major purpose behind China's decision to normalize relations with the industrialized world was to gain access to the scientific and technical assets of the United States, Japan, and Western Europe. As Denis Fred Simon shows in his chapter, foreign science and technology are viewed as a catalyst in China's modernization program. While the present leadership may vacillate in its openness to Western culture and non-scientific knowledge, it remains committed to preserving the country's open door to all forms of scientific cooperation and technology transfer. Scientific exchanges conducted through bilateral agreements signed between China and the respective industrialized countries provide a means for cooperative research activities, joint projects, and the reciprocal movement of scientific and technical data and literature between China and other nations.

As Gail Henderson's discussion of medicine and health demonstrates, these exchanges have been particularly important in acquainting China's scientific and medical community with the norms, values, and priorities of Western research and treatment. Yet, this does not always result in improvement for the overall population, as Henderson shows in her description of the move from the Maoist system of health care, which emphasized primary care, public health, and community medicine, to increasing emphasis on hospital care, imported technology, and professionalization of health-care workers.

Yet, in spite of the leadership's willingness to rely more on foreign technology and its efforts to become a more active participant in the global technology marketplace, China also wants to make sure it is not totally dependent on one source. The legacy of past dependency on the Soviet Union continues to weigh heavily on Chinese thinking, leading Beijing to diversify its technology contacts to reach Eastern as well as Western Europe and Latin America along with other countries in East Asia, including South Korea. Chinese concerns about having full and unincumbered access to a full spectrum of foreign technologies is most strongly reflected in its frequent admonishments of both Japan and the United States for not being more forthcoming in the export of technical know-how and advanced equipment.[7]

Having access to advanced technology is one thing; being able to assimilate it effectively is another. Roy Grow's chapter indicates that the Chinese experience with processing imported technology varies considerably, one good indicator of this variation being the extent to which managerial reforms have moved ahead in some enterprises but not in others. The difficulty in absorbing foreign technology is also noted by Baum in his discussion of China's attempt to enter the age of informatics by acquiring a large number of imported computers.

These multifaceted reforms—decentralization of research, introduction of the market, increased personnel mobility, and greater interaction with the international S&T community—have the potential for engendering a fundamental structural change in the Chinese political system because they undermine one of the Party's main instruments of power—its control over scientific and technological activity. Like the reforms in other areas, the S&T reforms obviate the Party's entrenched bureaucracy and aim at a radical transformation in the way government and Party manage the problems of invention and innovation. As the reforms enhance the position of newly rising technocrats, they necessarily lessen the authority of non-technocrats who have dominated the ranks of power. Consequently, because the recent changes threaten the bureaucracy's vested interests, Party bureaucrats have frequently obstructed, if not actually blocked, their implementation, particularly at the local level where Beijing lacks full control. Furthermore, at times, their antagonism to the reforms has been exacerbated by the increased opportunities offered to scientists and technically trained individuals to earn additional income and travel abroad, opportunities not readily available to the bureaucrats. Thus, while the reforms in science and technology are greeted with great fanfare in the press and in the outside world, much of what we see and hear about remains more in the realm of intent than reality.

If China's past experience with reform has taught us anything, it is that there is a difference between the enunciation of policies by Beijing and their effective implementation.[8] Throughout Chinese history, reforms have been thwarted by the very bureaucrats who are designated to carry them out.[9] Equally important, there are ideological factors that stiffen the bureaucracy's resistance. While the bureaucrats in the scientific and technological sphere are not against science and technology per se, they share some of the concerns Mao tried to redress in the Cultural Revolution—specifically a concern that faces all modernizing societies: As scientific and technological modernization develops, it becomes the exclusive preserve of apolitical experts, engaged in specialized activities that exclude the masses and inevitably increase inequalities. Mao tried to counter this trend by emphasizing science and technology practiced by the masses, whose innate creativity and revolutionary will, he claimed, bring forth new technological solutions. Perhaps, it is impossible to produce modern science and technology without a degree of exclusivity, but Mao, despite the destructiveness of his Cultural Revolution policies toward science and technology, was facing an important issue in the development of modern society: Can a nation truly be modernized scientifically and technologically if the masses of people are totally excluded from the process?

The apparent resistance of the entrenched bureaucracy is reinforced by the post-Mao regime's continuing use of political campaigns, some of which are directed against intellectuals. These campaigns lack the violence, fanaticism, and mass mobilization of the Maoist era, but they persist. Periodically, the Deng Xiaoping regime has launched political campaigns to stem the impact of Western political ideas and values that are perceived as threats to the Party's authority and its Marxist-Leninist orthodoxy. In these campaigns, the bureaucrats use the opportunity to discredit their intellectual rivals, even in the science and technology sector, and label them political deviants. The politicization and factionalization of the academic and intellectual community and the frequent abrupt changes of political line, while not as oppressive or as arbitrary as under Mao, continue to affect the research and educational environment.

The Campaign against the Democratic Movement in 1979, the one against the writer Bai Hua in 1981, the Spiritual Pollution Campaign in the fall of 1983, and the campaign against "bourgeois liberalization" launched in early 1987 do not directly touch on science and technology. In fact, the political leadership explicitly warned against carrying out these campaigns in the scientific and technological sphere. Such restraints, however, are unlikely to reassure people who from 1966 to 1976 were subjected to violence and humiliation for their Western orientation and expertise. Even if the campaigns' chief scapegoats are writers or a theoretical scientist (Fang Lizhi), the way in which these intellectuals are treated affects the scientific community whose help in the modernization drive is essential. It was Fang Lizhi's anger at the interference of Party bureaucrats, untutored in science, that led him to question the whole political system.

Writers are not directly relevant to science and technology modernization, but, if writers are prevented from learning from abroad or are subjected to arbitrary, repressive treatment, it may also inhibit scientific contact and experimentation. The ideologist Li Honglin points out that to exclude Western values that the leadership does not want may also exclude Western science and technology that the leadership wants. It is not possible to keep out Western values; they will accompany the import of Western science and technology. As he explains, "Decadent and degenerate ideology and culture will enter our culture along with science and technology and contaminate the air."[10] There is no way to let in one element of Western culture—its science and technology—and filter out the rest.

The political campaigns have a chilling effect not only on scientists in China but also on Chinese scientists and students studying abroad, especially when they hear, as in early 1987, that investigations are conducted in research centers to see if returnees bring back Western spiritual pollution along with their technical training. Under such circumstances, those studying abroad may be reluctant to return, finding it easier to remain where they are not vulnerable to political campaigns and their prospects for financial remuneration are greater. The fear and resentment these campaigns generate both at home and abroad may inhibit risk-taking, initiative, and creativity, essential for invention and innovation. Potential innovators may gradually withdraw as they did in the Maoist era as a form of passive resistance.

BALANCE BETWEEN STATE AND MARKET

China's reform of its S&T system must also be viewed against the backdrop of ongoing reforms in the industrialized nations and the Third World.[11] In the majority of cases, the role of government policy has assumed importance, whether through direct intervention aimed at targeting certain areas for technology generation by means of national funding programs and national research laboratories, or through "indirect" means aimed at market modifications that encourage innovation.[12] Despite the fact that much writing on the subject has emphasized the decisive importance of market forces, the strong role of government policy has become a central theme in the literature on technological innovation.

As the chapters of this book show, China's reforms are following this "two-pronged" strategy for bringing about fundamental improvements in its science and technology system. China has introduced "market forces" as tools for stimulating scientific advance and technological modernization. And, at the same time, it continues to rely on centrally directed control over research in key areas to promote rapid progress in important economic and military sectors. There are numerous historical examples in both the East and West that suggest the presence of inherently contradictory elements in such a strategy.[13] Nonetheless, Chinese leaders appear committed to simultaneous use of what Lindblom has called "authority structures" and "exchange mechanisms" to promote more rapid and sustained national scientific and technological progress.[14]

Thus, conspicuous in China's approach to science and technology is a focus on initiating improvements in the operating mechanisms of the existing centrally oriented R&D system. The leadership continues to look back fondly to some of its previous successes in science and technology, foremost among them the development of atomic weapons and ICBMs. Furthermore, the leadership refuses to allow the vagaries of the market to determine the outcome of future priority endeavors—preferring, instead, to reserve a number of key technology areas such as large scale-integrated circuits for nurturing by centrally directed organizations.

At the same time, China is pursuing a complementary strategy focused on the introduction of essentially new operating principles and institutions. The leadership recognizes that one of the most important elements in the post-war development of Western technological capabilities and vitality is the combination of entrepreneurial talent and market stimuli. Studies of the emergence of Silicon Valley in California and Route 128 in Massachusetts suggest that the "technological revolution" that occurred in these areas happened basically without the benefit of highly orchestrated efforts by universities, government, or other relevant groups.[15] In this context, an explicit attempt has been made to link economic reforms with the reform of the S&T system. As Deng Xiaoping has noted, "The new economic structure must be favorable to science and technology advancement, and the new R&D system should, in turn, be conducive to economic growth."[16]

China's strategy for S&T modernization, therefore, is to combine elements of state and market. This approach, in many respects, conforms to that of other newly industrialized Asian nations. India, Taiwan, and South Korea have used a mixture of state and market to establish a base for high-technology industries such as microelectronics and informatics.[17] While such "mixed" strategies may not always create major technological advances or long-term technological competitiveness, they can certainly influence the process in a positive, albeit modest, fashion by structuring incentives in the appropriate direction.

As several chapters in this volume indicate, the posture of the government in the post-Mao era is changing from the stultifying role it played in the past. Various types of policy instruments— including fiscal policy, trade policy, foreign-exchange policy, and macro and microeconomic policy—are being managed in a more sophisticated fashion to facilitate rather than inhibit long-term scientific and technological progress.

Yet, while the notion of launching a countrywide technological revolution is indeed attractive, it sometimes has led the Chinese to lose sight of the fact that incremental progress in science and technology, rather than quantum leaps, has characterized the success stories in the development process. Even in the case of Japan, whose rapid growth and technological advance have been held up as a potential model in China, progress was accomplished by taking existing technologies and improving on their performance through modest innovations.[18] While government did not always play a large role, it did play a critical role. The Japanese case as well as the case of the so-called "four dragons" in East Asia (Taiwan, South Korea, Singapore, and, to a lesser extent, Hong Kong) show that it is not the quantity of government intervention that counts but the quality and form of that intervention. This is an important lesson the Chinese are just beginning to learn.

Also, the effort to attain substantial levels of growth and technological advance can be accomplished only after a workable S&T infrastructure has been put in place. This includes both the physical elements (such as necessary materials and equipment) and an appropriate set of institutions. As the lessons of other Third World nations suggest, effective policies for science and technology need to be part of an overall institutional package, involving a variety of inputs ranging from finance to marketing, along with changes in the "culture for invention and innovation." The success of this effort will not come from adhering in a rigid fashion to catchy themes or by pursuing strategies that are based on the political fear of falling behind. Rather, the long-term viability of China's present "mixed strategy" depends on allowing the strategy to evolve in conjunction with the further changes that are needed in the cultural-sociopolitical realm.

This does not necessarily mean that China requires Western democracy to achieve substantial progress in science and technology. As studies of Third World development show, government policy can have its desired impact but only when the economic signals being sent to the various actors in the system are, on balance, logical and internally consistent, which has not always been the case in the Deng era. Nevertheless, Deng Xiaoping appears correct when he insists that China's technological revolution needs a "Chinese type" of modernization, one that is sensitive to both the size and complexity of the Chinese system *and* the "cultural" requisites for innovative behavior, which thrives in an intellectually diverse environment.

THE PROSPECTS FOR THE FUTURE

Undoubtedly, the drive to implement science and technology reforms could have a critical shaping effect on the Chinese political culture. But will the momentum and substance of this drive be sufficient to overcome the constraints of the traditional culture and the Marxist-Leninist ideology which inhibit inquiry, exchange, and behavior outside the established orthodoxy? China's opening to the outside world, exemplified by its expanded science and technology cooperation with both East and West, does not appear to be at risk in the short term, though the degree of openness may vary from time to time due to a variety of sociopolitical and economic considerations. However, the exaltation of Western science and technology as a panacea for all China's problems may prove difficult to sustain in the face of continued modernization problems as well as incomplete and uneven reforms.

As in traditional times and in the Mao period, China may make some scientific and technological progress. The chapter by Athar Hussain illustrates China's ability to upgrade general nutrition through improvements in agricultural science and technology in the turbulent decade of the 1960s. There is also good reason to believe that certain "pockets of excellence" will be created through a combination of trained manpower, local initiative, and nurturing by the central government. But will China develop the advanced levels of science and technology that it so desires? Will the structural reforms, which have run into resistance, be able to change a deeply embedded political culture that inhibits questioning, uncertainty, alternative views, and intellectual pluralism?

For a period in the mid-1980s, it appeared that structural reform and a loosening of ideological constraints were changing the nineteenth-century dichotomy between the *ti*–Chinese principle, and the *yong*–Western technology. It looked as though the *ti* was finally changing along with the *yong*. That was until the launching of the campaign against "bourgeois liberalization" in early 1987. The campaign imposed political limits on non-scientific areas, though it also explicitly excluded science and technology from those limits.

The ability of the reformist leadership to rein in this campaign after a few months is encouraging. What is most disconcerting, however, is that the same view of the world that plagued Chinese attempts at modernization in the late nineteenth century—a view that seeks to accept foreign science and technology while excluding the humanistic and political values upon which they are based— persists in the late twentieth century. The current orthodoxy, the *ti*, is not Confucianism nor is it as deeply ingrained as it was under Mao. Yet it still emphasizes the need for ideological unity and political control, and still relies, to a certain extent, on the Maoist campaign method in criticizing the differing ideas of the very strata of society who are to absorb and develop the new science and technology. Viewed from this perspective, the room for real "cultural" change appears to remain limited.

True, periodic campaigns of the post-Mao era against dissenting intellectuals and Western humanistic ideas, have had a less stifling effect on the country than under Mao. They have been limited to a few individuals and have been short in duration. But S&T modernization cannot succeed if it is carried out in abrupt spurts, as seen in late-nineteenth and early-twentieth-century China. It needs a sustained effort and extensive, ongoing interaction with the international scientific community. Unpredictable restrictions on the flow of experts, information, and initiative are bound to interrupt the continuity and contacts that transcend disciplines, institutions, and borders necessary for modern scientific development. Moreover, as China increases its participation in bilateral and multilateral scientific cooperation and technology-transfer activities, will it allow itself to become "interdependent" with the world community in ways that might mean a greater loss of "control" as a trade-off for certain scientific and technical benefits?

It is true that even with interruptions, China has been able to imitate and modify Western science and technology. It has also developed rocket and space technologies, made synthetic insulin, synthetic ribonucleic acid, and germanium crystal, and developed superconductors with temperatures about 100k. In this regard, China appears to have been able to follow in the steps of the Soviet Union, where, for the most part, the pattern of scientific and technological progress is conspicuous for its absence of change in the innovation climate. But China's aspirations are much higher than mere sporadic technological breakthroughs. It seeks to produce scientists at the frontiers of knowledge who can reach the world's most advanced scientific and technological levels. It aspires to win Nobel Prizes in the realm of theoretical science and become economically competitive by producing technologically sophisticated products for the world market. And, it seeks to catch up with and eventually surpass the West sometime in the next century.

Yet, modern science and technology, as defined by Sivin, require the establishment of an independent scientific community with its own norms and objectives. Its criteria for achievement are based on ability and scholarly attainment, which often conflict with the Party's criteria, based on seniority and political orthodoxy. It requires an ethos of free and wide-ranging scientific inquiry, tolerance of error, open communication, job mobility, and free exchange with the outside world on a continuing basis. Such an ethos needs institutions to protect it, but such institutions might undermine the Party's dominant political authority and control, as it might have undermined the Confucian political system. The very fact that the need for such a new operating environment was seemingly ignored at the reformist 13th Party Congress in October 1987 is a sign that the tensions remain and will persist for some time.

There is no question that China has the potential for developing modern science and technology. Even with the beginnings of structural reform in science and technology in the post-Mao era, the quality of its professional talent has improved very quickly. The problem is not with the talent or the age of the equipment in the laboratory, though these are important issues. The real problem is the overriding political system whose controls, while somewhat lessened by the reforms, still do not allow this talent the freedom of action and flexibility to develop beyond the physical manifestations of a modern research system. Unless this weight is lifted, China of the twenty-first century may achieve a modicum of scientific stature, but will be unable to achieve the global technological leadership to which it aspires. In addition to the tremendous amounts of energy and physical resources being devoted to science and technology modernization, China needs far-reaching changes in its cultural and ideological climate in order to produce its desired "technological revolution." Such changes, however, may lie outside the grasp of a Party leadership that wishes also to hold on firmly to its political power.

PART ONE

Historical Precedents

CHAPTER ONE

Technocratic Organization and Technological Development in China: The Nationalist Experience and Legacy, 1928–1953

WILLIAM C. KIRBY

Mao Zedong liked to stress the advantages of poverty and backwardness in China's economic development. His "poor and blank" thesis of 1958 implied that, because China was as "blank" as a "clean sheet of paper," it could forego its pre-revolutionary inheritance and undergo the most radical of transformations. Mao's thesis reflected his optimism in the early stages of the Great Leap Forward that rapid economic and technological progress could be achieved more easily through revolutionary virtue, self-reliance, and mass mobilization than by reliance on certifiable expertise, foreign technologies, and professional administrative planning. At the time, Mao was discarding much more than the economic model of the period of the 1st Five-Year Plan; he was rejecting long-term trends in modern Chinese scientific, technological, and bureaucratic development that were neither "poor" nor "blank" but rich in experience and precedents that still today help define the directions open to Chinese technological organization and development.

Since the mid-nineteenth century, Chinese governments have sought to apply modern scientific and technological methods to the political and economic purposes of the state. None has succeeded to the extent it would have liked. To various degrees, all have been uncomfortable with the reliance on nonpolitical "specialists" and foreign expertise that scientific and technological progress has seemed to demand. But Mao's periodic attempts after 1958 to return to China's traditions of technological insularity and bureaucratic omnicompetence are the great exception. When China's post-1976 leadership broke with the nativist and anti-specialist biases of the Maoist era, it returned to the idea, put bluntly by Deng Xiaoping, that talent and training matter[1] or, more specifically, that the development of a scientific and technological elite is essential for China's security and economic progress; and that this demands structured programs of foreign study, the use of foreign experts, and Sino-foreign cooperative ventures at home. Above all, it demands the institutionalization of scientific and technological expertise in government service.

Although it is seldom remembered, S&T development was an established part of China's economic and intellectual landscape before 1949. Certainly no government more identified with this "specialist" and "internationalist" approach to science and technology than the immediate predecessor of the People's Republic, the Nationalist regime that ruled China for over two decades before its relocation on Taiwan in 1949. The period of Nationalist rule witnessed the maturing of the first and second generations of self-consciously professional groups of Western-trained economists, scientists, and engineers, who shared the government's consistent commitment to what Sun Yatsen had called the "international development" of China. By recruiting leading specialists to plan and manage China's industrial development, the Nationalist leadership encouraged the creation of China's first technocratic civil service, the National Resources Commission of China, the principal focus of this chapter. In the process, the regime grappled with issues that remain of central concern to science and technology policy in the PRC today. How can China's technically competent manpower best be organized to serve national development? What are the virtues or limitations of centralized state planning for promoting technological advance? How are foreign technology and expertise best employed for Chinese purposes? What are the political costs of "expert decision making" in a one-party state? Historical parallels are never exact, though sometimes instructive. In this case, historical legacies may be more important. The Nationalists bequeathed to their successors approaches, institutions, and individual skills that were both prerequisites and models for early PRC efforts at technological and industrial development and Sino-foreign technology transfer.

TECHNICAL EXPERTISE AND INDUSTRIAL DEVELOPMENT IN THE NANJING DECADE, 1928–1937

An ambitious Nationalist regime assumed power in 1927, aiming to complete national reunification, reassert Chinese political and economic sovereignty, and rapidly industrialize the nation. The plan for Chinese industrialization drafted by Sun Yatsen, a leader of the 1911 Revolution and chief figure in the revived Guomindang as a political party in the early 1920s, emphasized the role of state planning, the participation of foreign capital, and the establishment of basic heavy industries and infrastructure (ports, railways, utilities).[2] This blueprint was followed by his GMD successors who established the Nationalist Government. They stressed to an even greater degree the precedence of industrial development over other sectors of the economy and shared Sun's dream of a modern, industrial, technologically advanced China. Industrialization became a primary concern for the government of Chiang Kaishek, which depended on the modern sector of the economy for much of its income, and sought security through the creation of a world-class armed force supported by state-of-the-art metallurgical, chemical, and munitions industries.[3]

The GMD's economic development plans have rightly been described as emphasizing certain technological preconditions for economic growth while resisting growth-inducing change in social institutions.[4] The many economic plans announced during the regime's first years stressed the government's commitment to industrial development with significant reliance on foreign technical assistance. This high-technology, internationalist bias was evident from the government's early emphasis in transportation policy on Sino-foreign civil air transportation companies, which, in the words of one historian, allowed the government to "soar above" the problems of local and interregional transportation,[5] down to its final, desperate years on the mainland, when it sought to erect under Sino-foreign auspices the world's largest hydroelectric dam in the Yangzi Gorges.

There were few useful precedents and no working institutions for the GMD's determination to mobilize China's Western-trained scientists, economists, and technicians. The traditional civil-service examinations of the old Empire tested intellectual mastery of literary texts deemed to be the embodiment of moral and political orthodoxy. Late-nineteenth-century "specialists" with training in Western science and mathematics were, for the most part, accommodated only in irregular, extra-bureaucratic positions. The end of the ancient examination system in 1905, the fall of the Qing in 1911, and the failure of early Republican governments to maintain central bureaucratic authority meant that China's slowly growing numbers of Western-trained specialists increasingly served regional militarist administrations and performed more as personal retainers than civil servants.[6]

The early Nationalist Government had no well-structured approach to talent recruitment. The Leninist political party offered the closest twentieth-century equivalent to the old mandarinate, being the defender of political orthodoxy in its role as overseer of a more specialized government. But the GMD was poorly linked to the community of scholars and new professional associations that comprised Republican China's intellectual elite. The formal organs of government were filled on the basis of political compromise and factional patronage. An attempt to reinstitute and modernize civil-service examinations was an embarrassing failure; the Examination Yuan, the government branch created to test prospective officials on relevant technical skills, chose perhaps 1 percent of government functionaries during the prewar decade.[7] The result was a series of overlapping and under-talented organizations competing to set economic development policies. The central institute of advanced study and research, the Academia Sinica, had no policy function. Consequently, there was little interaction between the government and China's most talented scientists, engineers, and planners.[8]

Only after the Japanese seizure of Manchuria in 1931, which aroused a sense of common danger among all patriotic Chinese, did the Nationalists seek greater expertise in the formulation of industrial policy. From the "Self-Strengthening" decades of the late-nineteenth century down through what Richard Suttmeier calls the "national-security state" of the 1960s and 1970s,[9] Chinese political leaders have periodically granted a degree of autonomy to specialists capable of enhancing national defense. The consensus of the 1930s between political and academic figures on the priority of a "national-defense economy" (*guofang jingji*) was particularly important, since it defined the economic priorities of the Nationalist Government and began an institutional process of historical consequence.

Leadership on the academic side came from members of a transitional generation of Chinese intellectuals with both classical Chinese and Western scientific educations. They shared the old literati presumption, now strengthened by "scientific" knowledge, that they necessarily had a role in national affairs. Thus, when two of China's most prestigious scientists, the geologists Ding Wenjiang and Weng Wenhao, were asked to advise on long-term national-defense strategy, they argued, in the international intellectual fashion of the 1930s, that military modernization was inseparable from basic industrial and technological development, which required the "scientific management" of a "controlled economy." As Ding put it, China needed a "new-model" developmental dictatorship of a technically educated bureaucracy.[10]

These arguments were well received by Chiang Kaishek, who dominated the political scene from his position as Chairman of the National Military Council. His basic policies after 1931 were to achieve internal unity and military-industrial development capable of resisting Japan. Although hardly willing or able to invest Chinese academics with dictatorial powers, he sought to recruit into government many of China's most gifted intellectuals, scientists, and technicians. The National Defense Planning Commission (Guofang Sheji Weiyuanhui), organized in 1932 under Weng Wenhao and the Cambridge-trained economist Qian Changzhao, was to direct this effort. Although born as a technical appendage to the General Staff in the tradition of informal "self-strengthening" secretariats of the late Qing, this organization, known after 1935 as the National Resources Commission (NRC, Ziyuan weiyuanhui), would become the central government research, planning, and managerial bureaucracy controlling state industries. When initially charged with the research and planning of basic industries and the preparation of economic mobilization, its responsibilities were not unlike those of the Economics Research Center and Technical Economics Research Center under the State Council of the PRC today.[11] By the end of the Nanjing decade, however, it had become de facto a ministry of industry and planning, already the forerunner of the State Planning Commission and several PRC industrial ministries. More immediately, its Three-Year Plan of 1936 put China on the path of a military-oriented economy in preparation for an expected war.[12]

In historical terms, NRC pre-war enterprises marked a major step in the direction of central-government control over industrial and technological development. During the late Qing and early Republican periods, industrial and technological advances had occurred largely under foreign, provincial, or private auspices. The Commission's activities were indicative of a return to strong Chinese traditions of central economic regulation and control of basic industries and also emulation of planned or interventionist economies in the USSR and several Western countries.

The NRC's major projects began a new era in Sino-foreign industrial and technical relations and in planned technology transfer. The Chinese government gained access to state-of-the-art industrial technology while retaining full state ownership over Sino-foreign projects, such as the new Central Steel Works, Central Copper Works, Central Machine Works, and Central Electrical Manufacturing Works. Credits for foreign equipment and technical assistance and the training of Chinese technicians were repaid through countertrading (*duixiao maoyi*) techniques set down in a series of barter/credit (*yihuo changzhai*) agreements. These agreements, made directly with foreign governments or industrial consortia, revolutionized the methods of Sino-foreign trade and circumvented existing financial obstacles to foreign investment in China. Similar arrangements guided contemporary joint ventures for China's first automobile and aircraft construction companies and for the expansion of state arsenals and chemical plants.[13]

The NRC also undertook domestic mining and manufacturing "joint ventures" with provincial governments, sponsoring internal technology transfer while enhancing central-government influence in areas not fully under Nanjing's military or political control.[14] In regions under more direct government supervision, the Commission gradually enforced the central government's legal claim to deposits of the strategic ores that were China's (and the NRC's) most valuable export commodities.[15]

Perhaps the most important feature of NRC activities during the Nanjing period was its development as a self-consciously technocratic agency. NRC enterprises were wholly state-owned and were operated by salaried state managers and engineers. They were not examples of "bureaucratic capitalism,"[16] a term that implies government officials as both planners and private investors. As conceived and administered by Weng Wenhao, Qian Changzhao, and Sun Yueqi, its three chairmen during the years 1932–1949, the NRC was not a political organization in any sense, but an organ through which China's scientific and technological elite could serve the nation in an institutionalized, planned fashion, reasonably free from political interference.[17] Its growth in size and reputation during the mid-1930s marked the high tide of "professionalism" in government service during the Nanjing decade.

WARTIME TECHNOCRACY AND POST-WAR PLANNING, 1937–1945

The NRC's strategy of planned, defense-related heavy industrial development was predicated upon the likelihood of war with Japan. But the Sino-Japanese War began before China's program could be completed and forced the government's retreat to one of China's least developed regions. The nation's internationally minded planners, scientists, and technicians adapted with difficulty and were increasingly cut off from international contacts. With the war's outbreak, many key industrial projects had to be abandoned or relocated.[18] Strategic ore exports went to buy arms rather than plant equipment, or to guarantee Soviet and American credits during the years 1938–1941.[19] But the war also enhanced government efforts at economic mobilization and control. The NRC emerged as the leading agency for planning the war economy and China's post-war economic agenda.

As the core of a new Ministry of Economic Affairs, which replaced a variety of moribund institutions, the NRC gained broad powers to "develop, operate, and control" the nation's basic industries and mining enterprises and to nationalize "other enterprises as designated by the government."[20] The Commission exercised its new prerogatives vigorously, expanding from 23 industrial and mining enterprises and 2,000 staff members in 1937 to 103 enterprises, 12,000 staff members, and 160,000 workers in 1944. By that time, nearly 70 percent of the total paid-up capital of public and private enterprises under Nationalist authority belonged to state-run concerns. Three-quarters of that amount was for NRC enterprises.[21]

Bureaucratization accompanied the NRC's growth. Before the war, it had recruited its central staff, technical experts, and managers through selective recommendations from the technical faculties of Chinese and foreign universities. By 1942, the process had become institutionalized with the routine forwarding to the NRC of the *curricula vitae* of university graduates in technical disciplines, and the use of the nearly forgotten Examination Yuan to rank potential and present employees. Individual referrals were still entertained, but were increasingly limited to Chinese studying abroad, who as a rule were not required to go through other channels. Although political pressure to find jobs for the sons of "refugee" officials was intense during this period, available documentation suggests that NRC hiring decisions continued to be made largely on the basis of scholarly credentials and technical expertise.[22]

This emphasis on certified competence, however, was not sufficient, and often not well suited to the challenges of wartime production. The goal of creating the industry to fight a modern war did not change after July 1937, but it had to be pursued in isolation from the most modern sectors of the Chinese economy and from China's foreign partners. Like a later generation of expert talent "sent down to the countryside" during the Cultural Revolution, NRC planners and engineers discovered that the most advanced training and prestigious university degrees were no preparation for the basic production problems of the underdeveloped hinterland.

There were some success stories. The NRC's Gansu Petroleum Production and Refining Administration performed important exploratory work, relying largely on local resources to provide the Nationalists' only domestic source of petroleum. The NRC pioneered the use of alcohol as a motor fuel, both in its own facilities for the production of industrial alcohol and by its encouragement of some 120 small, private plants using native wine as raw material. It mobilized an estimated 82 percent of practical alcohol-production capacity in Nationalist-held territory. Coal and electrical-equipment production also were sufficient.[23]

The record in other areas reflects the fact that the Commission, like the Nationalist Government, adapted poorly to its new environment. In steel making, for example, it had to exchange the Krupp-supplied Central Steel Works for a "refugee industry . . . located with reference to safety rather than with reference to raw materials supplies, transportation, or other industrial factors."[24] By November 1944, the iron and steel plants under NRC control were producing at less than 25 percent of capacity.[25] Many of the same factors affected the machine industry, which reached its production height in 1940 and declined thereafter.[26]

In these and other industries, the NRC's management style appears to have been ill-suited to wartime conditions in China's interior. The Commission, like the other bureaucracies in wartime Nationalist China, grew more rapidly than the economy it managed. By war's end, most NRC enterprises were overstaffed with recently trained engineers and understaffed with individuals who had practical experience. Overhead costs grew with the reluctance of most engineers to engage in actual production work, reflecting the traditional social divide between the degree-holders and those with less formal training. US observers of the American War Production Mission to China in 1944–1945 were unanimous in their complaint that recent college graduates were "cluttering up" NRC plants, which in some cases could be run just as well by intelligent and experienced foremen.[27]

The war was a frustrating interruption to the NRC's ambitious plans for national industrial development. It is not surprising that the Commission—originally founded as a research and planning agency—took the leading role after 1942 in the drafting of industrial plans for the post-war period. Although largely unrealizable because of the subsequent civil war, such plans provide evidence of the anticipated direction of Chinese development under Nationalist rule and of the large role accorded its central planning agency. These plans make it clear that the extension of government controls and the growth of the state industrial sector were not simply wartime "emergency measures" but were to serve as the basis for an even greater state role in peacetime.

Weng Wenhao, Qian Changzhao, and other NRC planners assumed that the post-war era would find not only China but all countries moving further in the direction of the "socialization" (*shehuihua*) of the economy. Weng Wenhao called Chinese policy "close to socialism though not entirely identical,"[28] while, according to Qian Changzhao, post-war China, in which, happily, capitalism would have no developed roots, would have to "follow the socialist road" even more than other nations.[29]

Whatever the terminology, there was a broad consensus among Chinese planners about post-war economic directions. The NRC's 1942 "Preliminary Enforcement Plan for Postwar Industrial Reconstruction" set out policies designed to ensure China's status as an "equal" and "modernized" country (*pingdeng zhi xiandaihua guojia*).[30] At a series of inter-ministerial conferences on post-war planning co-chaired by Weng Wenhao in 1943–1944, NRC plans gave continued high priority to state-controlled, defense-related development, now defined in the broadest terms. Detailed schedules were drawn up for industrial, mining, hydroelectric and communications projects in the first post-war Five-Year Plan at an estimated cost of at least US$10 billion.[31]

Such plans anticipated a resumption of Sino-foreign industrial and technical cooperation, which in fact began even before the war's end. China's logical post-war partner was its wartime ally, the United States, the only power likely to emerge from the war in any position to assist China's reconstruction. Accordingly, in the last years of the war, over 1,000 Chinese engineers and managers from all NRC divisions were trained in such American firms as Westinghouse, RCA, Du Pont, Monsanto, US Steel, American Cyanamid, and nearly a hundred others. In 1944–1946, the NRC also sent 400 of its top officials and managers for study tours of several months in US industries.[32] Immediately after the war, the NRC engaged American industrial consulting firms for detailed studies of Chinese industries and mines, the recommendations of which formed the basis for a series of technical-assistance and joint-venture negotiations between the NRC and US firms.[33]

The most ambitious joint effort of this period was between the NRC and the Bureau of Reclamation of the US Department of the Interior, for the planning and development of the Yangzi basin for hydroelectric power, irrigation, flood control, navigation, and other purposes. This "T.V.A.-like" project, led on the American side by

John L. Savage, the chief design engineer for the Boulder and Grand Coulee dams, was to end with the erection of the world's largest dam in the Three Gorges of the Yangzi River, capable of producing 10 million kilowatts annually for a region formerly limited to 50,000, all at an approximate cost of US$800 million. Between 1944 and 1947, the NRC expended considerable effort and funding on preliminary studies and subsurface explorations of potential dam sites. The Bureau trained 40 NRC engineers at its Denver headquarters.[34]

The Three Gorges project—whose tremendous scope and cost have made its completion a subject of hot debate in the PRC[35]—reflects better than any other project the size of NRC ambitions for the post-war period. Its demise "for urgent economic reasons"[36] in May 1947 was symptomatic of the fate of most NRC projects that had been planned during the war years.

POST-WAR FRUSTRATION AND GROWTH, 1945–1947

For Chinese technocrats of the NRC, the years following Japan's surrender in August 1945 were the best and worst of times. The Commission continued to expand with its nationalization of former Japanese and "puppet" industrial properties and grew in strength relative to Nationalist political and military interests. But, in truth, nothing went as planned for Chinese planners in the post-war years. Elaborate production charts and organizational tables, such as those for the first post-war Five-Year Plan,[37] were of little use as the magnitude of the wartime destruction of plants and mines and the necessity for a prolonged rehabilitation period became apparent. The Soviet expropriation of Manchuria's industrial plant as "war booty" and the onset of the Guomindang-Communist Civil War made a bad situation much worse.

In the first post-war years, assistance from foreign sources was even more necessary, but less available, than anticipated. Much of the industrial share of UNRRA (United Nations Relief and Rehabilitation Administration) aid went to the distribution of emergency goods;[38] and nothing at all was forthcoming in the way of Japanese industrial reparations, from which as much as US$5 billion had been expected.[39]

The greatest setback, however, was in the low level of American participation in post-war Chinese industry and the failure to maintain the momentum of economic and technical cooperation that had developed during the war. US industrial credits to China were a casualty of Sino-American political differences at the time of the Marshall mission (1946–1947) and, perhaps more important, of fundamental disagreement over the direction of Chinese industrial and technological development.

Chinese expectations that NRC industries would be recipients of US government credits and enter into a series of technical-cooperation ventures with American firms were not shared in Washington, which sought a leading role for private US enterprise in Chinese industrial development, unimpeded by Chinese government regulation. Although the US Export-Import Bank "earmarked" US$500 million in credit for China in April 1946, funding was forthcoming in very few cases, since projects in potential competition for private US capital were excluded. The protracted negotiations (1943–1946) for the Sino-American commercial treaty replacing the system of extraterritoriality and frequent American complaints about China's "trend toward nationalization and government monopolies"[40] revealed the gap between Chinese and American expectations.[41]

American pressure was insufficient to reverse what were now long-standing and institutionalized trends in Chinese economic and technical organization. Even as the Nationalist Government entered a period of nearly total dependence on the good will and material assistance of the United States (the source of nearly 60 percent of China's imports and the destination of almost 40 percent of its exports in 1946),[42] the NRC resisted American demands to "privatize" Chinese industry and liberalize the terms for private foreign investment. On the contrary, the state industrial sector and this technocratic agency continued to grow at a rapid rate. At its height in August 1947, the NRC's industrial empire had 11 major divisions extending from power plants to paper making, employed 32,791 staff members and 228,247 workers, and accounted for over two-thirds of China's total industrial capital.[43]

The NRC's continued expansion, although a logical outgrowth of wartime plans, was beyond the limits of available Chinese managerial and technical manpower. When the political decision was taken to repatriate all Japanese technicians in Taiwan and Manchurian industries in 1946, Chinese technical talent was stretched even thinner. The Commission's archives contain numerous reports on problems encountered in the "reconversion" period, which was so confused that the Commission's own Nanjing headquarters was handed over by the Nationalist army to US forces.[44] NRC organizational and management inefficiencies—what one memoir writer has called its "yamenization" in this period—[45] combined with a belated recognition that the NRC's central production targeting of its various industries created as many problems as it solved and led to a wholesale reorganization and rationalization of decision making in late 1946. The NRC was raised to ministerial status, its chairman given cabinet rank. Most units operated by the NRC or affiliated with it were established as independent corporations with greater managerial autonomy. Unprofitable operations were to be allowed to close.[46]

It is faint praise to say that NRC operations compared favorably with the activities of other Nationalist Government bodies in these years. Yet, the NRC maintained its reputation for nonpartisan integrity and relative honesty. At the least, its continued recruitment of university graduates on the basis of education and skill rendered it immune from increasingly anti-government student agitation.[47] This did not mean, however, that this technocratic and managerial agency could remain sheltered from the political storms that swept around it.

TECHNOCRACY, POLITICS, AND TRANSITIONS TO COMMUNIST RULE, 1948–1953

"When I was in Chiang Kaishek's government," former NRC chairman Weng Wenhao wrote in Beijing in 1961, "I devoted myself completely to my work. I paid no attention to factions and politics."[48] This cannot be entirely true for an individual who served in a succession of ever-higher government posts in the Nationalist regime: Secretary General of the Executive Yuan (1935–1936); Minister of Economic Affairs (1938–1945); Vice-Premier (1946–1947); and Premier (1948). At the least, Weng and his longtime associate Qian Changzhao had tactical skill enough to keep the NRC out of the organizational grasp of the factions of Song Ziwen (T. V. Soong) and Kong Xiangxi (H. H. Kung), the groups most properly associated with the term "bureaucratic capitalism."[49] Weng and other NRC figures were themselves often identified with the Nationalist-era "Political Study" faction, a loose grouping of mostly Western-trained bureaucrats and scholars. That implied no independent political power, but authority derived from a holder of political or military power, in this case Chiang Kaishek, who, in the 1930s, provided the NRC with the political insulation needed to establish its organization.[50] Yet the NRC was not simply an appendage to a "Chiang Kaishek clique," nor were its state enterprises his economic secretariat in the manner of the informal bureaucracies of the late Qing.[51] Rather, in the later years of Nationalist rule, the NRC became an increasingly autonomous part of the state structure. It had no independent political power, but it proved capable of serving any political group that could provide peace, order, and prospects for economic development.

The wartime and post-war periods of NRC expansion witnessed a precipitous decline in Nationalist military and political fortunes. Even more than in the 1930s, therefore, NRC enterprises constituted an important component of Nationalist Government influence in the many areas where its rule was weak. Inevitably, the NRC came under intense political scrutiny from the more "ideologically oriented"[52] organizations in the GMD. The Party's role in policymaking had never been adequately defined, and had declined steadily since the late 1920s, as both the army and civil government rose in relative importance. To reverse this trend was the aim of GMD "renovationists" of the mid-1940s, who resented the "usurpation" of authority by non-Party professionals.[53] Weng Wenhao, in particular, was subjected to abusive attack in what might be termed a struggle between "blue" (the GMD color) and "expert," between the demands for political orthodoxy and the need for technical competence.[54]

Unlike later struggles between "red" and "expert" in the Mao Zedong regime, in this case the experts prevailed. Not only was Weng not removed from his posts, as the "renovationists" had demanded; he assumed new, more significant positions during the Civil War years. This was certainly in part because Chiang Kaishek knew that Weng was not seeking a political base of his own. Most important, the services of the state industrial sector were indispensable to the survival of the regime. When the regime faced economic, military, and political collapse in May 1948, Chiang made Weng Wenhao his premier. The highest point of "expert decision making" in the Nationalist period thus occurred at the lowest point of the political leadership. Weng's tenure is chiefly remembered for a spectacularly unsuccessful currency reform that put another nail in the Nationalist coffin. By year's end, this unpolitical fellow was being blamed for the imminent Nationalist defeat, while branded a "war criminal" by Chinese Communist forces.[55]

Within two years, however, Weng Wenhao was in the service of a new Chinese government, having been encouraged by the new, Communist premier, Zhou Enlai, to return from a self-imposed exile. More important, much of the organization that he had built and most of his former colleagues were also in service to the new state.

With few exceptions, NRC industrial facilities were transferred peaceably from GMD to Communist control in 1948–1949. Assurances were given that NRC employees could remain at their posts. The vast majority did, at all levels, befitting the NRC's professionalism and commitment to national service over political allegiance. Of course, for most industrial workers at NRC plants, fleeing with the GMD regime was not a realistic option. What is surprising is how few top NRC personnel left with the GMD,[56] and how assiduously NRC leaders assisted the transfer of power, often in direct contravention of Nationalist orders.

The pattern began in August 1948, with the public pronouncement of Chinese Communist Party (CCP) policy toward state-run industries in the northeast, and its refinement in January 1949 into general policy for all China, emphasizing the need for continuity and order.[57] NRC managers and workers alike joined "factory protection teams" in a widespread action that became more regularized as Communist armies moved south. While workers guarded installations from sabotage or looting, management negotiated the "reception" (*jieguan*) of NRC property by the People's Liberation Army.[58] Although the success of this policy has sometimes been attributed to the underground work of Communist cells,[59] the transition was largely coordinated by top NRC leaders, who planned their activities even as Communist policy was in formulation.

In October 1948, NRC chairman Sun Yueqi met with over 40 leaders of NRC industrial groups throughout the country at a clandestine meeting in Nanjing. Sun had ample experience with botched transitions of power as NRC special commissioner for Manchuria in 1945–1946. He had served as Weng Wenhao's vice-chairman after 1947, and, when Weng assumed the premiership, became the Commission's head. His message to this tightly knit group, most of whom had worked together for over a decade, was straightforward: Their tasks—to maintain production and work toward Chinese industrial development—did not change with political circumstances. If it were not possible to do so under the GMD and with American assistance, then it might well be possible under the CCP, and with Soviet aid.[60] Sun implied the need to resist GMD efforts to sabotage or remove factories. In the next several months, he himself deflected Chiang Kaishek's directives to transfer 5 factories from the Nanjing area to Taiwan.[61]

The peaceful transition of NRC units in Nanjing and Shanghai in the spring of 1949 bore witness to the success of NRC-CCP cooperation. Like all other Nationalist Government agencies, the NRC as such ceased to exist. But, despite a formal ruling by the new authorities that, "since the Nanjing government was a bureaucratic clique, its enterprises therefore constitute bureaucratic capital to be confiscated," a huge exception was made for NRC operations, which were classified as "public enterprises" (*gongying qiye*), and allowed to continue operations as before.[62] Moreover, at the behest of the new regime, the NRC's main industrial-planning committee met some 35 times as a unit from May to October 1949, helping to coordinate the CCP takeover of NRC enterprises to the south, and advising on industrial issues nationwide.[63]

LEGACIES

After 1949, both the People's Republic and the Republic of China on Taiwan sought to put historical distance between themselves and the discredited period of Nationalist rule on the mainland, emphasizing the innovations of the "New China" of the PRC and the "reformed" Guomindang on Taiwan, respectively. Certainly both governments have presided over social and economic changes that can be called revolutionary. But in some sectors there was more continuity than change under both regimes. Although the NRC was unable to fulfill its ambition of planned, internationally oriented industrial and technological development during the period of Nationalist rule on the mainland, it contributed personnel, experience, and precedents (positive and negative) to subsequent modernization efforts on both sides of the Taiwan Straits.

When the Chiang Kaishek government moved to Taiwan in 1949, it inherited the upper- and middle-level NRC managers and engineers already in place in the 18 NRC-run state corporations on the island. These provided an important portion of what the economist Simon Kuznets, a former NRC consultant, has called the "experience and human capital" needed for Taiwan's subsequent progress.[64] Although the NRC on Taiwan was dissolved as an institution after 1952, when its activities were reincorporated into the Ministry of Economic Affairs, its "alumni" made significant contributions to the island's remarkable development: as heads of state corporations and government economic planning commissions; as 8 of the 14 ministers of economic affairs; and, in one case, as premier. Its tradition of technocratic planning—now more in the realm of "guidance" planning—remained strong, even as Taiwan's economy became more truly "mixed" between public and private sectors, while Chinese state planners, under now irresistible American pressure, created an environment conducive to private as well as public entrepreneurship. Yet, even with these developments, Taiwan's economy retained features directly associated with pre-1949 industrial planning. Emphasis on state-led, import-substituting, defense-related heavy industrial growth dominated until 1956, while strong protection of key import substitutes continues into the present. A cautious, if not hostile, attitude toward private foreign investment lasted until the late 1960s. Despite the denationalization of 4 former NRC corporations in 1953–1954 as part of the land reform program, the government refused to divest further, and continues to dominate areas of original NRC monopolies (steel, shipbuilding, petroleum, engineering, heavy machinery, and so forth) in a persistent *étatisme* that, as Alice Amsden correctly notes, is in the "tenacious tradition of Sun Yat-sen."[65]

The new political order imposed in the People's Republic also carried several trends begun under the NRC on to logical, if more extreme, conclusions. The first concerns the growth of central-government control over industrial and technological development. The NRC's accumulation of state capital "laid the foundation" for the PRC's even more extensive state monopoly.[66] In 1947, the Commission's enterprises controlled 33 percent of China's coal output, 45 percent of cement, 63 percent of electric-power production, 90 percent of iron and steel, 90 percent of sugar, and 100 percent of petroleum and nonferrous metals, accounting for 67.3 percent of total industrial capital. The PRC's incorporation of this industrial and human capital was a precondition for the early and rapid growth of the socialist state sector.[67]

Second, although the PRC's 1st Five-Year Plan assumed far more extensive state control of domestic resources, including agriculture, than NRC planners had ever dreamed of, like the NRC's industrial programs under the Nationalists, it was predicated on a close but controlled trading and technological relationship with an advanced industrial power. The PRC's alliance with the Soviet Union in the 1950s provided China's state enterprises with industrial goods and technical training on credit in exchange for raw materials (*yihuo changzhai*). This exchange was the logical extension of NRC approaches, with a foreign partner far more compatible than the United States of the 1940s. In post-Mao China, retired NRC officials who continued to serve under the PRC look back on the period of the 1st Five-Year Plan as "golden years" in which, finally, unencumbered by war, revolution, incompetent politicians, or the competing demands of foreign capitalists, they could do their work.[68] If, in the 1980s, the Chinese government has introduced rules of foreign investment more liberal than those countenanced in the last years of Nationalist rule, foreign investment still continues to be recruited according to the needs of state plans. The long-term viability of Western private investment in an historically inhospitable bureaucratic culture remains to be proven.[69]

Third, the NRC's devotion to central economic planning and research was taken much further in the PRC than on Taiwan. It found expression (as some NRC officials found employment) in the regional industrial bureaus of the first years of the PRC and, later, in the State Planning Commission, various industrial ministries, and research organizations such as the Chinese Academy of Sciences and the State Scientific and Technological Commission. The NRC experience showed some of the strengths and weaknesses of a centrally planned, nationalized industrial sector. NRC strengths ultimately were greater in research than in organization; the scholarly studies of its Economics Research Institute are now among the best available sources for Republican-era economic history. The NRC's approach to industrial organization suited the purpose of harnessing technology and expertise for national purposes. But, as the state sector grew, it created problems for which there were no easy "technical" solutions. These included a serious lack of managerial skills, a result of its primary concern with technical training in hiring and promotion decisions; adherence to planning guidelines too rigid for a nation of China's size and diversity; and toleration of unprofitable enterprises. The NRC was only beginning to address these issues when the Civil War broke out in 1946. Its successor organizations in the PRC have tried to address them under a variety of socialist models, but with indifferent success.

The NRC's ability, however, to maintain its technocratic outlook and apolitical status throughout the period of Nationalist rule make it different from its successor organizations in the PRC. It was able to do this in part because its activities early on bore the legitimacy of "national defense enterprises" at a time of a national emergency. But the NRC remains the high point of "expert decision making" in modern China for reasons that go beyond the immunity of a national security clearance. In Nationalist China there was still room for the patriotic but apolitical expert, committed more to a discipline or methodology than to a political "ism" to make a contribution to national development. This was possible in an institutional sense because the need for technically competent industrial planning created an organization that Nationalist-era political authorities could not effectively oversee or control.

Before the founding of the NRC, there was no working mechanism for recruiting technical expertise for government service. As China's first attempt at a technocratic civil service, the NRC had, by 1942, institutionalized its selection of China's "best and brightest" in the technical disciplines. Through its growing empire of industries, the NRC became an essential part of the Nationalist rule, though increasingly separated from military and political authorities. It was, to be sure, never a "purely bureaucratic" entity in the Weberian sense; the varied nature of its recruitment policies and the extraordinary disruptions of war and civil war precluded such a development. Nor was it a "nonpolitical" civil bureaucracy created for coherent political and social reasons, as in the cases of early nineteenth-century Prussia, or Meiji Japan. The early marriage of convenience between the NRC and the military rule of Chiang Kaishek may well be comparable to what Guillermo O'Donnell has called the "bureaucratic-authoritarian" regimes of post-1964 and post-1966 Brazil and Argentina, though not to later periods of military rule in those countries. More accurate comparisons, particularly after the NRC's restructuring in 1946, may be made with the public holding companies for state corporations in a variety of developing economies from the 1930s on, such as the IRI and ENI of pre- and post-war Italy, respectively; CORFO in Chile; the National Development Corporation in Tanzania; and public firms in Korea and India.[70]

In its relationship to the Party-state, the NRC was potentially not unlike what Kendall Bailes calls the "technical intelligentsia" of the early Soviet Union, which prospered under Lenin's commitment to Taylorist "scientific management."[71] But, in the Nationalist case, the creation of a technocratic arm of government did not by itself narrow the gap between politics and expertise. Instead, it institutionalized it. Without effective Party oversight or control,[72] political supervision of economic policy weakened. As the Guomindang's decision-making capacity decreased after the late 1920s, the Party's main function increasingly was to assert a vague ideological authority. Because the NRC was the agency charged with implementing Sun Yat-sen's industrial agenda, it was ideologically defensible, while the GMD had little capability of judging the expertise of the "experts." Over the course of Nationalistic rule, and particularly in the 1940s, important economic decisions were made more and more by individuals with little if any political standing and holding only titular Guomindang membership, if that (Weng Wenhao, Qian Changzhao, Sun Yueqi). Although this could have positive results, as in the economic mobilization of the mid- and late 1930s, the continued separation of expert talent (even in government service) from political authority seems to have encouraged a lack of realism in the former and ever more incompetence in the latter. How else can one explain the NRC's "Great Leap" mentality in post-war planning, particularly its dogged and expensive pursuit of the vast Three Gorges project for three years at a time of general economic dislocation and civil war? How else could Chiang Kaishek allow Weng Wenhao's desperate (and, as it turned out, unworkable) currency reform of 1948 that sealed the political fate of the Nationalist regime? This combination of consistently weak political institutions[73] and increasingly autonomous technical and managerial elements in the government was a major, and still understudied, feature of Nationalist disorganization and disintegration in the 1940s. Such autonomy allowed the NRC, whose professional responsibilities and national loyalties ultimately transcended the political interests of Nationalist rule, to shed one set of political authorities for another.[74]

Certainly the Nationalist experience provided some justification for the skepticism toward a "specialist" and "internationalist" approach that marked Maoist policy toward scientific and technological development. But, at the other extreme from the Nationalists, the Maoist period saw the imposition of political controls stringent enough to call into question "the very existence of professional life"[75] for China's scientists and technologists, and at significant cost to China's economic development. The technocratic structure of the Nationalist period by and large survived well into the 1st Five Year Plan, albeit greatly expanded and systematized.[76] Initially, the political controls were bearable, but, from the late 1950s until Mao's death in 1976, the extreme insistence on political purity and denigration of expert talent reduced technocratic contributions seemingly to the point of no return, as many Chinese with advanced technical training were placed in conditions of internal exile in some ways more harsh than those endured by their NRC predecessors of the wartime period. But institutional patterns have been known to survive tyrants as well as incompetents. Many veterans of the Nationalist technostructure and their protégés have resurfaced under the post-Mao return to policies of scientific, technological, and economic development in a global context.[77] Political controls still restrict their contributions, however. Mao understood, but feared too much, the historically apolitical nature of Chinese technocracy. Today, such fears still put limits on the current Chinese leadership's desire to grant greater autonomy to a new generation of internationally minded technocrats, and complicate its task of finding a new role for a Communist Party whose control of China's scientific and technological development during the last decades of Mao's rule yielded results that were truly "poor and blank."

CHAPTER TWO

Learning from Russia: Lysenkoism and the Fate of Genetics in China, 1950–1986

L. A. SCHNEIDER

Today, genetics enjoys a high priority in China's science-modernization plans.[1] Pleased as they are with this situation, China's life scientists cannot help recalling with a sense of irony and grief the "twisted path" genetics has traveled in China since 1949. They remember that, during the 1950s, their work was repressed by the Soviet biological doctrine named for its creator and ruthless enforcer, the agronomist T. D. Lysenko (1898–1976). Lysenkoism denied the validity of classical genetics from Mendel to Morgan and promoted its own theory of heredity, based on the belief that acquired characters could be inherited. Lysenko claimed that spectacular improvements in agricultural productivity resulted from applications of his theory. Starting from 1949, Lysenkoism was imported from Russia and established in China. With the aid of Soviet advisors, it quickly gained hegemony in the biological sciences. Even after it lost this position in China, its debilitating effects were felt for decades.

The relationship of Lysenkoism to genetics in China is a case study of the effects of government and Party policies on the practice and development of a science. The field of genetics is a useful example because so many basic and divisive issues have converged and persisted there. Some of the issues concern scientific thought and practice, while others address the relationship between science and social class or science and production. Still other problems arise from the motivation to follow foreign models of science and technology. Periodically, there have been attempts to resolve such problems through the liberalizing reforms known as the Hundred Schools policy which first occurred in 1956 and the first half of 1957.[2] Where the natural sciences were concerned, the implications of the Hundred Schools policy were first explored in the conflict between Lysenkoism and genetics. In 1986, when the 30th anniversary of the Hundred Schools policy was celebrated, the policy's origins and intent were illustrated with this classic conflict. China's Lysenkoist experience has become a cautionary tale for the redevelopment of science in the post-Mao era. It warns against Party and government interference with scientific institutions and practice, and opposes the imposition of any philosophic system or doctrine on scientific thought.

The initial victims of Lysenkoism and beneficiaries of the Hundred Schools policy belonged to a group of about fifty scientists who received their doctorates from American universities. Most were trained in the joint program for plant genetics and improvement sponsored by Nanjing and Cornell Universities. Others studied and collaborated with T. H. Morgan's groups at Columbia and Caltech. Gradually, the chilling effect of Lysenkoism spread to the rest of the bioscience cohort inherited by the PRC. This larger group also included a number of distinguished scientists trained at the best American and European centers. Their fields ranged from botanical systematics to embryology to biochemistry. Generally, they were superbly trained and in close touch with international schools and trends.[3]

Unlike its role in the Soviet Union, Lysenkoism in China did not result in death, jail, or even unemployment for any of these scientists, but it prevented or delayed for decades the realization of their great potential to develop science education and research, and to improve agricultural productivity.[4]

THE ATTRACTION OF LYSENKOISM

What are the basic technical elements of Lysenko's self-styled new biology? Most important is its transformist impulse. Lysenkoism hinges on the belief that nature, especially the part that feeds humanity, is infinitely malleable and infinitely responsive to the appropriate human efforts to direct and reshape it for practical purposes. Lysenko built his career on claims that he could improve food plants so that they would grow in previously hostile environments, or grow more abundantly in native settings. His "agrobiology" consisted of a set of transformist techniques to exploit what he called the phasic or stage-like development of plants. Lysenkoists argued that, by manipulating the temperature or available light at the appropriate time, a plant could be directed to change in a desired direction, even to change its "hereditary nature." Winter-ripening wheat could be changed into the spring-ripening variety, and vice versa. Particular characters of a food-grain could be developed so that a variety of wheat that originally required a moderate climate could be made to grow in harsh northern climates.

Lysenko legitimized and enhanced the appeal of his transformist programs by coopting the nativist and populist aura that surrounded the peasant horticulturalist I. V. Michurin (1855–1935). Lysenkoism's success in China depended on the Michurin imagery: the earthy, grizzled farmer whose personal involvement with production is the source of endless creative innovation. Lysenko followed Michurin's trial-and-error technique, eschewing experimental controls, pure lines, and statistical evaluation of results.

By the time Lysenkoism reached China, it had surrounded its utilitarian, transformist ideas with a theoretical structure countering classical genetics—Mendel's laws of inheritance and T. H. Morgan's chromosome theory. Lysenkoist theory denied the reality of the gene and the importance to heredity of the chromosome. The notion of a basic hereditary material of fixed position, structure, and function was inimical to the transformist fantasy. Instead, Lysenkoism emphasized "outer," environmental sources of organic change. "Heredity" implied a conservatism that the transformist wanted to overcome and redirect. The Lysenkoist doctrine argues that, by manipulating an organism's environment, one can alter its characteristics and also alter its heredity, and so cause new characteristics to be inherited by the organism's progeny. This doctrine had special appeals: It bypassed the complexities of modern genetics with a scheme that was, in the words of the Chinese practitioners, "easy to explain to the masses, and easy for peasant minds to grasp." It promised quick, inexpensive results.[5]

The Chinese who were first attracted to Lysenkoism were CCP cadres involved in plant-breeding and improvement programs in the Beijing area during the late 1940s. Their leader, Luo Tianyu (1900–1984), used the prestigious new Beijing Agricultural University as his base. He was appointed its first president when it was established in October 1949. Luo achieved preeminence in his field in Yanan during the 1940s when the CCP had hopes of attracting scientists from the GMD-controlled areas. To that end, a science program was launched and eventually a university was established in Yanan, with Luo Tianyu as head of its biology department. In 1942, he played a significant role in the Zhengfeng (rectification) campaign. His job was to implement the "mass line" in science and education. This line emphasized populist and nativist values which resonate strongly with Lysenkoism. Perhaps this was a factor in the continuing failure of Yanan to attract scientists to its fold.[6]

Luo Tianyu's Yanan experiences help to explain how and why he was able to launch and establish Lysenkoism with such speed and efficacy. In its full-blown form, Lysenkoism provided Luo and his colleagues with a complete ready-made doctrine, not merely an agricultural technique. Though the Chinese Lysenkoists always touted the practical results of their doctrine, Luo's campaign propaganda initially stressed its ideological correctness. For Luo and his colleagues, Lysenkoism in 1949 provided the means for achieving what they had dreamed of in 1942, that is, establishing guidelines for scientific education and research within the context of socialist dialectical materialism. Generally, Luo understood this to mean that scientific creativity and education began in production, among the workers. Now they had the organizational means to carry out this goal, and the imprimatur of the Soviet Union itself. Even though the CCP did not issue any policy statement on Lysenkoism until 1952, Lysenkoism's quest for monopoly and the status of orthodoxy had already been achieved, de facto, with the license and encouragement afforded by the CCP's comprehensive pro-Soviet policy of "learning from the advanced experience of the Soviet Union."

As early as October 1949, Lysenko's representatives were in China, helping Luo Tianyu's campaign. First came V. N. Stoletov, Minister of Higher Education and Chair of the Department of Genetics at Moscow University. He lectured at Luo's University, and was soon followed by other Lysenkoist luminaries, such as N. I. Nuzhdin and A. P. Ivanov. These visiting experts traveled extensively throughout China, giving introductory lectures which were rapidly transformed into college textbooks in the Chinese language. The Russians also preached to key scientists about Lysenkoism's scientific validity and effectiveness.[7]

Luo Tianyu initially guided the Lysenkoist campaign through the nationwide network of agriculture colleges and extension programs under the jurisdiction of the Ministry of Agriculture. Grass-roots Michurin Societies were established for instructing local educators and farmers in the new biology. At the colleges, Michurin study groups disseminated literature and monitored the censorship of biology textbooks wherein Lysenkoism was supplanting classical genetics.[8]

At the outset of this successful campaign, Luo harassed and intimidated American-trained geneticists who worked in the Beijing area. An international scandal resulted when he drove C. C. Li, a young Cornell graduate, out of the country in 1950.[9] Nevertheless, Luo's campaign continued for almost two more years, during which his heavy-handed treatment of scientists exacerbated tensions between Party and non-Party intellectuals at a time when the CCP was strenuously recruiting intellectuals, especially scientists, into the Party.

The Lysenkoist campaign was almost impossible to assail from any solid institutional base. There were not yet any centralized science-planning agencies nor any science plan to which anti-Lysenkoists could appeal. Leaders of the Ministry of Agriculture, while not necessarily themselves Lysenkoists, nevertheless were strongly attracted to the doctrine because of their empirical and pragmatic bias. The Ministry of Higher Education was preoccupied with teaching, almost to the exclusion of research, and so ignored the fundamental questions raised about the validity of Lysenkoist science. The leadership of the CAS was sufficiently subservient to the CCP to accept the Lysenkoist juggernaut as one more element of the pro-Soviet policy. Units of the CAS, however, acted as havens for biologists fleeing

Luo Tianyu's abuse, and as safe harbors for anti-Lysenkoist sentiment.[10]

It was Luo's heavy-handed treatment of scientists, not his Lysenkoism, that brought CCP action against him during a major "thought-reform" campaign in June 1952. Although he was condemned, the status of Lysenkoism was regularized by making it an official CCP policy. Luo was subjected to fierce criticism at a joint meeting of the CAS and the Party, simultaneously made public in *People's Daily* alongside a defense of the "Michurin line." His primary responsibility, the Party said, should be to help intellectuals reform, but, instead, he sought to enhance his own status by "ingratiating himself with the backward masses." Because of his "sectarianism" and his "individualistic heroism" he was unable to "rally together non-Party intellectuals." Luo's job had been to persuade non-Party scientists that Michurinism was appropriate for new China and Morganism was inappropriate, but he was unsuccessful because he himself had little grasp of Michurinism and even less of natural science outside his narrow agronomic training.

The most serious errors of which Luo was accused were "empiricism and dogmatism" in his view of science. He over-emphasized the minutiae of practical productive experience while neglecting systematic agronomic theory; and he stressed study rooted in production while ignoring laboratory work. He had written that "the fields are a laboratory; nature is a classroom." Moreover, he set mass science above that of the intellectuals, saying that the People are genuine materialists, who grasp the answers to things directly, because, if they do not understand and live by the laws of nature, they die. Since scholars do not operate this way, they are idealists.[11] This vintage Yanan rhetoric embarrassed elements in the CCP and CAS who, in the early 1950s, were responsible for organizing and deploying natural science personnel during the forthcoming 1st Five Year Plan (1953–1957).

Luo drew political conclusions about scientific matters; he labeled Morganism and the scientists who used modern scientific techniques idealistic, metaphysical, and reactionary. The Party, however, would not endorse this tactic and expressed its official opinion in *People's Daily*: "If one says that the old genetics is reactionary, one cannot say that it is true that the practitioners of the old genetics are consequently definite members of the politically reactionary element."[12]

The Party's dilemma was clear: It did not want to alienate any more scientific talent; but it could not yet bring itself to compromise its pro-Soviet policy. No Party official was ready yet to point out that Russia was providing contradictory directions for development. On the one hand, there were the populist, voluntarist, anti-specialist values of Soviet Lysenkoism. On the other, there was the "Stalinist model" Five-Year Plan for urban, heavy industrial development, dependent on an elite of technical specialists, guided by a centralized planning bureaucracy.

Luo Tianyu immediately lost all his administrative posts, and the classical geneticists were promised that one would not be labeled a political reactionary because of one's scientific beliefs. The price for these concessions was at least passive compliance with the Michurin line: One need not espouse it, teach it, or follow it in research; but one must not criticize it nor continue to teach or do research in any aspect of the "old biology."

When Lysenkoism first came to China, C. C. Tan (b. 1909), China's most accomplished geneticist, was Dean of the Science College at National Zhejiang University, renowned for its outstanding natural-science faculty, assembled under the presidency of Zhu Kozhen. Tan with Zhu Kozhen and the faculty strongly criticized and resisted Lysenkoism. In 1950, Nuzhdin and other Russian Lysenkoists tried with conspicuous lack of success to convert Tan. The following year, Tan was a prominent spokesman for the "old biology" at a hearing held by the central Ministry of Higher Education on the subject of reforming the biology curriculum. At that hearing, Tan sided with the Ministry's biology division against representatives of the Ministry of Agriculture, which proposed the adoption of a Lysenkoist curriculum on the subject of evolution and genetics. Tan stubbornly continued to teach and do research in classical genetics until the Michurin line was promulgated[13] and the "old genetics" was proscribed in July 1952. Tan's public "self-criticism" in August 1952 was the political ritual that formally legitimized Lysenkoism.

The CCP's final acceptance of Lysenkoism stemmed from three considerations. The Party believed that natural science was "superstructural." Therefore, a science developed in the capitalist West could not have universal validity or be compatible with the needs of a socialist society. Second, there was a compulsion to see natural science, properly conceived, as an integral stage of the production process. Sciences like genetics, especially when viewed through the lens of Lysenkoism, seemed self-serving, remote from production, and concerned only with esoterica. Finally, on the eve of Russia's massive assistance with China's 1st Five-Year Plan, the putative successes of Soviet science, including Lysenkoism, seemed to be the surest guide. The CCP's virulent anti-American, anti-Western attitudes guaranteed that Soviet science would be the only guide.

THE 1ST FIVE-YEAR PLAN PERIOD

Soviet contributions to China's science and technology were comprehensive and of varied significance. In the urban, heavy-industrial sector, the Soviets provided everything from organizational models for planning development to close supervision of the construction of new or expanded industrial infrastructures. They supplied construction blueprints and sometimes even prefabricated structures. Most important, thousands of specialized consultants were sent to supervise construction projects. Although the idea of centralized S&T planning, planning agencies, as well as CAS and the higher-educational system, existed in China before Soviet contact, over the course of the 1950s, they were developed, rationalized, and tinkered with along lines provided by Soviet experience and advice in order to have them better serve the goals of the Five-Year Plan and Party control.

Soviet-inspired changes affected the content as well as the form of S&T. By the mid-1950s, thousands of advanced Chinese students were studying in a variety of disciplines in various Soviet universities. Hundreds of Soviet scientists came to Chinese universities to offer introductory courses in their specialties and to give state-of-the-art colloquia to researchers in the institutes. Their lectures were regularly converted to Chinese textbooks. Still greater amounts of technical literature were translated from Russian and circulated widely throughout China in periodicals, monographs, or textbooks.

As yet, no thorough assessment has been made of this Soviet legacy to China's S&T. Some generalizations can, however, be ventured. First, Soviet experience and advice led Chinese S&T research to be compartmentalized, even fragmented, within a number of organizations—the institutes of CAS, various ministries, and the universities. Horizontal communication and cooperation across organizational boundaries became difficult, if not impossible. Given the scarcity of properly trained and experienced personnel, transferring them from one organization to another was unthinkable. Second, the Soviet tendency was to separate research and education in CAS and the universities. Lack of personnel, however, prevented the Chinese from achieving this separation to the degree encouraged by the Soviet example. Third, Soviet technical education was narrow and highly specialized. Consequently, curriculum reforms in China resulted characteristically in over-specialization. In practical terms, this quickly produced large numbers of S&T personnel to carry out preplanned tasks, but this tactic limited the scope and duration of the personnel's usefulness. Finally, the Soviet pattern planned S&T research in coordination with production goals. This contributed to a Chinese preoccupation with immediate and demonstrable production applications to justify research. Often, basic research was denigrated. Party officials without S&T expertise played a substantial role in the planning process, thereby assuring continuing confusion about the difference between applied and basic research, and pervasive ignorance about the essential roles played by basic research in the developmental process.[14]

Throughout the 1st Five-Year-Plan period, as Soviet-inspired S&T reforms were being implemented, the "Michurin line" flourished. A constant stream of Soviet experts shaped the biology curricula of colleges and universities. In cytology, the crackpot work of Olga Lepeshenskaya was used to support Lysenko's transformist beliefs by supposedly showing that living cells could be created out of nonliving organic material. A. V. Dubrovina lectured to huge audiences on Lysenko's version of Darwin, and introduced the notion that there was neither rivalry nor "mutual aid" among organisms within a species. I. E. Glushchenko, leader of Lysenko's international program, and other agricultural experts gave extensive lectures explaining the relationship between Lysenko's "New Darwinism" and plant cultivation programs.[15]

Except as the target of nasty epithets, classical genetics disappeared from podium and print. Genetics research was usually impossible. The exception was a plant geneticist like Bao Wenkui (b. 1916), who had the support of local Party cadres. Bao (PhD, Caltech, 1950) had worked before 1949 with successful plant-improvement programs. After 1949, he continued to be involved with applied research. But, in 1956, when he lost his cadre sponsors, his "old" techniques were criticized and Lysenkoist zealots destroyed his experimental plots.[16]

Most biological scientists did not leave themselves open to this kind of harassment. Like C. C. Tan, they typically carried out their routine duties, kept quiet, and tried to keep informed about their special fields through colleagues returning from the West and from Western scientific literature which somehow managed to get into China. C. C. Tan moved to Shanghai in 1952 and founded the biology department of Fudan University, where he assembled a superb American-trained group in human genetics and in biochemistry. Under the ban, this group was permitted to teach only general biology courses and an occasional course on evolutionary theory.[17]

Visiting Russian biology experts often exacerbated the tension and hostility between the "two schools" by publically criticizing unconverted scientists, especially older, prestigious ones with distinguished foreign pedigrees. This may have been encouraged by Chinese Lysenkoists as a tactic to get at their enemies indirectly. In any event, it was perceived by many as an arrogant interference by outsiders, and it was the start of a profound resentment of the Russian presence.[18] This sentiment was in no way assuaged by events such as the joint celebration of Michurin's centenary in 1955. Great amounts of time, energy, and precious resources were spent on marathon conferences, ponderous festschrifts, and elegant editions of Michurin's writings. Chinese and Russian Lysenkoists staged this dramatic celebration not only to promote themselves further in China but, more importantly, to bolster Soviet Lysenkoism, which was in trouble.[19]

During this period when the Michurin Line was an exclusive CCP orthodoxy, other imported Soviet science doctrines also affected the development of natural science in China. Chief among these was Pavlovian physiology. Of lesser impact were doctrinal controversies on resonance theory in chemistry and quantum and relativity theory in physics. Although these controversies stirred passions, none had any of the debilitating force of Lysenkoism, for no doctrine had Lysenkoism's combined cultic qualities and organizational power. Moreover, by the mid-1950s, genetics was a scientific field where biology, physics, and chemistry came together. Thus, Lysenkoism in China, as in Russia, affected the natural science community well beyond genetics.

THE HUNDRED SCHOOLS AND THE QINGDAO GENETICS CONFERENCE

By the end of 1956, the Michurin Line was abrogated and Lysenkoism lost its monopoly in China. The context for these developments was the CCP's comprehensive reevaluation of the role of S&T education and research in the 2nd Five-Year Plan. The Party raised questions about the ability of China to produce a sufficient number of its own qualified experts and, by extension, lessen dependence on Soviet expertise. To answer the questions, in the fall of 1955 the Party charged appropriate people to begin drawing up a 12-year science plan. This in turn led to a January 1956 national symposium to reappraise the treatment of intellectuals in China. The key issue was whether the best conditions existed for the development and use of their talents.

Seizing the moment, leaders of the CAS and the Ministries of Education prevailed on CCP leadership to reconsider the problem of genetics. Mao Zedong himself was apprised of a parallel problem and its resolution in the German Democratic Republic, where Hans Stubbe, a prominent biologist, had been permitted by the GDR Communist Party to challenge and then reject Lysenkoism. Events in the Soviet Union also stimulated action in China, beginning with Khruschev's speech on "de-Stalinization" in February 1956 at the 20th Party Congress of the Soviet Communist Party. In April, after many months of criticism from the Soviet science community, Lysenko was finally forced to abdicate his presidency of the Lenin Academy of Agricultural Sciences. The Chinese immediately learned of this from the Soviet biologist N. V. Tsitsin, who had been invited by the CAS to be an advisor for the 12-year science plan. Although they had kept a close watch on the criticism of Lysenko which had been mounting in the Soviet Union since 1952, they were stunned by the news.

Mao and the Central Committee now decided to give the "genetics question" special attention. At the end of April, Lu Dingyi's Propaganda Department was charged with making this question an example of the Hundred Schools policy. Yu Guangyuan, Director of the Propaganda Department's Science Division, and one of the leaders of the 12-year planning commission, was to use the medium of a national genetics symposium to communicate and develop the charge. Preparations began immediately for the symposium, scheduled to be held in August. In the interim, Yu had informal discussions with Party, CAS, and Ministry of Education personnel. They determined the basic line on the natural sciences and the "genetics question."

The first of these decisions was that the CCP would not specify the detailed meaning and application of the Hundred Schools policy; that would be left to the scientific community. Second, science orthodoxies and monopolies were impermissible. The CCP, Yu said, would not do as the CPSU had done when it "created the Lysenko faction and gave it its special place." The CCP would not rule on scientific controversies, and the ideas of minorities would be protected ("scientific questions cannot be decided by majority vote"). Third, political attacks under the guise of philosophical labeling must cease (for example, "Morganism is idealistic"). Finally, only scientists should decide if and when philosophy is of any aid to science. The genetics symposium was asked to consider these four points and how they could be implemented through educational curriculum reform, and through reforms in research, including the creation of a CAS Institute of Genetics.[20]

The Genetics Symposium was held in Qingdao from 20 through 25 August, jointly sponsored by CAS and the Ministries of Education, notably excluding the Ministry of Agriculture from its planning and administration. Fifty-four scholars, representing every branch of the life sciences, all educational and research institutions, and a broad spectrum of opinions, accepted invitations to prepare reports or remarks on special technical topics and to participate in the free-form discourse. Seventy-three others, from central and local CCP and government organs, accepted invitations to be observers.[21]

The Symposium's general chairman was Tong Dizhou (1902–1979), a distinguished European-trained experimental embryologist, head of the CAS Biology Department, and CCP member. He outlined the general goals and ground rules. First, "a comprehensive analysis of the recently developing international conditions of genetics." Second, proposals to the ministries of secondary and higher education for a new genetics curriculum. And third, exploring new directions for research which would break down the barriers erected by Lysenkoism and permit the Chinese biological sciences to meet international scientific standards. Tong urged all participants to speak their minds and not be inhibited by courtesy or fear of reprisal. He explained that the symposium's purpose was not to determine who was right or wrong, nor to reach any conclusions about particular scientific issues, but to establish open discourse.

The Qingdao Symposium accelerated the Hundred Schools policy in the sciences. The proceedings of the Symposium reveal that it was an unrestricted forum for lifting bans. For the first time in seven years, biologists were able to describe the nature of modern genetics free of the distortions and misconceptions purveyed by Lysenkoism. Before a distinguished audience, "Morganists" confronted their Lysenkoist antagonists. Representatives of both sides agreed upon a future of academic freedom, mutual understanding, and cooperation. Granted, the Lysenkoists had no choice in the agreement, since that was the Party line, but the Symposium was not a pro forma political exercise. Only practicing scientists spoke; there were no speeches by or for the government or Party.

Virtually all the Symposium's reports and discourse were technical. The Morganists presented accurate, detailed, state-of-the-art discussions of each field of biology related to genetics, interwoven with sensitive comments on scientific epistemology and method. These discussions were meant to be the first step in reintegrating China's life sciences into the international community. Through government and CCP auspices, Symposium materials were printed, distributed to specialists, and used as means for reeducating the broader public. They were expected to provide a foundation for new biology curricula in secondary and higher education wherein the "two schools" were to co-exist.

There was (and remains), however, some ambivalence about Qingdao's significance. At the Symposium it was bitterly remarked that criticism of Lysenkoism was permitted in China only after it had erupted in Europe and the Soviet Union. When the Soviets changed their policy toward Lysenkoism, so did the Chinese. Symposium participants argued that Chinese science had become a "colonial science" in submitting itself to Soviet guidance. Was Qingdao anything more than another, albeit salutary, imitation of Soviet policy?

Despite these criticisms, there is no question about Qingdao's immediate and long-range consequences. Lysenkoism steadily declined, and its influence was more circumscribed. Chinese Michurinists (as they now preferred to call themselves) increasingly distanced themselves from Lysenko, whose tactics they repudiated and whose scientific errors they acknowledged. Into the 1960s, leadership of the Michurinists continued in the hands of Luo Tianyu's closest associates: Zu Deming, Liang Zhenglan, and Hu Han (the only one of this group to receive education in the Soviet Union). The headquarters of Michurinism were divided between the CAS Institute of Genetics, established in 1959, and some Beijing-area agriculture schools. An isolated pocket of die-hard Lysenkoism persisted at Wuhan University. In the context of the extensive, Soviet-inspired S&T reorganization of the latter 1950s, these bases ultimately acted as constraints upon Lysenkoism, limiting its influence to a corner of agricultural science.

The CAS Genetics Institute was one product of the new S&T reorganization as well as a result of the bioscience community's formal recognition of the science of genetics. Technically, the Institute was the product of consolidating a number of small, fragmented CAS units which purportedly conducted genetics research. All these units, however, were controlled by Lysenkoists. Qingdao Symposium participants surely would not have recommended the new CAS institute had they known it would conduct Lysenkoist research.

Dominance of the CAS Genetics Institute by Zu Deming and his Lysenkoist colleagues demonstrated the influence and support they still had within the Party and CAS. Other organizational changes, however, worked against them. For example, from 1956 there evolved at the highest level of government and Party a series of increasingly more powerful and comprehensive S&T planning organs which attempted to make S&T organizations more efficient in implementing policy. To this end, an Academy of Agricultural Science (AAS) was established in 1957 under the Ministry of Agriculture, which oversaw the bulk of agricultural research. Its main organ for conducting genetics research was under the direction of non-Lysenkoist scientists, trained by Cornell University,[22] who hastened the rehabilitation of plant genetics research.

At the highest level, such changes gradually transferred policy formation from CAS to a powerful organ, the State S&T Commission, which was meant to transcend all other S&T units. At the same time, the units themselves became more numerous, specialized, and separated into a number of administrative hierarchies. The result for Lysenkoism was its increasing restriction to the Beijing area and the CAS

Genetics Institute. Within CAS, the Lysenkoists became an anomaly, since the research they conducted was not genetics but green-thumb exercises in plant selection and breeding, which should have been conducted, if anywhere, under the auspices of the new AAS. With no knowledge of genotypes, the Genetics Institute's research was still trying to achieve earlier ripening and greater yields and hardiness using standard Lysenkoist techniques based on vernalization, photoperiodism, or wide crosses. The only new departure was the introduction of anther cultures and haploid breeding techniques, which had the potential for satisfying the transformists' impatience to create new plant varieties.[23]

Between 1957 and 1961, leaders of the "Morgan" school, like C. C. Tan, proceeded cautiously and diplomatically to bring genetics education and research up to date without provoking the Michurinists, who still had a considerable following. As long as the Russians were still in China, serious attempts were made to impede the Morganists' progress. Tan and others reinstated basic genetics and evolution courses, wrote a series of textbooks for audiences of various ages and sophistication, and translated fresh editions of Mendel's and Morgan's works, and—with a special sense of irony—the works of Lysenko's rival, Dubinin. Moreover, their writings were published extensively in CAS journals as well as in the journals of national and local agricultural-research institutes. By 1960, they were also able to publish a number of works for general audiences on the subject of molecular biology and the genetic code.[24]

Although the field of genetics had undergone a transformation in the years since Chinese geneticists had last been able to do research, plant genetics and breeding research quickly recuperated and showed significant progress. Hybrid-maize research was a case in point. In the early 1950s, the Chinese had followed Lysenko's proscription on the use of inbreeding techniques which were proving enormously successful in American maize production; they lifted it only in 1956, after the Qingdao Symposium. Nevertheless, some of China's Cornell-trained geneticists secretly performed inbreeding experiments during the early 1950s, and, by 1957, performed them openly, extensively, and successfully. Contrary to Lysenkoist assertions, their success demonstrated the power of modern genetics and its close link to production.[25]

Although the Great Leap and the Anti-Rightist Campaign of the late 1950s did not directly affect bioscience personnel or institutions, the Michurinists, taking advantage of these campaigns, tried to intimidate the Morganists back into passivity and silence.[26] The potential chilling effect of their efforts was dissipated during the post-Leap economic collapse and the withdrawal from China of Soviet advisors in 1960. All areas of genetics education and research enjoyed a spurt of development from 1961 to 1966.

During the early 1960s, the Hundred Schools policy reemerged. Various government and professional agencies sponsored national conferences on all aspects of science and technology, with the general purpose of developing plans to facilitate research and education. The Qingdao position on genetics was reconfirmed at the Guangzhou S&T Conference in the spring of 1962 and at subsequent professional meetings. This was paralleled by a substantial public-education campaign on modern genetics.

At Fudan University, C. C. Tan established a Genetics Institute that was to become the counterforce to the CAS Genetics Institute. Its small but brilliant and energetic staff assumed a leading role in reestablishing in China a professional network of scientists working in various areas of genetics, particularly in radiation, microbiological, and medical genetics. Because Lysenkoist misrepresentation had anathematized human genetics as a form of "Facist eugenics," special care and imagination were required to introduce this subject.[27] Genetics research at Fudan and elsewhere in China developed slowly, focusing on limited, manageable projects that reflected current techniques and questions being explored by geneticists abroad. The CAS Institute of Biochemistry, for example, kept abreast of and became involved in DNA research. C. C. Tan led a radiation-genetics group at Fudan which was linked to Dubinin's distinguished radiation studies in the Soviet Union.[28]

Genetics research and education were facilitated by an extraordinary effort to create an accurate public awareness of modern genetics in the hope that Michurinism would be permanently eliminated as an inhibition to the further development of genetics. Where Qingdao had been aimed rather exclusively at the professional community, this new effort was carried out in the national media. Between the autumn of 1961 and the end of 1963, scores of detailed articles about genetics filled the pages of *People's Daily*, *Guangming ribao*, and *Wenhui bao*. They were written by senior geneticists from all over

China, but the Fudan group was especially prominent. Unlike Qingdao, the tone of these articles was aggressive and uncompromising. They stated conclusions and made judgments that openly contradicted the Michurinists: Modern research showed that genes were real; there was no evidence— acceptable by international standards— that Michurin biology had ever improved a plant or increased a yield; but there was abundant evidence that Morganists genetics had played an essential and productive role in such applied areas as agriculture and medicine.[29]

The Michurinists' response was weak. When they bothered to respond, it was to defend some esoteric point or concede that recent advances in genetics had laid to rest their doubts about the basic Mendel-Morganist positions. By 1966, the "genetics question" had all but disappeared from Chinese science publications. New biology textbooks, even those concentrating on genetics or evolution, mentioned Lysenkoism only in passing. Sometimes it was not mentioned at all.

The outcome of the "genetics question" had no effect on the Cultural Revolution. Geneticists suffered as much and in the same fashion as all intellectuals. C. C. Tan labored on a farm; Tong Dizhou was beaten, paraded with a dunce cap, and put under house arrest. Teaching, research, and publication virtually stopped. Though CR zealots often sounded very much like Lysenko himself, even Chinese Michurinists were not safe from attack: The CAS Genetics Institute was criticized and forced to undergo reform because of its alleged failure to link research and production, theory and practice. C. C. Tan observes sardonically that the only good thing to come out of the CR was the fact that the Michurinists finally got a taste of their own medicine.[30]

THE POST-CR RECOVERY OF GENETICS

After the height of the CR in the late 1960s, there was an effort, led by Zhou Enlai, to resuscitate the intellectual community, particularly the scientists so crucial to economic modernization. In 1972, as the bioscience community began to revive, the CAS Genetics Institute was the only significant institutional base for latter-day Michurinism, but even it included important non-Lysenkoist personnel. That year, the government sponsored a national biology symposium, which addressed the need for basic research, especially in areas such as biophysics and molecular biology. While these were new areas for the CAS Genetics Institute, they had already been introduced into a number of CAS institutes as well as at Fudan. In 1972–1973, science journals started reappearing, and Tong Dizhou's CAS Experimental Biology Laboratory published the first reports on new genetics research.[31]

Potentially, the most important genetics publication in the country was the journal of the CAS Genetics Institute. When it last appeared in 1966, it was the exclusive organ of the Michurinist-controlled Institute. In 1974, when it reappeared, it was an ecumenical forum with an editorial board representative of the entire genetics community. Its early issues were devoted mostly to reports on plant-improvement projects conducted in previous years, especially those employing anther cultures for haploid breeding. More remarkable, however, were state-of-the-art essays which went out of their way to cite current international literature as evidence for the validity of gene theory and to refute theories like the inheritance of acquired characters. Even bland experiment reports made efforts to say a few words in passing about the unreliability of Lysenkoist techniques and data.[32]

It was not until the autumn of 1978, however, with the return of Deng Xiaoping to full power and after the March 1978 National Science Congress, that the genetics community as a whole had an opportunity to achieve professional solidarity. A National Genetics Society was established with China's most senior geneticist, Li Ruqi (b. 1896), as its first president. Li had been a student of T. H. Morgan and was the mentor of C. C. Tan. The Society publishes a national journal, *Yichuan* (Hereditas), sponsors annual conferences, and participates in international scientific exchanges.[33]

Although the CAS Genetics Institute in the post-Mao era continues to be directed by Russian-trained plant-breeders, among whom are still a few unrepentant Lysenkoist associates of Luo Tianyu, its greatly expanded staff includes many younger scientists trained in the West since 1978 or trained in Chinese schools like Fudan. The scope of the Institute's research has broadened into many areas of contemporary genetics. Like most CAS institutes, it has developed relations with the international science community, especially in the United States.

The most significant developments in genetics, however, occur outside the CAS Genetics Institute. Resources have been shifted to new centers for genetics and "biotechnology," as seen in the successful efforts of Tong Dizhou and the Chinese-American geneticist M. C. Niu of Temple University in obtaining funds from the United Nations and the Rockefeller Foundation for genetics research relevant to "population studies." The result is a new CAS unit, the Developmental Biology Institute, which opened in Beijing in October 1983.[34]

In response to long-range science plans, CAS also established a Genetics Engineering Laboratory in Beijing, adjacent to the CAS Genetics Institute. Fully equipped and operative since 1984, this unit is administered by younger members of the Genetics Institute who were trained in the United States. Their policies, budget, and staff are determined by a special CAS committee, independent of the institute.[35]

More significant are developments in Shanghai, which is expected to be the chief national center for the study of genetics. At Fudan University, C. C. Tan's group has been awarded long-range funding for the construction of a major biotechnology institute outside the CAS structure. By the end of the 1980s, Fudan will have a unit that comprises the full range of teaching, research, application, production, and marketing in biotechnology. On the other side of Shanghai, Wang Zhiya and Luo Deng are directing the first-stage developments of the massive, comprehensive Shanghai Center of Biotechnology. This is a CAS unit which is drawing its key personnel from various Shanghai CAS institutes, such as the outstanding Biochemistry Institute.[36]

All these organizations are benefiting from and, in the case of those based in Shanghai, owe their existence to new, centralized agencies for science policy and funding. The so-called National Science Foundation (or Fund, NSF) for example, allocates considerable research funds, using a system of competitive grant proposals and peer review.

Largely because of the funding policies of agencies like the NSF, successful efforts are being made in genetics to overcome some of the major problems inherited from Soviet-style science organization. While Soviet-style institutes are primarily research centers, all genetics research units also engage in teaching. The CAS units offer MS degrees and some will soon offer the PhD. Cooperative teaching and research projects penetrate former barriers, whether between CAS institutes, or the CAS and the universities, or Beijing and Shanghai.

If these trends continue, it will be to a great extent because the NSF and other science-policy agencies have continued to be successful at linking science research and education to "market forces." For example, by 1986, all the aforementioned genetics research units were able to contract for the training of personnel from any organization in China; and all could do contract research for any organization inside China or abroad. The CAS units can also lease their facilities to qualified scientists or agencies who lack their own research wherewithal. Future operating and development budgets for all these units are supposed to depend on their ability to sell these contractual services.[37]

Competitive grantsmanship and marketing strategy have enlivened the biosciences in China and stimulated a process of complex organization building that was hitherto unthinkable. Some scientists and policymakers are concerned, however, that these salutary developments will be paid for by sacrificing basic research to applied research, because many officials without science expertise are still involved in making science policy.[38] Even if most bioscientists share this concern, it is unlikely they will express it publicly, as long as they benefit so handsomely from current policy, which makes substantial funding available for building biotechnology institutions and training scientists for them.

Paralleling these institutional developments, careful efforts have been made to document the Lysenkoist experience in China and to explain its negative impact on China's scientific development. While many scientists and scientist-historians have contributed to these efforts, one group of scholars engaged in science policy and the history, philosophy, and sociology of science have taken the initiative. Some of them are based in the CAS History of Science Institute. Characteristically, all these scholars have some training and experience in natural science.

Their most important forum is the *Dialectics of Nature* journal, begun in 1979.[39] Its first issue published an accurate, detailed chronicle of Lysenko's career in the Soviet Union, detailing its destructive effects on the science community there. This was the first time such a statement had ever been made for the general Chinese public. Subsequent issues published critical pieces on various areas of genetics harmed by Lysenkoism. The journal has also published documents relevant to the Qingdao Symposium and has presented the first historical surveys of the "genetics question" in China.[40]

On the 30th anniversary of the Qingdao Symposium, in December 1985, a new edition of the Symposium proceedings was published. Intended for general circulation, this well-annotated edition is the most detailed statement yet on the role of the Hundred Schools policy in resolving the "genetics question." This kind of scholarship is "using the past to instruct the present" and dramatically supports current science modernization.[41]

Between April and July 1986, for the 30th anniversary of the Hundred Schools policy, dozens of national newspaper and periodical articles cited the experience with Lysenkoism and the ensuing Qingdao Symposium as major factors in the formulation of the policy. They argued that natural-science education and research can flourish and be productive only if free of government and Party interference.[42] An article in *People's Daily* complains that government and Party are still on the backs of scientists and must be removed if the scientific community is to progress.[43]

On 29 May 1986, at the conclusion of a national symposium commemorating the original Hundred Schools policy, an unexpected note was sounded: Xu Liangying, a veteran philosopher of science, ridiculed the symposium for celebrating a phantom. Dramatically citing facts from a pile of old documents he had brought with him, he argued at some length that the Hundred Schools policy had never come close to fulfillment in the past because it had repeatedly been betrayed by China's highest leadership; and it was far from fulfillment in 1986. When he finished he received a standing ovation. Later, some of the symposium speeches were recorded in the press, but his was not among them.[44]

PART TWO

The Reorganization of Science and Technology

CHAPTER THREE

Reform of China's Science and Technology Organizational System

TONY SAICH

The current Chinese leadership sees S&T as providing the key to China's modernization. The failure of successive movements to provide a policy program capable of launching China on a path of self-sustaining growth has caused Deng Xiaoping and his supporters to search for new sources to stimulate economic growth. They have identified technological change as the main mechanism for improving economic performance and maintaining economic growth. They argue that long-term increases in productivity depend on the capacity of China's R&D establishment to provide the innovation to make the industrial and agricultural sectors realize more of their potential.

The importance China attaches to S&T has brought it into line with the rest of the Communist world, where it is recognized that increased supplies of land, labor, and capital are simply not enough to expand production. These factors must be accompanied by an increase in technical know-how. This has resulted in greater emphasis on the central role of technological advance in policymaking.[1] In March 1978, Deng Xiaoping confirmed that S&T were productive forces, giving them pride of place in the Communist lexicon.[2] China's leaders found, however, that the S&T system they had inherited was incapable of meeting the new demands placed upon it.

PROBLEMS IN THE S&T ORGANIZATIONAL SYSTEM

Following the arrest of the "Gang of Four" in October 1976, important measures were taken to redirect S&T policy. In particular, a more positive attitude was adopted toward the role imported technology could play; scientists were encouraged to air their views in the press and through conferences;[3] and, most important, top-level political endorsement was given to the importance of developing the S&T system. China's leaders, while paying lip service to the need to be both politically aware (red) and technically competent (expert), made it quite clear that the latter was the more important.

In the organizational sphere, China's leaders looked back to the "golden era" of the mid-1950s. As a result, they set about abolishing the innovations of the Cultural Revolution and began to revive the structures that had, in their view, served China well in the past. Thus, the S&T commission system was revived, the Chinese Academy of Sciences (CAS) was restored to its former strength,[4] and the Chinese Association for Science and Technology (CAST) was revitalized. This organizational structure was based on the Soviet system and shared many of the same problems.

Very quickly China's leadership became aware that not all the current difficulties could be blamed on the CR but that deep-rooted structural problems in the Soviet system itself were holding back China's progress. The party leadership unveiled a new development strategy at the 3rd Plenum of the 11th Central Committee (December 1978), further energizing the need for reform of the organizational system. The key task for all future work was to be economic modernization, not class struggle. Deng and his supporters decided than an increased role should be given the market in allocating resources and greater freedom given producers in deciding what to produce. Yet, the increased use of economic levers, rather than administrative mechanisms, to regulate and stimulate economic activity was frustrated by China's over-centralized and segmented organizational structure. This meant that new measures had to be introduced to make the system more flexible, including, where possible, decentralizing powers of decision making. More links were to be developed to join vertical structures. However, to make sure that such decentralization and increased use of market forces did not lead to dislocation, the state apparatus was strengthened to improve policymaking and coordination at the macro level.

Reformers highlighted a number of problems in the S&T sector that required solving. The centralized system had shown its value in mobilizing scarce resources to focus on designated problems, but its inability to provide consistent linkage between research and production sectors was conspicuous. Centralization had not solved the coordination of work within the research sector. The organization of the sector into vertically distinct units created strong barriers to cooperation and caused wasteful duplication of work. Poor coordination between the military and civilian sectors has been particularly noticeable. This derived from the fact that military and civilian S&T work came under different commissions. The Party identified the inability to turn research results into production as the key problem in the post-Mao era. The then Premier, Zhao Ziyang, attributed this difficulty precisely to the vertical structure of the S&T system, with research institutes being responsible to higher authorities in their own command structures and not developing horizontal links with society and individual production units.[5]

Not surprisingly, the production units and the research institutes blame each other for their inability to bridge the divide. Production units had little incentive to adopt new technology before the reform program started in 1979. In some cases, innovation could even be detrimental. The lack of production demand meant that there was very little incentive for China's scientific research institutes to focus on projects that could be swiftly applied to the production process.[6]

Lack of appropriate incentives and knowledge of the market meant that, in many instances, products were developed without sufficient guarantees that they were actually needed. Such problems were compounded by inadequate knowledge about the production system. Thus, industrial departments complained that research institutes did not consider product competitiveness in the market when planning research on new products.[7] Such complaints about the separation of the research and development sectors are not unique to China. In China, however, industrial organization tended, on the whole, to inhibit the development of new technologies. There was not much finance available for such development within the enterprises themselves and very little incentive for them to seek out new technology from the research institutes.

Even where relevant, there was little incentive for researchers and their institutions to popularize the results of their research.[8] There was no system for a fair, paid transfer of research results, which were treated as common property that could be used and adopted without any form of recompense. The free use of invention had a disincentive effect on the conduct of research and led to the hoarding of information for fear that research results might be "pirated" by other institutions. As one writer noted, "Our manure shall not fertilize other people's land." Thus, even when an institute or a researcher could not make use of the results, they were not passed on to those who could.[9]

Not surprisingly, this attitude has led to the duplication of research. Insufficient funds and channels for the dissemination of research results have exacerbated the problem. One significant change in the post-Mao era is the mushrooming of scientific journals and conferences which provide an extensive system of communication and dissemination of information to the scientific community. This will lead to less duplication of effort. While duplication may decline, the question of research application to industry is still far from being resolved, as a survey of 1984 indicated.[10] The survey, covering over 3,500 scientific institutes throughout the country, disclosed that fewer than 10 percent of what the state terms "scientific achievements" had been applied to production.

One further major problem was the insufficient contribution to research made by institutions of higher education. In the 1950s, when China opted for the Soviet model of development, it adopted a system that de-emphasized university-based research and concentrated research in institutes of the various academies. Consequently, the management of education was based in the Ministry of Education and, to a lesser extent, in the trade committees of the various industrial ministries. This model reinforced China's pre-1949 system in which research was carried out in special research institutes outside the formal education sector.

Working within a Soviet-style system has given many college administrators a tunnel vision which they find difficult to expand. A member of Xiamen University noted that the leaders of institutes of higher education were not geared to the new demands placed on them. He felt that some viewed scientific research as in opposition to teaching and saw no role for research in teaching.[11]

While such problems were identified in the late-1970s, it was only in 1981 that China's Party and state leaders introduced substantive reforms for the S&T sector. The new policy called for the forging of much tighter links with the production sector and for each part of the R&D network to increase its input to economic modernization. In 1985, Zhao Ziyang called for "countless organic links between scientific research and production units" to be created "on a regular basis and in different forms."[12] Clearly, more attention was to be paid to application and development. Basic research was not to be stopped altogether but was to be made more relevant to China's needs. One of the most notable casualties of this reorientation was the high-energy physics program.

THE STATE S&T ORGANIZATIONAL SYSTEM

The increased use of market mechanisms does not mean that China's Party and state leaders want to relinquish central control over S&T policymaking. They are trying to find a better balance between government control and the role of market forces in the research sector. The Party, albeit through its control of state organizations, will still decide on research priorities. It will not allow the "whims" of the market to deflect it from pursuing what it sees as key research areas.

In all policy areas, along with more flexible policies and decentralization of some decision making, there has been a strengthening of the state administrative system. In the S&T sector, this is clearly shown by the restoration of the SSTC and especially by the formation of the S&T Leading Group (STLG).

The decision to create the STLG in January 1983 testified to the increased importance accorded S&T within China's development strategy and the need to coordinate more effectively work within the

The Organization of China's S&T System (Simplified)—1986

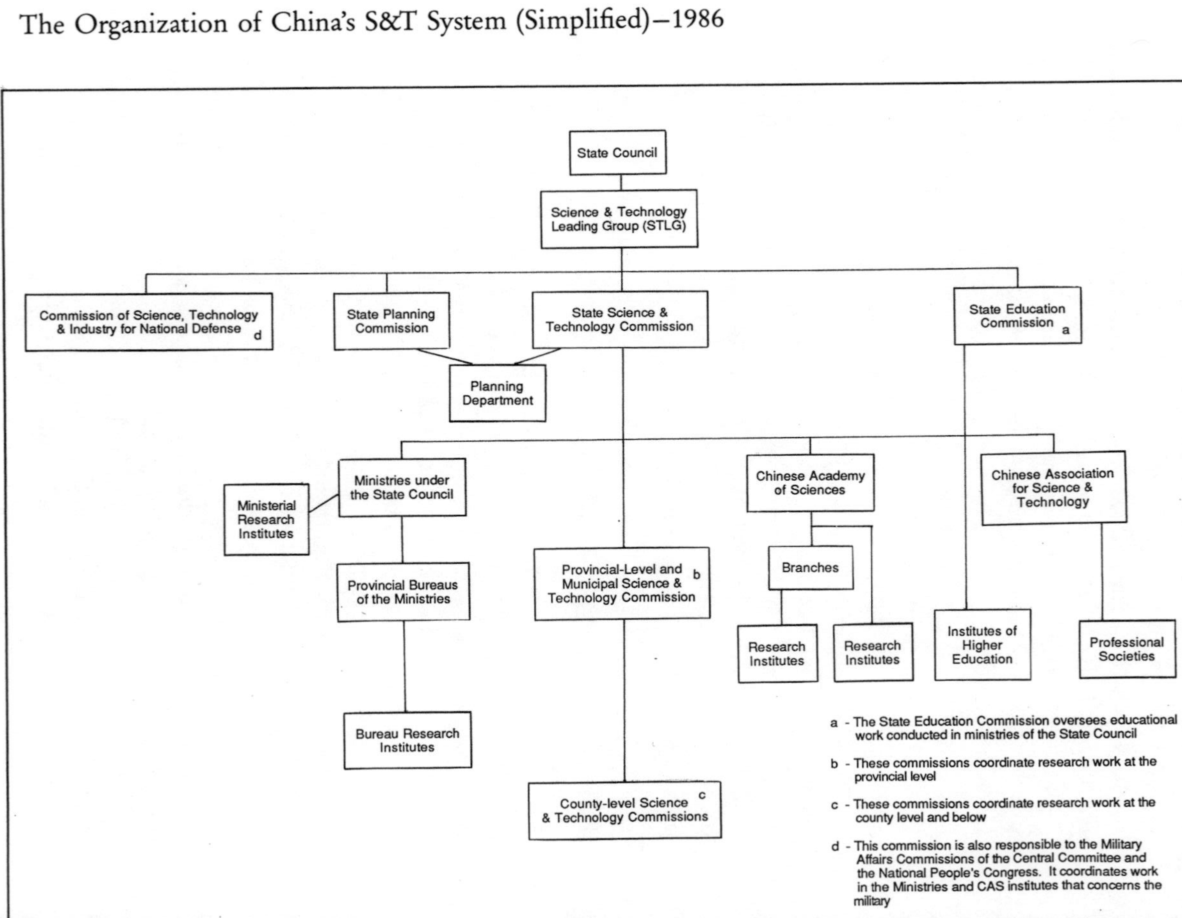

sector. It signaled that the SSTC by itself was not sufficiently powerful to break down the barriers between the different vertical hierarchies that make up the S&T system. Because, administratively, the SSTC was on the same level as the Planning and Economic Commissions, it was unable to coordinate properly work in the S&T and economic sectors.

Under these circumstances, China's leaders decided that a body with more political muscle was necessary within the governmental framework. The STLG is expected to provide this muscle. It comes directly under the authority of the State Council, the highest organ of the state administration. The STLG was headed by Premier Zhao Ziyang, a clear signal of the political priority given the modernization of the S&T sector. Members of the STLG are drawn from all the key organizations involved in work relevant to S&T development including the SSTC, the State Planning and the State Economic Commissions, The Commission of Science, Technology, and Industry for National Defense, and the Education Commission, a composition designed to break down the compartmentalization of the system by bringing together key figures in one body.

The STLG is responsible for general long-range S&T planning and the targeting of key S&T areas for national development. Clearly, it is expected to act as a broker within the S&T system, cutting through the vested interests of the different sectors to ensure that national policy goals are met. The establishment of the STLG has led to a decline in the preeminent role of the SSTC in policy formulation. However, the SSTC remains the principal body for gathering information on the S&T system, which gives it considerable influence still over policy formulation. Moreover, it remains the most important organization for coordinating policy once formulated.

One of the STLG's major tasks is to ensure that policy for the civilian S&T sector is properly integrated with that of the military sector; the need for better integration between these two sectors was a major reason for the STLG's establishment.[13] Although the military sector is generally thought to be more efficient, there were problems with coordinating R&D within the sector. As a result, in February 1985, the Commission of Science, Technology, and Industry for National Defense was set up through the merger of the NDIO and the NDSTC.

THE RESEARCH AND DEVELOPMENT SYSTEM

It is in the R&D sector that market influences are expected to play a growing role in determining the kind of research to be undertaken. To facilitate this, research institutes have been given more autonomy in deciding the kind of research to be pursued so that they can respond to the expected "pull" of the market.

China's specialized research institutes are divided across five sectors: CAS; institutes of higher education; institutes and departments under the State Council; institutes under local government; and national-defense sector institutes and departments. In theory, a division of labor exists, but, in practice, the distinctions are blurred. Both basic research and applied research are mainly conducted in large-sized comprehensive R&D institutes such as those under CAS. Seventy-five percent of basic research funding, excluding the military sector, is spent within CAS. Some 50 percent of developmental research funding is invested in the institutes under the jurisdiction of the various ministries. R&D institutes under provincial-level jurisdiction are the most important for carrying out such activities as engineering and design, dissemination and service, and production activities.[14] Now that CAS institutes and university departments are encouraged to undertake contract research projects, their boundaries will become less distinct.

CAS was originally styled after the academy system in the Soviet Union, but post-Mao reforms have made it necessary for it to reassess its role.[15] As one Chinese writer has pointed out, it had to change with the shift away from a "completely planned economy under central control."[16] The institutes under CAS no longer simply wait for research projects assigned by the government but must seek out contracts with local government and business. In the future, they will be expected to win funding for central-government research projects on the basis of competitive bidding.

Although CAS is China's premier center for basic research in the natural sciences, such research is the least important for CAS in terms of both funding and personnel. Only 10 percent of personnel are engaged in basic research; 30–40 percent do key, long-term application projects and large-scale projects, and another 30–40 percent engage in developmental research.[17] At present, CAS is trying to reduce its developmental research. With enterprises being encouraged to develop their own R&D institutes, demands on CAS for this kind of research are expected to decrease.

The Party clearly sees a change in the role CAS will play in the future. In its decision on the reform of the S&T system, it suggested that institutes mainly engaged in developmental research which have already developed extensive contacts with production ministries, enterprises, and so on be moved out of the CAS sector into another administrative system.[18] This shedding of institutes is resisted by the CAS leadership. While they acknowledge it as a policy about which they have no choice, they deflect the policy's thrust, noting that it has not yet occurred and interpreting it only as a very long-term development.[19] The Party's intention is to use the "magic of the market" to redistribute institutes more rationally. The CR policy of reallocating CAS institutes by decree to ministries or local government is criticized for not having made research more relevant to production needs, since there was no economic spur toward linkage.

The nearly 5,000 research institutes under the jurisdiction of the ministries and the provinces are expected to carry out applied and developmental research.[20] Research institutes within these two sectors have the most to gain from the new policies to promote contract research, provided that they have staffs that are sufficiently competent and research facilities that are adequate.[21]

China's top leadership has realized that, far from being an advantage, the separation of research from teaching can have a detrimental effect on promoting and disseminating the latest research results. However, the low level of educational funding leaves little money over to finance the purchase of equipment necessary for advanced scientific research. China spends only 3.1 percent of its gross national product on education. The median for other developing countries is 4–5 percent.[22] Though China's attempts to develop education on the cheap have been remarkably successful in creating a basic framework, they have hindered the development of the more advanced research sector.

Nevertheless, three measures have been introduced in an attempt to promote scientific research within the higher-education sector. First, the creation of the State Education Commission (May 1985) is expected to lead to better coordination and promotion of research. Previously, such guidance as did exist was fragmented. In the former MOE, an S&T department was supposed to take care of the planning of scientific research and its financing, but other ministries also had an education department that handled S&T work in institutions within their jurisdiction. A measure of the fragmentation is that, just before its abolition, the MOE directly administered only 36 of the institutions of higher education; 250 were attached to various ministries and commissions, with the remainder administered by provincial-level authorities.[23]

Second, available funding for research is going to be directed to selected institutes to ensure better returns on investment. Thus, in November 1981, when the World Bank loaned China US $200 million, it was to be used for purchasing instruments and computers and for training at 26 "key universities."[24] In early 1985, MOE selected 5 key universities as sites for government-funded laboratories, based on their track records and the quality of their staffs.[25]

Finally, to ensure that more research for production is undertaken, institutes of higher education are being encouraged to forge links with local communities and the production sector and are being given greater autonomy to facilitate such contracts. Institutions with a technical focus will be able to take greater advantage of these opportunities, but even comprehensive universities have been able to diversify their work to take advantage of the new reforms.

CREATING HORIZONTAL LINKAGE: THE USE OF THE CONTRACT SYSTEM

The major organizational problem within the S&T system has been lack of linkage across vertical structures, particularly between the research and production sectors. Insufficient innovation takes place in industry to bring about the necessary technological advance, and it is difficult to transfer to the production sector the innovation that takes place in the research laboratories. The relatively autonomous and isolated situation of the research sector is a problem in most countries, but it is particularly acute in developing countries. Research must be properly integrated with the broader socioeconomic structure for optimal use to be made of scientific discovery and technical development.

One way to foster this integration is to give the market a greater role in deciding what kind of research is necessary. Now China's Party leadership defines science, technology, and knowledge as commodities that have a price and can be exchanged in the market-place. Initial reforms to encourage the transfer of technology followed those in the Soviet Union and Eastern Europe. Faced with the same problems of the lack of integration between research institutes and the production sector, in 1961 the Soviet Union introduced regulations instructing research institutes to derive part of their funds from economic agreements signed with enterprises.[26] Similar reforms were introduced somewhat later in the East European nations, and this system became common in Hungary after the economic reforms implemented in 1969. For example, the Budapest Steel Research Institute derived 80 percent of its annual research expenses from contracts concluded with factories and only 20 percent from government organizations. When introducing the contract system, the Chinese leadership made a careful study of such reforms.[27]

In China, experimentation with this reform began in the late 1970's, but it became national policy only in mid-1984 when the State Council announced that it had approved a remunerative contract system for all China's developmental research institutes to be implemented within three to five years.[28] To compel institutes to switch over to this new system, the state authorities are gradually cutting allocations. To persuade research institutes of its efficacy, the Chinese press is filled with examples of success stories under the new system.

The contract system not only opens the way for diverse funding channels but helps overcome the uneven distribution of research institutes within China. With the reduction of the organizational barriers, it is expected that technology will flow not only into the production sector and across the different organizational systems but also across the different regional systems. Thus, paid contracts, consultancy, and also increased mobility of personnel[29] are seen as mechanisms to transfer research results and expertise from advanced centers to the "more backward regions." The previous method for the dissemination of new technology relied on enterprises' emulating the experiences of "advanced units" or models. While this method highlights a particular experience, it is a very hit-or-miss approach to ensuring that individual enterprises acquire the appropriate technology for their particular needs.

Introduction of the contract system and the use of market mechanisms, apart from changing the focus of existing research institutions, have produced a variety of new organizational forms. China's top leaders have given their seal of approval to bodies that integrate scientific research and production. Sometimes they may even include a teaching organization. Zhao Ziyang supported the formation of these research and production alliances as early as May 1982,[30] and the Party adopted the policy that they should evolve from developmental research institutes "on a voluntary and mutually beneficial basis" in its decision on reform of the S&T system.[31] To stimulate their development, such organizations are permitted to set aside a certain percentage of the "newly increased profits" as technology-development funds.

The growth in the number of these integrated alliances creates the potential to deal with research and production in a coordinated manner. The existence within one organization of research and production processes will speed up the transfer of research results; it is hoped not only that development time for a new product will be shortened but that product quality can be improved.[32]

Zhao Ziyang also approved the evolution of developmental research institutes into technology development centers (TDC). In October 1982, when speaking at the National Science Awards Conference, he commented favorably and advised that their number be small, well-managed, and effective. A TDC differs from the research and production alliances in that it does not include all the steps in the research-production process.

The TDC's primary task is to concentrate on technical transformation of the industries within its own specific trade and assist medium- and small-scale enterprises in the development of new technology and products. It is also expected to help absorb the import of new technology and ensure that what is imported is relevant and can fit smoothly into China's production sector.[33]

To match those looking for technical knowledge with those who wish to sell their research results, S&T information and service centers have been set up. They operate under a variety of names such as an S&T development exchange corporation (*keji kaifa jiaoliu gongsi*) or a technology service corporation (*jishu fuwu gongsi*). To enable sellers to advertise their wares and buyers to see what is on the market, technology trade fairs have become regular events. The scope of such fairs has expanded over time. By mid-1985, the "First National Technology Products Fair" was convened in Beijing. Trade fairs are also held to transfer military industrial technology to the commercial sector and encourage the military to produce more for the civilian sector. In March 1985, the national trade fair held for this purpose carried out business transactions to the tune of 1,120 million yuan and concluded transactions on more than 3,200 projects, 44.3 percent of all projects offered at the fair.[34]

MARKET REFORMS IN AN IMPERFECT MARKET

Ultimately, the success of the S&T organizational reform depends on the reform of the urban industrial economy. While rural reform has resulted in remarkable increases in both output and productivity, industrial reform has been a stop-go affair. The success of the rural reforms has provided ammunition for those who wished to introduce more wide-ranging changes into the urban sector. In fact, as the American political scientist Tom Bernstein has pointed out, the centralized industrial system has not been able to meet properly the needs of an "increasingly commercialized, decentralized agriculture."[35]

It is too early to judge how successful the urban industrial reform will be, but the experience of other state-socialist regimes is varied and not encouraging. Tremendous political will plus good economic results are necessary to sustain the reform momentum. While a reversal of direction is highly improbable, the process to date has been erratic, with spurts of activity followed by temporary halts as results are assessed and ways are sought to deal with the problems that arise.

A major weakness in China's economic system is an irrational price structure. The current government has committed itself to revising the price system and the potentially explosive subsidy system. It is recognised that the effort will be slow, complicated, and for some possibly painful. To date, a system of state-administered pricing has been a central component of the central planning system adopted in the mid-1950s. Prices are not set by market forces but by government authorities. As a result, wholesale prices do not provide the main basis for decisions made by production units.[36] At present, the PRC is in a transitional phase where central control has been relaxed somewhat but where market discipline has not taken its place. The inflation that has accompanied each attempt at price reform has panicked China's leaders. Therefore, it was no surprise that, just after the 13th Party Congress (October–November 1987), it was announced that yet again price reform was being delayed because of the rising inflation in the urban areas. Instead, industrial policy was to focus on giving enterprise managers a freer hand.[37] In the absence of thorough price reform, however, the effect of greater freedom for enterprise managers will be limited.

The inconsistencies in the pricing system may mean that some developmental research institutes may not become profitable through no fault of their own. At present, the true market value of research that is being done by a particular institute may not be reflected in the price they can charge the enterprises. Indeed, in the absence of a market, how does one know what the true market price is? At the same time, it may be uneconomical for an enterprise to introduce a new technology. In May 1984, the *Guangming ribao* carried a report on an investigation of 472 factories conducted by the Shanghai Joint Section for the Development of New Products, in which it found that introducing new products generated lower profits than old products. For example, the average profit rate on new pressurized vacuum flasks was said to be only 10 percent as compared with 25 percent for older products. Moreover, the higher the output of new products, the lower the average profit.[38] Not surprisingly, many enterprises resist introducing new products and adopting new technology.

Consequently, the government has proposed certain measures to encourage innovation, such as preferential treatment in terms of loans, reduction in taxes, trial sale prices, and submerging funding for S&T achievements and technical transformation into production costs. The crippling effect of industrial and commercial taxes on new product development is illustrated by a Tianjin research institute that found itself faced with industrial and commercial taxes of 2.06 million yuan in 1982 for the new eddy-spinning technology it had developed, although, in 1981, the new technology had been exempted from tax. After mediation of officials in the Tianjin local government, the tax was reduced to 980,000 yuan, but, in 1983, a further bill of 1.2 million yuan was presented. Not surprisingly, this seriously hampered the development of the technology and finally, after numerous appeals, it was agreed that no taxes would be levied beginning in 1984.[39]

The fact that production units are now responsible for their own profits and losses does not mean that profit-making units will introduce new technology to upgrade their productive capacity. Adoption of new technology and processes disrupts production for a time, and some factory directors opt for a certain short-term gain rather than gamble on a potentially larger gain over the long term. In the earlier years of the reform program, this tendency would have been reinforced by uncertainty about how long the program might last. Despite continual reassurances by the leadership about the continuity of policy, even after Deng Xiaoping's death, China's people has seen too many abrupt changes over the last thirty years to take any chances.

Some directors who do upgrade the technological level of their enterprise may rely not on domestically produced technology but on imports from abroad, because, in many instances where technology is imported, the cost is borne in some form by the state instead of by the factory. Finally, whether an enterprise takes advantage of new technology depends on the calibre of its leadership personnel. In this respect, the policies of appointing those with technical skills to leadership positions, increasing enterprise autonomy, and encouraging the independence of the enterprise director from the enterprise's Party committee have been important.

As yet the paid contract system for developmental research institutes has not been extended fully throughout the country. In the initial phase, adoption of the system has not been mandatory. Only those institutes that have actually applied operate with the contract system. By the end of 1985, 32 percent of China's R&D institutes, excluding the military sector, had adopted the contract system. A total of 317 institutions were said to be financially self-supporting; and 640 were receiving less financial support from the state.[40] Presumably, those institutes that have already adopted the contract system are those that have most to gain from such a system. This helps explain the high percentage of success stories that have been reported. Many of the remaining developmental research institutes, however, have neither the facilities nor the necessary personnel to make a success of the contract system.

Do the authorities have the power or the political will to make those institutes that prefer the status quo adopt the new system? Will those institutes that are not financially viable be closed down? Chinese officials answer enquiries about the latter question by saying that such institutes will be amalgamated with other research institutes or industrial enterprises or that their personnel will be redeployed. It seems unlikely, however, that successful institutes or enterprises will willingly absorb unprofitable institutes that might undermine their new-found economic success. Moreover, if institutes are really to be given greater powers over the employment of personnel, it is difficult to see how extra staff can be forced upon them.

The introduction of the contract system does not necessarily mean that the barriers between research institutes will be broken down. Certainly, if there is money to be made, various research institutes may pool their knowledge and facilities for particular projects. But institutes may jealously guard their innovations in order to maximize their own gains and prevent others from profiting from their work. Cases have been reported of plagiarism, the pirating of technological developments, and the sale of unreliable technologies. One remedy is the establishment of an effective legal framework to govern the S&T system. Already, problems with the technology market have led the SSTC to strengthen its controls and codify behavior.[41] The degree of efficacy of the new Patent Law will be of great importance in regulating the process of sufficiently rewarding innovation.

THE PARTY AND THE S&T SYSTEM

As in other areas, Party dominance has been singled out by reformers as a key impediment to the proper functioning of the S&T sector. As in the Soviet Union, the heavy hand of Party control obstructs an atmosphere amicable to scientific discovery and technical innovation. The British political scientist Peter Kneen has remarked that, in the Soviet Union, the Party's centralized structure and authoritarian style "renders it incompatible with many of the aspects of the working situation it claims to supervise."[42] For the current reforms to work effectively, the Chinese Communist Party must not only create an atmosphere conducive to scientific research but sustain it consistently by decreasing its direct involvement in decision making and the administration of research institutes. As the Party now gives the highest priority to the development of S&T, it has acknowledged that it should withdraw from its over-dominant position, allowing scientists and technicians greater freedom within their areas of professional competence. However, the Party is still uncertain about how to strike a balance between freedom and autonomy for the individual scientist and institute and guidance and control from the Party.

The post-Mao reform leadership is caught in a contradictory situation. The Party professes to be the only organization capable of defining the correct strategy for the attainment of communism. Yet, it has chosen a development strategy that relies heavily on the advice of professionals whose skills are conspicuously absent within the Party. One way in which the Party seeks to resolve this contradiction is to acknowledge that the official ideology cannot explain every aspect of life, especially in the rapidly changing world of the "new technological revolution."[43] Moreover, it recognizes that questions of enquiry in the natural sciences have a universal character. S&T policy options, however, operate within the broad guidelines laid down by the Party. During the years 1979–1981, the Party allowed scientists and technicians more leeway in setting the S&T agenda, but, in 1981, the Party firmly decided that some guidance was necessary. Furthermore, in a general trend of re-centralization as the Party prepared to launch the next Five-Year Plan (1986–1990), it issued a series of key documents summing up the reforms and laying down guiding principles for the subsequent phase. The Party Central Committee decision on reform of the S&T system was published in March 1985.[44]

Nevertheless, emphasis on a more empirical approach to decision making on technical issues had led to placing greater value on consultation and discussion; attempts to draw key groups of experts into the decision-making process on a more systematic basis have resulted. As in other periods when economic modernization has been pushed to the fore, scientists and technicians have been called on to increase their contribution. Thus, they were closely involved in the drafting of the 1956 Twelve-Year Plan for Science and Technology, the 1975 plans to support Zhou Enlai's and Deng Xiaoping's modernization program, the ambitious 1978 Ten-Year Plan for Science and Technology, as well as the 1985 Party Decision on the Reform of the S&T Management System.[45]

To stimulate the scientists' enthusiasm for participation, the Party has upgraded their position ideologically and materially. Gone are the references to the "stinking ninth category"; intellectuals are now defined as an integral part of the working class. This "ideological upgrading" has been accompanied by attempts to improve their workplaces, housing, and salaries. To facilitate their contribution, intellectuals have been given greater freedom within their fields of professional competence. There has been an explosion of professional journals and convening of congresses to enable the exchange of views. Professional societies have been mushrooming as a further forum for the exchange of ideas and as channels through which the Party and state can obtain expertise. Of these societies, the most important for modernization are the 138 professional bodies under the umbrella of CAST.

The role of the Party within research institutes is to be changed. The tendency in all Chinese organizations is to push decision making on even the most trivial matters to the highest levels. The potentially adverse consequences of making an incorrect decision have far outweighed the potential gains of displaying initiative. In practice, this has meant referring all decisions to the Party committee, so that Party committees have become overloaded with minor matters and called on to make decisions for which they have not been trained. Consequently, the Party has begun to appoint more experts to run research institutes and enterprises.

In 1986, China introduced a trial system of administration within research institutes, part of a more general overhaul of enterprise administration. This new system referred to as the "institute director responsibility system" was to replace the "institute director responsibility system under the leadership of the Party committee." A clear demarcation is to be made between Party and administration. The Party will look after questions of ideological and political work; the "expert" director will take charge of management and scientific matters. Certain major questions on the direction and plan for the institute and important personnel management matters will still be considered by both the director and the Party Committee.[46]

Not surprisingly, these attempts at reform have met stubborn resistance. Within research institutes and other work units, they have led to conflict between cadres who owe their position to their political and administrative skills in the old system and those who derive their power from their detailed technical knowledge. While those with technical skills push for greater autonomy, many administrative cadres push for greater control over the enterprise in order to maintain their own positions. In discussions of the new enterprise law at the Standing Committee meeting that preceded the 6th Session of the National People's Congress (March 1987), the new form of management was a key point of disagreement. Opponents of change argued that it would undermine the role of the Party in the enterprise. As a result of disagreements the Standing Committee did not put forward the draft of the law to the National People's Congress session.

Even in situations where those with professional knowledge

achieve leadership positions, they do not necessarily have the power to implement their policies. A Party secretary can always invoke the ultimate authority of the Party to ensure getting his or her way. To help combat such resistance, the Party is trying to upgrade the educational level of its membership and recruit more intellectuals, in the hope that it can exercise more appropriate supervision over technical matters and appear more capable of managing the modernization program. The situation at CAS is ideal; the President is both a highly respected professional *and* a party member. This serves the dual function of subordinating the leading functionary of CAS to the Party and of ensuring that the Party receives professional advice in its decision making. However, this policy of recruiting intellectuals has also been resisted by some Party members who fear that, "if intellectuals are promoted to leading positions and then join the Party, they will have all the best jobs."[47]

The essentially unreformed Soviet-style S&T organization system as it existed in the late 1970s was a highly rigid, bureaucratic structure. The reforms undertaken subsequently are intended to introduce flexibility so as to make innovation possible. The decentralization of decision making and the attempts to create horizontal links in an essentially vertically structured system should enable research institutes to engage in more relevant research, which is increasingly defined by a mixture of central control and market forces, or imperfect market forces.

China's reform leaders realize that the S&T sector cannot be overadministered; at the same time, they do not wish to leave it to its own devices. They are searching for an appropriate balance between central control and the delegation and decentralization of powers. Thus, the scientists and research institutes have been granted greater autonomy over their own work than at any other period in the PRC but, as Deng Xiaoping and his supporters made clear in 1981, this autonomy exists within a general framework laid down by the central authorities. Consequently, while powers for specific areas of implementation have been decentralized, the new central bodies such as the STLG and the Leading Group for Development of Computers and Large-Scale Integrated Circuits are to devise and monitor policy implementation in key areas.

Reform of the S&T system is not, however, taking place in isolation. Whether the S&T system can deliver what is expected of it depends on whether reform of the urban industrial economy and the rural economy can provide the necessary "pull" to encourage appropriate research and the development of necessary technologies. The Party must ensure that an atmosphere conducive to scientific research is created and maintained. If the Party finds itself faced with a scientific group working outside direct Party control, it remains to be seen whether the Party will continue to grant this group autonomy, especially if the Party's monopoly of power and its theoretical monopoly of truth are undermined.

CHAPTER FOUR

Reforms and Innovations in the Utilization of China's Scientific and Engineering Talent

LEO A. ORLEANS

As we all know, one of the fundamental stumbling blocks facing China's quest for modernization has been and is the shortage of high-level scientific and engineering personnel. China's post-Mao leadership immediately recognized the obvious and initiated steps that would lead to expanded and upgraded institutions of higher education, but that is a long-term solution. In order to ease the serious manpower problem in the present, the leadership has sought to use existing professional personnel more productively. Virtually all China's top leaders have expressed views on how to stimulate the enthusiasm and the creative spirit of China's scientific and technical talent. Among his many statements on the subject, Deng Xiaoping pointed out that, "in restructuring the economic system, what is most important and what I am most concerned about is talent; in restructuring the science and technology system, what I am most concerned about is still talent." Premier Zhao Ziyang has reiterated the two often-stated primary goals of the current reforms in the scientific and technical system: "One is that it should be advantageous to the integration of research with production; the other is that it should promote the fuller utilization of scientists and technicians."[1]

To utilize existing talent it was first imperative to change the tarnished image of the intellectuals. After so many years of abuse, which culminated in the violence of the Cultural Revolution, negative attitudes about intellectuals were ingrained in the psyche of the masses, who had to be reeducated into believing that brain work was as productive as manual labor. At the same time, it was necessary to heal the deep scars of the intellectuals and to assure them that they are indispensable to China's goals of economic modernization. Not all intellectuals have profited to the same degree from the new policies. Since the post-Mao reforms focused primarily on scientists and engineers, considerable (albeit uneven) progress has been made in improving their working conditions, status, and privileges. As for the humanist intellectuals, policies toward them continue to fluctuate even in the 1980s and, according to Merle Goldman, writers and ideological theorists constantly have to remind the Party that "the way they are treated affects the scientific intellectuals" and the ultimate goals of modernization.[2]

Yet, even for those with scientific and technical talent, stories of mismanagement have become legion. Efforts to improve their lot and thereby to "liberate their productive forces" start at the top of the governmental pyramid, which includes not only the State Education Commission and the Ministry of Labor and Personnel but also the State Planning Commission, the State Economic Commission, and the State Science and Technology Commission (SSTC). These government agencies have been experimenting with reforms relating to more rational assignment of college graduates, more mobility of professional manpower, and greater incentives to maximize a professional's effectiveness.

JOB ASSIGNMENTS FOR COLLEGE GRADUATES

Ever since 1949, Chinese planners and educators have been struggling with the problem of matching college graduates with national needs, but with little success. In the 1950s, Chinese authorities complained that "mathematics students were assigned to teach Russian in factories, physics students to serve as proofreaders in printing establishments, electrical engineering students to install electric light bulbs;"[3] in the 1980s, they are still complaining that history graduates are driving trucks, graduates of cryogenic technology are working in hotels, and graduates of applied physics are working in warehouses.[4] The reason there has been no visible progress in this matter is systemic. The Soviet Union has spent many more years than China trying to improve the state assignment of college graduates, but with little more success. Even with the best of intentions (which cannot be taken for granted), and even if all the necessary data were available (and they are not), it is impossible for the central-government planning institutions to forecast the enormous and diverse needs of every economic sector for specialized professional manpower and then to achieve a precise match between those needs and the current crop of university and college graduates, who have already been in the "pipeline" for three, four, or more years. The process is further complicated by the many hierarchical levels through which an assignment must pass. For example, graduates of universities under the State Education Commission are assigned to production ministries, which then appoint them to their own bureaus, which place them in companies, which finally assign them to a specific job in a factory or some other enterprise. Given these procedures, it is easy to see how graduates find themselves in jobs unsuited to their training.

A system that forces enterprises and institutions to vie with one another for the annual graduates of institutions of higher education creates its own irrationalities. At each administrative level, employers strive to hire as many as possible of the best and most qualified graduates, but, in an environment of shortage, they hold on to anyone who is assigned. To paraphrase one factory manager's remarks on this subject: "When we need 25 engineers, we request 50 in the hope of getting 15." Everyone is aware of this ploy, and it tends to be perpetuated. An article in the *Jiefang ribao* (Liberation daily) blamed many of these problems on the "iron-rice-bowl" system: "Units which demand specialized personnel needn't pay a penny for college graduates because of the distribution system. They pay no attention to what kind of talent they actually need and talent often goes to waste."[5]

The problems of misassignment and misuse of graduates are accentuated by two traditional prejudices common to many of China's old-line managers. The first is a bias against youth, who must go through the apprenticeship process no matter what their qualifications. One unit that was assigned about 300 university graduates in the past few years "put more than half of the graduates to such work as dealing with mail, copying documents, buying office materials, and cleaning equipment. Those who complained of under-use of abilities were criticized for being arrogant."[6] The same source tells of a young scientist who returned from abroad with two masters degrees and was appointed to head a research project, but "some of his leaders felt he was too young to be qualified for the leadership position."

The other bias relates to the "longstanding and traditional prejudices and the force of old habits" toward women graduates. In Henan province "over the past two years, about 70 to 80 percent of the units in need of personnel have selected only male graduates, giving various pretexts for not selecting females. Some even openly refuse to recruit female graduates."[7] Females also suffer inequities in the distribution of housing and land. Even though employment prejudices against women are not as prevalent in the large municipalities, such as Beijing and Shanghai, women everywhere anticipate impending employment problems and tend to select traditionally female professions, such as teaching and medical care.

Experiments and Reforms

As the graduates of revitalized institutions of higher education started to enter the labor force in the early 1980s, Chinese leaders realized that the country could no longer afford the extravagance of irrelevant job assignments. But, although they called for reforms in the planned assignment system existent since 1949, reforms were slow in coming. The main reason for delay was expressed at the March 1982 national graduation-assignment conference, which concluded that it would be too risky to implement changes rapidly, as long as demand for graduates continued to be much higher than supply. In 1983, for example, there were 570,000 requests for graduates, while the graduating class numbered only about 280,000; in 1986, the requests were 700,000 and the graduates 321,000, or a decrease in supply from 49 to 46 percent.[8] A freer flow of graduates would not only effectively negate any guarantee that priority projects and activities would receive the necessary personnel, but would also increase the likelihood that, in their efforts to obtain the necessary professional manpower, enterprises would subvert the existing regulations by resorting to "back-door" tactics.

Nevertheless, some initial and tentative steps were recommended and approved by the State Council in 1983.[9] First, an effort would be made to develop assignment plans a year in advance, "so that college and employing work units will have more time to contact each other and make job assignments better suited to the students." Second, the assigned graduate would have a one-year probationary status in order that his competence could be judged and to make sure that the assignment was relevant to his studies. And, third, employing units would be permitted to interview and test graduates before accepting them. Although this last reform was first introduced "on a trial basis" for students completing postgraduate studies in 1984, interviews were held not between graduating students and their prospective employing units but between the college officials and the units. Moreover, it was stressed that, once assignments were made, neither the graduates nor the units would be permitted to alter or reject them.[10]

The next phase of the reforms—still experimental and modest—was announced by Vice-Minister of Education Huang Xinbai in July 1985 and applied to the 1985 graduates of only two of China's most prestigious universities, Qinghua in Beijing and Jiaotong in Shanghai.[11] This phase provided for employment of graduates based on recommendations by the universities and exams by the employers. Graduates with the best academic records would be allowed to choose their own employers. Graduates who did well in school but were not employed (or felt dissatisfied with the prospective employer) would get assistance from the university. Students with poor academic records would have to find their own jobs. Graduates who volunteered to work in the border and backward regions of the country would be allowed to choose their region and work unit and would be able to transfer elsewhere after five years (seven years, according to some sources). And, finally, if after three months a graduate still had not registered with the work unit to which he was assigned, he would have to refund all the subsidies or scholarships received in the course of his studies. Similarly, if a work unit hired a graduate outside the work-assignment plan, that unit would be required to pay the university the total cost of the individual's education.

Huang Xinbai also announced some modest changes with regard to most (but not all) of the universities and colleges under the Ministry of Education (MOE). (This has since become the State Education Commission.) In the case of 73 percent of the graduates, their institution (not the students themselves) would have more say as to the final job assignment. One would assume that the 27 percent directly assigned by the state (without the involvement of the institution) would be graduates of the less prestigious specialized colleges. As for the remaining 160,000 graduates in 1985 attending about 800 locally run universities and colleges, they were to be assigned as in the past by the relevant provincial, municipal, and other local authorities.[12]

There are now two types of students who bypass the difficult state job-assignment procedures altogether. Since the mid-1980s, some production units have contracted with colleges and paid for the training of a certain number of specialists, who on graduating are then employed by that work unit.[13] In this way an enterprise can assure itself of getting the personnel it requires and have a voice in determining the content of the curriculum offered to its future employees; the institution of higher education, for its part, can supplement its income. This arrangement is more common in the specialized two- and three-year technical colleges than in the four-year university, because paying for four years of education would be too expensive for most production units; also, from the practical perspective of the manager, a four-year curriculum is likely to include too much "superfluous" knowledge.

The other group of college graduates who fall outside state job assignment are students who pay their own college tuition and all other expenses. Upon graduation they can either seek their own jobs or ask the school to recommend them.[14] So far their numbers are relatively small, but, with continued economic growth and perhaps access to funding from overseas relatives, the proportion of self-sustaining students may increase.

The authorities were serious in their desire for reforms and their belief that, "with little freedom of choice, graduates are bound to be disappointed with their work if it has little relevance to their hopes and aspirations."[15] But the rapidly introduced changes were short-lived. Although, in 1985, only 23.6 percent of university graduates were assigned jobs by the state and 63.8 percent found work through contacts between schools and employers, it did not take long to discover what opponents of change argued all along, that the new system "did not ensure that key industries and government units, especially those in remote and backward regions, received much-needed graduates." In 1986, the figures were almost reversed; 69 percent of the graduates started work in state-assigned jobs.[16] To reinforce the assignments, a 30 June 1986 circular warned that "local public security bureaus are not allowed to provide residential permits to those graduates who refuse to work in their assigned places. And anyone who tries to use 'back-door' connections to place graduates will be punished."[17] It should be stressed that the drastic reduction in the proportion of graduates who were assigned jobs in 1985 did not necessarily mean that the graduates themselves had more say in selecting jobs. Nevertheless, the greater involvement of the institutions from which they graduated could have been a first step. If the professors and administrators are of good will, they should have some idea of the preferences and strengths of the students in question and thereby may be able to prevent the types of blatant misassignments of graduates so prevalent in the past.

Shifting policies suggest that the authorities responsible for drawing up nationwide annual plans for assigning college graduates are still undecided about how to mesh the desires of the graduates with the needs of the country. This uncertainty is intensified by the fact that the less-developed provinces cannot adequately expand and upgrade enrollment in their educational institutions; relatively small numbers of their secondary-school graduates are able to pass the college-entrance examinations with adequate scores. It is therefore up to the state to "subsidize enterprises located in remote and backward areas and assure that a certain number of college graduates work for them."[18] Furthermore, the quality of the individuals who are sent out is likely to get worse because the modest reforms already introduced and still in effect (allowing some preferences by graduates and the involvement of universities in matching graduates with jobs) will undoubtedly skim off the cream of the crop for the major municipalities and the economically advanced coastal provinces.

Another problem has also emerged which may explain the government's renewed stress on the overall assignment plan. Apparently it is not unusual for organizations to "raid" a graduating class by offering higher salaries, better housing, and other privileges to individuals already assigned elsewhere by the state. Sichuan province, for example, planned for 1,593 of the 1985 graduates to take up teaching positions, but 360 failed to report for duty, while, of those assigned to the town of Guangyuan—located in the mountains of northern Sichuan and therefore undesirable—only 58 of 103 graduates arrived to take their posts. Moreover, some of those who did report immediately informed the authorities that they wanted different jobs and/or wanted to stay in the city and not assume positions in the countryside. A circular from the Sichuan People's Government urged that "organizations stop recruiting graduates who have been assigned to other jobs and return those who already have left. But some organizations ignored the government's request."[19] Obviously such actions would have been unheard of in the Mao years.

Because the cumbersome planning process places many young people in positions that do not make use of their skills, it would be wrong to infer that the "man-bites-dog" examples favored by the Chinese press are representative of the nation as a whole. According to a "random sample" taken in 1984, only 12.3 percent of the graduates were assigned to jobs that did not fit their specialties.[20] Of course there is no way to know how large or how random the sample was, but, if it was representative, then the problem may not be quite as widespread as some of the students would have us believe. It is significant, however, that the most vocal complaints come not from the graduates of such technical institutions as the Wuhan Water Conservancy and Electric Power College or the Harbin Ship Engineering College, but from the most prestigious universities. It's almost one-third of the Nankai graduates, about 11 percent of the Lanzhou graduates, and almost 27 percent of the graduates from Jiaotong that are reported to be misplaced.[21]

Why is it that, the more elite the educational institution, the more likely complaints about job assignments? The prestigious key universities enroll students with the highest grades on the college-entrance examinations and are mainly urban-bred offspring of parents in the intellectual or government hierarchy; in the freer atmosphere of post-Mao China, these graduates with high expectations are more likely to complain about a displeasing assignment or locale. Many of these graduates would insist they were misplaced if assigned to a factory or to an enterprise or institution located in a smaller city or in the interior of the country. Most would probably prefer to go to graduate school. Barring that, preference would be for a position in an institute of one of the academies of sciences. Graduates from key universities have, in fact, complained that their talents were wasted because they were not assigned to "key employment organizations,"[22] which presumably refers to top-level state institutions or enterprises. In a different context, one of the top officials of the Chinese Youth League complained: "Many young people, especially students, tend to be arrogant and exaggerate their own abilities. They should be made to realize what shortcomings they might have and properly evaluate themselves."[23]

Another aspect of this problem was discussed in Shanghai's *Jiefang ribao*:

> Some students have had limited experience with personal relationships. . . . At work they are confronted with the complicated relationships between colleagues of different ages, experience, and background. They often find it difficult to cope. Sometimes their forthright manner offends their colleagues or leaders. Their suggestions for improvements may not be encouraged by the leaders, which means the students mistrust them and lose interest. Consequently they ask for transfers.[24]

Obviously a situation like the one described above could develop whether or not job assignments were improper, which obfuscates efforts to distinguish between graduates actually placed in unsuitable jobs and those simply dissatisfied with their assignment; this further complicates the task of improving the job-allocation system.

Although students with degrees from prestigious universities—especially advanced degrees—may have unrealistic professional expectations of what their country can do for them, there is something to be said in their defense. With the shortage of scientific, engineering, and other higher professional personnel, agencies responsible for job assignments should be especially careful not to waste skilled manpower through faulty job allocations. The new graduates may well be mindful of a line from a Qing dynasty poem: "A wise man cannot display his resourcefulness if no scope is given to his abilities."

MOBILITY

Restrictions on mobility make the first job assignment of vital importance in the careers of young graduates. Until recently, such assignments were for life. The Party leadership is acutely aware that restricted mobility of China's professional manpower—and especially of scientists and engineers—is detrimental to modernization. Lu Jiaxi, President of the Chinese Academy of Sciences, expressed this view when he said that "scientific and technological communities suffer from a lack of mobility, a condition not suited to modernization. . . .(but) conducive to mental ossification, conservatism, and bureaucratism."[25] As in other areas of post-Mao China, however, there is a great gap between the recognition of the problem and doing something about "the practice of lifelong profession being decided by one job assignment"—a longstanding traditional policy.

Any discussion of mobility has to start with the *danwei*, or the work unit to which all workers and employees are assigned. These assignments have been akin to lifelong indentures, where the individual is literally "owned" by the unit and is unable to move without the approval of those in charge. Unit ownership of employees gives rise to such abuses as hoarding and misassignment, and, the higher the qualifications of the individuals, the more prevalent and serious the abuses. Until China improves its current assignment policies, it is especially important that increased mobility alleviate some of the injustices of centralized assignment of college graduates. Since they are China's scarcest resource, scientific and technical personnel are greedily hoarded. Even if they are not immediately utilized, a few extra engineers can be likened to a bank savings account—something for a rainy day. As one cadre explained in denying a transfer to an unused specialist: "We maintain an army for a thousand days to use it for an hour. We will use you some day."[26]

Theoretically, a hardworking employee should find it easier to "float," but actually it is often the reverse. The tragedy, according to a *China Daily* commentator, is that dedicated, hardworking people—even if misplaced—find it almost impossible to transfer jobs, while the troublemakers and lazy employees are much more likely to be released by the *danwei*.[27]

To make matters stickier, if an individual has worked in a unit for some time and has attained special skills and knowledge, the unit is likely to demand up-front compensation for the investment made in "property improvement" and for the loss of services. Some units demand as much as 2,000 yuan for the release of a specialist, while others say outright that "no college graduate will be permitted to leave."[28] To circumvent these obstacles, some production units in need of scientific and technical personnel resort to undercover methods. For example, the Dalian Haiyan Bicycle Corporation picked up 12 technical personnel from the Dalian Rubber and Plastics Machinery Factory by promising higher wages, two-room housing, and more prestigious professional titles. The journal *Gongcandang yuan* warned that this type of action is "a trend toward anarchy" and that mobility "does not mean that an enterprise may recruit people in any way it wants to, nor may an individual go to work anywhere he wants to."[29] However, the message on this point is somewhat contradictory. Because small enterprises are unable to obtain assignment of any college graduates, the government does not discourage the flow of personnel in the direction of such enterprises,[30] as in the case of the bicycle plant. In recent years, however, many small enterprises and especially production cooperatives have prospered and can now pay higher salaries than can the large controlled state-owned enterprises. They are accused of "robbing" some "large enterprises and scientific research units of their key scientists and technicians," and the departing specialists are reprimanded for not "going through the necessary transfer procedures, thus affecting the normal work of the units."[31] The *Gongcandang yuan* commentator concludes that, since "state-owned enterprises do not have as much sovereignty as the collective enterprises," they require protection to maintain the stability of their scientific and technical personnel. Obviously, "freeing talent from the fetters of departmental ownership" is more difficult than an outsider might imagine.

Conflicts between the seekers and the owners of talent is further exacerbated by the practice of advertising for specialists in the mass media, which started a few years ago. *People's Daily* lists three advantages of "inviting applications for jobs." First, advertisements list the necessary qualifications, conditions, and procedures and eliminate secrecy, which "creates unhealthy work styles." Second, they help employers and applicants to "meet in person and understand each other." And, third, the voluntary nature of this practice meets the needs of the enterprises and the wishes of the individuals. As the article points out, publicly advertising for personnel does indeed bypass the bureaucracy, and this is precisely why "some comrades" strenuously object to the practice as being "divorced from party leadership" and a "practice of bourgeois liberalization."[32]

In addition to increasing the mobility of nearly 7 million scientists and technicians, reform leaders also want to ensure that their flow is in the direction of China's needs and priorities. In 1983, the State Council pointed out (and not for the first time) that "efforts should be made to allow scientists and technicians to flow from the city to the countryside, from the large cities to the medium-size and small, from the developed coastal regions to the frontier, and from overstaffed organizations to those urgently needing employees."[33] But even higher salaries, better housing, and other incentives are unlikely to lure specialists voluntarily to leave the Beijings and Shanghais and head for smaller towns and distant provinces. On the contrary, most of the voluntary movement takes place between inter-city organizations and coastal municipalities or in directions opposite from those intended. This last concern has been expressed by a number of provinces, as seen in the blunt statement of the Sichuan Party leadership: "It is necessary to prevent people from moving to better places, big cities, and higher organs under the pretext of mobility of talent."[34]

Despite these difficulties, policies and procedures are being introduced to facilitate the recruitment of scientists and engineers in less desirable localities. The Governor of Yunnan, for example, publicized measures taken by his province in *People's Daily*: Special delegations are sent to other areas to recruit scientific and technical personnel; provincial units may hire "undocumented specialists," investigate afterwards, and, if necessary, make placement readjustments at a later date; the recruiting departments will recognize the individual's pay scale, position, seniority, grain ration, and residential status; and, if a scientific or technical worker who is anxious to "support the construction of the border areas" is impeded in this desire by his old *danwei*, "the recruiting departments can make the necessary arrangements directly with the departments concerned."[35]

It is impossible to say how successful the undeveloped province of Yunnan has been in luring specialists by offering better salaries, more spacious housing, and greater professional recognition to an area where professionals are scarce. These incentives may be further enhanced, however, by implementing "fixed working periods" in frontier and hardship areas (currently 8 years in Tibet) with a guaranteed return to a more desirable locality after the specified service.[36] Another effective incentive is the promise to unite spouses who are living apart. The city of Changzhou (Jiangsu province), for example, managed to get 7 researchers from Shanghai to fill key positions by promising to bring the husbands and wives together.[37] Some of these measures may prove successful, but the number of individuals involved will not be large, certainly not as large as Beijing would like to see.

An important innovation to facilitate mobility of personnel was the establishment of the Talent Exchange and Consultation Service Centers by the Ministry of Labor and Personnel. The first center was opened in June 1984 in Beijing. By December, the Ministry reported that similar centers had been established in every province, municipality, and autonomous region, "forming a national cooperative network for talent exchange."[38] These experimental centers are not only to move people from jobs where their specific talents are not being utilized to places where their expertise is needed. They are also to act as intermediaries between units that lack skilled personnel and those with a surplus and to arbitrate differences between units.[39] Individuals wanting to change employment can register at a center and indicate their desires. If no resolution is reached between the employer and employee, the employee can resign his position and be recommended to another unit by the center. Similarly, the employer can register at the center for a new employee.

Since so many professionals believe that they are improperly utilized, the centers have been popular. Within two weeks of the opening of the Beijing center, more than 1,000 persons registered in the hope of finding an opportunity to better apply their talents.[40] The first exchange meeting in the auditorium of the Ministry of Labor and Personnel was attended by some 500 people from the personnel departments of various ministries of the State Council and their subordinate units in Beijing. By December, it was reported that 180 units in Beijing had registered with the center, asking for more than 2,100 professionals in all fields.[41] Other provinces and municipalities are also reporting widespread interest in the talent-exchange centers. However, although it is reported that talent-exchange organizations "began to open wide the door to rational exchange of talent," the number of people who manage to get transfers is relatively small—only 10 percent in the first six months of operation.[42]

The problem continues to be that "localities and units have placed major obstacles in the way of talent exchange"—in shorthand referred to as "indigenous policies."

Some of the provinces have held "personnel-exchange conferences" to increase the mobility of professionals. In Guangzhou, for example, the Guangdong Provincial Personnel Bureau convened a conference at which 148 work units announced an urgent need for 23,000 specialists, whom they hoped to pick up from units that were overstaffed or that wasted scientific and technical personnel.[43] A similar conference was held in Changsha, Hunan province. Every type of unit appeared to be represented at the six-day session: personnel departments, employing units, state enterprises, collective enterprises, township and town enterprises, and even individual households. Of the 2,700 professional and technical personnel attending, 2,323 exchanged positions, although supposedly only "592 have gone through the formalities of transfer."[44] According to the Guangdong report, under this experimental program the unit that will suffer the personnel loss "can urge them to stay or advise them, but it cannot withhold them." If the unit stands in the way of the transfer, the individual can resign "with the approval of the Guangdong province or Guangzhou city personnel department" and then accept the position that was offered.

Reports from Jiangsu province indicate more success in implementing the talent-exchange program than other areas. Eleven municipalities and 56 counties have established organizations for the exchange of talent and "transferred thousands of professional technical personnel of all kinds from ministries, provincial units, schools of higher learning, scientific research units, whole people ownership enterprises, and military and industrial enterprises."[45] The province also held a seven-day talent exchange conference in Nanjing in the course of which "10,717 scientific and technical personnel filled out forms requesting transfers, 2,540 persons were initially recruited by recruiting units, and 567 persons were transferred, loaned, hired, given concurrent jobs, or contracted." It is difficult to imagine, however, that some type of pressure was not applied on the "many talented scientific and technical personnel" who "spared no effort to quit the near in favor of the distant and to quit the soft in favor of the difficult." This skepticism is supported by another article, entitled "Why Can't Certain Places in Northern Jiangsu Keep Their

Talented People?" because it lists the very same localities to which people were recruited at the Nanjing conference. It also seems to suggest that many people who are recruited for undesirable localities do not stay long and sometimes "leave without saying a word."[46]

Shanghai, with its high concentration of scientific and technical personnel, is confronted with serious problems and has introduced functional innovations to facilitate mobility. From the following example it would be easy to agree with a *China Daily* reporter that the situation is hopeless: "When an electronics company advertises for 250 workers, gets 1,000 applications but just one person is allowed to leave his present job to take up the new appointment, questions need to be asked about the present personnel system."[47] Yet, Shanghai is trying to make sure that the personnel departments of industrial bureaus, administrative districts, and other employing institutions coordinate their personnel-exchange services under the umbrella of the municipal personnel bureau and meet periodically to discuss problems, compare experiences, and exchange information. The idea is to combine the planned assignment procedures with the free mobility of personnel. The example of the Shanghai Light Industry Bureau illustrates how the new plan is supposed to work.[48] Before anyone from outside the bureau is assigned to any existing vacancy, the personnel people of each of the companies under the bureau are to examine the lists of specialists currently on the payroll who have requested a transfer. Since the transfer would take place between units under the jurisdiction of the bureau, those involved can be easily consulted and a quick decision made. Once this obvious approach was introduced, half the personnel requesting transfers were accommodated within the various work units of the bureau. Units that continue to block the transfer of personnel are subjected to a pincers movement: serious investigation and patient persuasion. Without speculating on how typical this example might be, there is no doubt that it is much easier to implement personnel transfers within a large bureau with scores of large work units under it than in smaller organizations and independent work units.

In January 1985, the Science and Technology Talent Development Bank—"the first of its kind"—opened up in Shanghai.[49] Although it has achieved "pleasant results," the only obvious distinction between it and some of the other institutions involved in improving the use of scientific and technical personnel is that this talent bank also offers training classes in enterprise management, contracts, finance and taxation laws, and other subjects that would supplement scientific and technical backgrounds.

Personnel-exchange activities cover a wide variety of professionals at all levels, but special and independent measures are taken to increase the mobility among the top high-level scientists and engineers. Yang Jun, Vice-Minister of the State Science and Technology Commission (SSTC), announced the establishment of its own national exchange center for scientific and technical personnel and urged provincial and regional science-and-technology commissions to set up similar centers.[50] This was followed by the SSTC authorizing its local branches to "look into overstaffing and improper use of talent and intervene if necessary" in order to help the mobility of scientists and engineers.[51] In addition to correcting imbalances, the transfer of these high-level personnel would also "promote the transfer of technology and the results of scientific research."

In addition to restructuring the personnel management system for high-level scientific and technical institutions, the government is also gradually trying to replace the old system of planned assignments, organized transfers, and administered management of personnel with a "new" system of hiring. A September 1985 article in *Forum of Science and Technology in China* discusses some of the benefits and problems associated with this "new" system.[52] The hiring system will overcome the abuse of the "iron armchair" (the scientists' version of the "iron rice bowl"?); encourage the movement of talent; and increase the scientists' contribution by giving them some say in where they will work and what they will do. Presumably, the work unit and the individual will negotiate from "completely equal positions," and the final agreement will stipulate the responsibilities, authority, and benefits that go with the job. The author argues that hiring procedures are entirely compatible with the socialist system, but also warns that, lacking experience in a hiring system, "we must proceed cautiously, make measured experiments, and gradually perfect it."

Nevertheless, despite China's efforts to reform her employment system by allowing scientists and engineers to have more opportunities to choose their jobs and work units, the most authoritative verdict was given in a July 1986 Circular of the State Council; this pointed out that, despite some progress and good prospects, the phenomenon of "overstocking, wasting, and misusing scientists and technicians has not been fundamentally eliminated."[53]

NEW INITIATIVES AND NEW INCENTIVES

In today's China (and to the chagrin of the old Mao followers), many of the complaints still expressed by scientific and technical personnel can be corrected by an increase in income. By decreeing that egalitarianism is now obsolete and by assuring professionals that it is not just permissible but desirable for them "to become well-off through their scientific and technical related activities,"[54] China is introducing the universal stimulant. Indeed, monetary incentives have been most effective in maximizing the utilization of the scientific and technical personnel, improving their living conditions, and increasing their job satisfaction. In addition, their numbers are growing. According to a special Science and Technology Census, started in November 1985, civilian research institutes employed 770,000 people, including 231,000 scientists and engineers and 121,000 technicians. China's 760 colleges and universities of science and engineering, and agricultural and medical sciences now have "a technical force" of 481,088 people, including 356,088 scientists and engineers.[55]

Contracting for Research

One of the reforms which not only facilitates the integration of research with production but more fully utilizes scientists and engineers is the "paid contract system." This system, at best, allows many of the scientific research institutes to become economically self-sufficient and gradually decrease their total dependence on state funding for research. Such outside funds are expected to provide new incentives, stimulate new enthusiasm and creativity of the individual researchers, and, ultimately, eliminate the unhealthy practice of everyone's "eating from the same big pot." The new independence of the research institutes was described in a *China Daily* editorial. The director of an institute is in complete charge of all professional and administrative work; he has the power to hire and discharge according to requirements; and he can spend funds without external interference. Furthermore, the institute can

> forge cooperative ties in various forms with enterprises, designing units and colleges. And it can enter into contracts with other units in terms of technical development, transfer and application, . . . and obtain income through these channels.[56]

When the initial steps were taken in 1979 to broaden the decision-making powers of research institutes, some experimentation with contractual agreements between institutes and enterprises was attempted, but it was not until the early 1980s that the practice became policy. In April 1983, with the approval of the State Council, the SSTC and the State Commission for Restructuring the Economic System issued a circular urging units involved in research and development "to put operational expenditures under a contract system."[57] The first step in this process was to select "experimental point units" within each institute which were best able to perform contractual research either for departments within the institute's own hierarchical (vertical) structure or for outside enterprises. In theory, the government expects gradually to reduce and finally stop allocating operating funds to institutes and departments engaged in applied science and technology that are expected to make profits.[58] Already, in 1983, 3,536 research institutes directly under the control of the central government reportedly were receiving 36.6 percent of their operating expenses from contracts—not an insignificant proportion, which undoubtedly has increased since then.[59]

Contracts, however, are not the answer for all institutes. Institutes doing basic research or those involved in medicine, public health, family planning, environmental sciences, agricultural sciences, and others which are not doing primarily technological research may find it impossible to get outside support from production-oriented entities, and therefore will continue to get funds from the state.[60] Nevertheless, to encourage institutes and scientists involved in basic research to extend their work into applied research and development, the Vice President of CAS, Zheng Haining, stated in an interview that "we must select some scientific and technical personnel and train them in business and management so that they can handle developmental work (and) correct the mistaken notion that only people unfit for scientific research will go in for development."[61] At the same time, institutions and individuals who have expressed concern about the cuts in state appropriations for research have been assured that the new system of financing will actually increase funds for scientific research; "only the method of managing the funds will change."[62] Furthermore, in order to encourage scientific research units to undertake contractual obligations, the Ministry of Finance has exempted them from taxes on income derived from "technological transfers, advisory activities, services," as well as from "product taxes and appreciation taxes for new products they produce as a result of their research work."[63] The regulations include some exceptions, but funds accrued as a result of tax exemptions "should be used exclusively for technological development."

Clearly, the thrust of the post-Mao policies is to encourage the scientific and technical establishment to engage in entrepreneurship and risk-taking. But how does the contract system translate into earnings and job satisfaction? The most frequently cited advantage is that the contract system will "abolish egalitarian general awards and uphold the principle of remuneration according to labor, directly linking awards with the performance of tasks."[64] In other words, the more successful the unit in performing its tasks, the more money it will make, and the more there will be to distribute to members. Logically, then, the greater the contribution of the individual, the greater his remuneration. This is not necessarily true. Most of the money received by the contracting research unit is expected to go for operating expenses, new equipment, and capital improvements. The leading cadres must determine allocation of these funds as well as the proportion of the income to be distributed to the participants. Furthermore, they must decide the relative contribution of each individual. These subjective decisions have created fresh problems, among them abuses unthinkable in the past: "illicit and abrupt promotions, illegitimate issuances of bonuses, and unauthorized increases of wages."[65] The SSTC is attempting to correct these wrongs, blaming not only the "lack of experience and insufficient legislation," but also the "new unhealthy tendencies," which are not limited to the scientific establishment and which are much more difficult to remedy. The benefits, however, outweigh the abuses. Contracts as well as economic incentives are expected to stimulate freer exchange of information and innovations. Before contractual agreements between research institutes and enterprises, there was a tendency for units to guard the results of their research jealously, to make sure that "our manure shall not fertilize other people's land."[66] Now, of course, the institutes stand to gain from any research results that may make a contribution to production and will therefore seek to publicize their results to potential customers. Organizations involved in research now participate in fairs and exhibitions to advertise the results of their work in the hope of selling them to a production unit. Also, at these events, research institutes "are invited to bid for technical problems raised by enterprises and factories."[67]

Spare-time Work for Intellectuals

A major new liberalization in China's employment practices has been the introduction of spare-time and part-time employment for intellectuals. This practice, which started in the early 1980s, has been growing rapidly, especially among tens of thousands of scientists and engineers who use their skills in their spare time to supplement their incomes. But, again, as the government actively encourages these activities, there is considerable opposition and jealousy on the part of management cadres. Some who were providing after-hours technical services not only have been accused of being engaged in work labeled "not proper" and solely to "pursue money," but "have been implicated in 'economic crimes,' brought to trial, investigated, and dealt with." According to an article in *Guangming ribao*, "such a large number of scientists and technicians being made into 'economic criminals' has exerted a negative influence on the efforts to invigorate the economy and to implement the policy toward intellectuals."[68]

In June 1985, the SSTC issued a circular calling attention to some "unhealthy tendencies" in the personnel practices of the S&T establishment, some of which related specifically to spare-time work. The policy is clear: "We shall permit scientific and technical personnel to take on proper concurrent work or after-hours intellectual work, so that the potential that we have in our scientific and technical personnel is fully brought into play."[69] It then calls on the managers to make some important distinctions. First, the responsible cadres must distinguish between those who complete their primary functions and therefore do not hamper the technical or economic interests of the units by accepting outside responsibilities and those who "unscrupulously neglect their principal work to seek personal profits elsewhere." And, second, a distinction must be made between personnel who accept outside contracts with the prior approval of their principal employers and those who "simply run off without taking leave and grab extra income by hook or by crook."

Some reports suggest that the SSTC circular may have had an adverse effect on spare-time employment by scientists and technicians. According to an article in *People's Daily*, precisely since mid-1985 it became more difficult to obtain approval to do outside work from the leadership of the *danwei* to which the individual belongs. Moreover, at least in Shanghai, the taxation rate on income from spare-time work has been raised on three occasions.[70]

The basic concern about spare-time employment is how to compensate scientific and technical personnel and at the same time limit their earnings so as to diffuse the jealousy and antagonism of colleagues and cadres. The authorities in Hebei province have given much consideration to this question. One of the "opinions" issued by the province states that remuneration received for after-hours work is "legal income and should not be treated as improper income"; nor should it be returned to the work unit. However, if "remuneration is too great (over 1,000 yuan greater than the average per capita income) or if equipment or materials from one's own unit are used, the individual should hand in an appropriate part of the amount to his unit after both sides have discussed it."

Not only are Hebei scientific and technical personnel permitted to hold outside jobs, but units with large numbers of such specialists are urged to arrange for spare-time work by organizing them "to carry out after-hours' technical service and sign contracts that combine responsibilities, rights, and benefits." Since it is unlikely that work units will voluntarily seek outside employment for their personnel, an incentive is offered. Under contractual arrangements in which the employing unit is involved, the proceeds would be shared between the unit and the personnel assigned to these activities. Scientists with an entrepreneurial bent probably find the involvement of their own *danwei* intrusive and limiting, but, because of past experience, there must be a good number who are reluctant to "go it alone" and prefer consulting arrangements made by the unit. Even if the home unit does not initiate outside contractual relationships, provided individual contracts are made through prescribed channels, the units "may not use any pretext to refuse, obstruct, or create difficulties" for anyone wishing to work in his spare time. In an effort to put an end to some of the unfair situations that have plagued professional personnel in the past, Hebei prescribed a new practice: "The department seeking service should explain the situation to the concerned units and bear the responsibility, and the scientists and technicians should not be blamed."[71]

Solutions and policies may be somewhat different, but the issues and problems of Hebei are surely not unlike those of other provinces. In the final analysis, what the authorities throughout China are striving for is a more systematic and a more formal approach to spare-time work by scientists and engineers. They prefer agreements between units, rather than between the scientist and secondary employer; and they prefer written contracts with responsibilities carefully spelled out for both parties, rather than ad hoc arrangements.

Other Contractual Innovations

Other innovations have been introduced to stimulate imagination and flexibility in the use of professional manpower. These can be seen in the new role for institutions of higher education and in the private or cooperative scientific and technical consulting organizations that have recently "sprung up like flowers in a spring rain."

Modeled after the Soviet system, Chinese institutions of higher education were virtually divorced from the research establishment, which was centered in the institutes of the academies of sciences.[72] Moreover, limits on such research were perpetuated by inadequate student qualifications, narrow specializations with an insufficient theoretical background, and inadequate research funds and facilities. Consequently, institutions of higher education were isolated from production—a condition now surprisingly blamed on "leftist influences," which, if memory serves, attempted to integrate education with production.

As academic standards rose and research became part of the graduate curricula of institutions of higher education in the early 1980s, it became possible for the universities to get a piece of the consultative action. This new role was advocated by Premier Zhao Ziyang in May 1982, when he urged the establishment of associations for teaching, scientific research, and production in universities.[73] By now, many of the universities and colleges are indeed "providing independent departments and enterprises with research results, scientific consultancy and technical services, signing long-term contracts or setting up centers combining teaching, research and production."[74] The 1985 reform of the educational system reconfirmed this function, stating that, as part of the expanded decision-making powers now vested in institutions of higher education, they "can accept projects from, or cooperate with, other social establishments for scientific research and technological development, as well as set up combines involving teaching, scientific research, and production."[75]

The better universities, especially those with strong faculties in the areas of science and technology, have been obviously in a stronger position to take advantage of the new policies. Close behind, if not on par, are the engineering colleges, which can more readily become involved in production problems.[76] In Shanghai alone, which of course has the highest concentration of scientific and technical personnel, 39 universities and colleges signed some 2,500 technical-service contracts with factories and enterprises—many of them with enterprises in other cities and adjacent provinces.[77] As early as 1983, Jiaotong University's technical-service department earned 1.5 million yuan through research projects, technician training, technical consulting, and translation work—and 176,000 yuan of the total profit went to "teachers and technicians" who participated in these projects.[78] In general, over the past several years, all the lines on the graph have been on the ascent in the number of participating institutions, number of contracts, amount of money made by the universities, and the amount of money saved by the enterprises. And, in the process, faculties have been able to supplement their relatively meager academic salaries.

As usual, however, while the policy makes great sense, it is not without inevitable problems. With the likely exception of the most prominent universities, most of the institutions of higher education are at a disadvantage when competing for outside contracts. Large state enterprises seeking outside assistance for research and development think first of the institutes of CAS or, as an alternative, turn to the research capabilities within their own (ministerial, bureau) vertical structure. This is because, for so many years, the best (and often the only) research was associated with CAS institutes; the notion of university research is new. Moreover, most managers still mistrust outside research agreements, believing that "rich waters do not flow to the houses of others"—any significant achievements in science and technology will not be passed on to another institutional entity. Furthermore, most institutions of higher education are still at a disadvantage with regard to advanced technology and equipment and therefore are limited in jobs they can undertake. Only 28 key higher institutions in science and engineering fields have been lucky enough to be part of the World Bank's Chinese University Development Project, which not only raised the quality of education but provided them with specialized laboratories, computer centers, and other equipment to strengthen research. Most universities don't have funds for importing foreign technology and equipment, and industries and enterprises are not likely to share technology or turn to the faculty of a university for help in assimilating it. When it comes to advanced technology, institutions of higher education "can only frustratingly chase after what is left." Therefore, most universities and colleges find themselves working for small local enterprises that require a lower level of scientific and technical development or work on problems that no one else wants to handle. As one college-based specialist put it: "It is always the case that only ungnawable hard bones and inedible chicken ribs and tails are turned over to the higher institutions."[79]

Faculty working within the university structure generally share only 14 to 17 percent of the income from outside contracts. Compare this with a 30-percent share for individual remuneration within scientific societies, 60 to 80 percent in private development companies, and remuneration that need not be shared with anyone when a professor takes on a spare-time job on his own[80]– perhaps taking a chance by bypassing the institutional bureaucracy. This gradation understandably creates a tendency for faculty to search out more lucrative research arrangements independent of the university. Many professors prefer to participate in outside research as members of scientific societies, which have played a significant role as consultative bodies in recent years. At the same time, since "intellect and knowledge are a formless valuable product residing within the minds of teachers" and since teaching schedules are flexible, university administrators find it difficult to keep track of the faculty's outside activities to determine whether they are abiding by the established rules for acceptance of outside contracts.[81] If outside jobs are not approved by the university, some faculty members end up "being on the job but with their minds elsewhere." The result is that a professor who finds part-time outside employment may neglect teaching responsibilities. There is undoubtedly somewhere a happy medium which some institutions of higher education may have found. But, for most, these forward-looking policies, which have so expanded the horizons of the teaching personnel, will continue to create a variety of problems.

Chinese publications repeatedly point out that "the scientist qua entrepreneur is a well-established phenomenon in foreign countries." They cite scientists who founded many of the successful knowledge-intensive enterprises in the West.[82] It is not surprising, therefore, that some of the most efficient and flexible R&D organizations emerging in China's new environment are also entrepreneurial in nature. They include the various S&T development companies and centers operating in the technology markets with their "daily bustle of both buyers and sellers." They also include science associations and technology associations, which, by offering "generous economic remuneration," can call on their own network of professionals. Because these types of organizations are "horizontally aligned and non-governmental," they are least hampered by institutional constraints and can be most responsive to market needs.

CAS reported in 1986 that the institutes under its jurisdiction had established more than 100 technology development corporations.[83] These corporations emphasize the "conversion of scientific results into commodities" and have entered into joint ventures not only with domestic but also with foreign companies.

Several large cities have S&T centers run by individuals and cooperatives ("run by the people"). In Nanjing, for example, 210 scientific and technical-development consulting service companies have been established, employing 4,600 people on a full- or part-time basis.[84] The activities of these service companies are said to have been so beneficial to the production units and the individual participants, that some "large factories give leave to scientific and technical personnel to run economically independent scientific and technical service companies." Certainly not a common occurrence.

One of the most successful S&T centers run by an individual or a collective appears to be the Huaxia Technology Development Center in Beijing. It was opened in 1983 by a research fellow from the Institute of Physics (CAS) with a 100,000-yuan loan from a local industrial company.[85] Within a year, the Center grew to 70 staff members and dozens of paid part-time technicians, entered into cooperative agreements with some foreign countries, and earned a 600,000-yuan commission from just one contract with a radio factory.

Another example of a somewhat different operation is the Jiusan (3 September) Society, one of China's eight "democratic parties" and primarily composed of intellectuals. In fact, most of the members are scientists, engineers, and professors—one-quarter from CAS and many of the academy's retirees. The society organized 78 teams to offer scientific and technical consulting services throughout the country and, in the past five years, it has provided consultation on 3,670 projects in many cities. In addition to research and consultation, "members also give lectures and classes to train other scientific and technical personnel."[86]

Although some phony consulting companies have been established and some have been accused of "swindling, buying, and selling for profit, and trafficking in state-controlled goods and materials,"[87] the authorities tend to be more tolerant of these abuses because consulting services are still considered to be a "newborn baby." Furthermore, as long as the government not only tolerates but even encourages a market economy in science and technology, privately run S&T consulting and development companies will continue to function. For, as the Chinese journal *Kexue yu jishu guanli* points out, they have great advantages:

> They are not limited to a certain field or a certain discipline, but do whatever the market needs. They do not need a state-allocation system, nor do they need state-allocated expenses; for, on the one hand they take on research topics that are needed by the market, and, on the other, organize their scientific and technical capacities. When things have been sized up, they act; when projects are completed, they let them go. Scientists and technicians get paid according to their efforts, and profits are distributed according to the degree of contribution toward completion of the project.[88]

PROSPECTS

The post-Mao economic spurt, the introduction of incentives through various types of responsibility systems, and the ingestion of foreign capital and know-how have created a dynamic economy. Can the reforms in the utilization of scientific and technical personnel ameliorate the shortage and maximize the contribution of specialists so vital to sustaining China's modernization?

It is difficult to judge the long-term effects of the numerous and sometimes radical changes introduced by the post-Mao reformers. Given China's current predisposition to "tell all," one can easily substantiate successes and failures in almost any sector of development. Judgment, however, is complicated by the fact that many of the new policies are still experimental, often eliciting simultaneously positive and negative commentary in the Chinese media. As a result, the following sequence has become rather familiar: Reforms are tested in one province, city, enterprise, or other appropriate entity; they are gradually introduced nationally; unforseen problems arise and solidify resistance to the changes; Beijing eases up on implementation, performs some fine-tuning of the policy, and then pushes forward again, but in a slightly modified direction. This is not an unreasonable way to introduce reforms, and some aspects of this process have been evident in efforts to utilize scientific and technical personnel more effectively and to ensure that both individual and institutional decisions become more responsive to the country's needs.

In the case of student assignments, we have some liberalization in allowing greater employment options for graduates, followed by some backtracking, because, given their choice, students will not fill the scientific and technical positions in border provinces and in more backward enterprises. Similarly, despite the vocal support of scientists and political leaders in Beijing, there is continuing opposition to increased mobility among scientific and technical personnel because the work units do not only resent the loss of qualified people but also because unrestrained mobility likewise deprives the less desirable localities of skilled manpower. One can speculate that those in favor of increased mobility point to the United States and other Western nations; those opposed can point to the Japanese successes under conditions of minimal labor mobility. The bottom line, however, is that, despite the creation of "over 100 organizations for the transfer of trained people," of the scientists and technicians who attempted to relocate only 2 percent were successful in 1983 and 3 percent in 1984.[89] There is also no unanimity on having scientific and technical personnel involved in spare-time work in various private consulting organizations or in other privately initiated arrangements outside the work unit. Some of the opposition stems from simple jealousy, but there are also more legitimate reasons, such as that part-time activities tend to diminish an individual's commitment to the primary occupation. Moreover, since the most energetic and best-qualified specialists are the most likely to become involved in some form of private scientific and technical activities, there is inevitable loss to state enterprises and institutions.

An important peripheral issue has a direct bearing on the effectiveness of the utilization of engineers. In contrast to the industrialized nations, which have 3 to 5 technicians and 10 to 15 mechanics as assistants to one engineer, the ratio of engineers to technicians is reported to be 3 to 1 in China's machine industry.[90] A similar disparity between engineers and supporting technicians probably exists in other industries as well.[91] In order to provide adequate support personnel and to assure that engineering talent is not wasted in less demanding activities, the educational system has been emphasizing secondary technical education. But progress has been slow, because the education in vocational and secondary technical schools has been admittedly poor, and because China's urban youth seek to enter secondary general schools which prepare them for college. Only a few, however, are selected to fill the limited number of slots available in institutions of higher education. Most are left with an education that has not provided them with the practical skills necessary to perform as middle-level technical personnel—where the shortage is. China is hoping that at least some of the gap in demand will be filled during the 7th Five-Year-Plan (1986–1990), which calls for the training of 8 million young people in vocational technical middle schools.[92] If the goal is met, it may to some degree compensate for the shortage of engineers by increasing the on-the-job efficiency of senior professional personnel.

By expanding and improving the educational system at every level, China has made a clear commitment to invest in people. But, since the educational cycle is a long one and scientists and engineers are still at a premium, it is imperative that they not be wasted. Not only must they be properly placed; they must also be given the incentives and opportunities to perform at capacity. Changes in China's manpower policies are an integral part of the overall socioeconomic transformation that has taken place in the 1980s. Thus, with regard to scientific and technical personnel, as with the economy in general, the very basic problem that concerns the political leaders is how to combine central planning with free market forces and how far to take this new hybrid system. With regard to professional manpower, they have come to realize that, although cracked, the "iron rice bowl" cannot as yet be entirely discarded. Egalitarianism and the principle of "eating from the same big pot" were indeed dampers on people's enthusiasm and obstacles to the development of the country's productive forces. But, until some semblance of equilibrium is achieved in the supply and demand of scientific and engineering personnel, China will have to pursue a policy that combines a degree of personal mobility and private initiative within what must, of necessity, remain a system of control over professional talent. It will be a difficult balance to maintain.

CHAPTER FIVE

China's Industrial Innovation: The Influence of Market Forces

WILLIAM A. FISCHER

Throughout its period of governance, the PRC has been concerned with the role that science and technology (S&T) should play in promoting economic development.[1] In fact, the levels of that concern, the accompanying debate, and the attempts at organizing science and technology have been considerably more intense than in most other developing countries. With the exception of the Cultural Revolution period, 1966–1976, there has always been open acknowledgment by China's political leaders that S&T can be an important contributor to the building of a modern economy in areas ranging from factory automation to increased industrial productivity to biotechnological approaches to improved agricultural performance. Although frequently there has been little national consensus on how best to develop the linkages by which S&T would support economic achievement, or to what extent "appropriate" or indigenous technologies should be the focus of such activities, the interrelationship between S&T and economic development has been recognized and deemed important enough to warrant sustained government attention at the very highest levels.

In the post-Mao era, the Chinese government's policies of economic reform include the significant introduction of market influences into economic decision making, which represents an abrupt change from the strong reliance on central planning that had characterized the PRC for much of its existence. These policies also continue to embrace a commitment to advanced S&T, and industrial innovation, across the whole spectrum of the Four Modernizations. By stated policy and by implication, the two streams of reform, market influences and technological advancement, should intersect and reinforce each other. Given the size and state of the Chinese economy, however, it is no exaggeration to state that the magnitude of this effort is immense, and the sudden shifting to the use of market mechanisms to induce innovation on this scale is probably unprecedented. In a very real sense, the present attempt to employ market forces in China to stimulate industrial innovation is still a great experiment whose ultimate outcome remains unpredictable.

CHINESE INDUSTRIAL INNOVATION IN THE PRE-REFORM PERIOD

In order to understand where the Chinese are today in terms of their ability to stimulate industrial innovation, it is necessary to understand where they were in the period prior to the post-Mao economic reforms.[2] Philosophically, China's leaders believed that socialist economic systems, which theoretically reduced competition and the resulting R&D redundancies and possible "hoarding" of knowledge, offered the best environment for marshaling and making full use of the nation's creative resources. Removing incentives that encouraged the privatization of knowledge, they believed, would make their economic system most effective in stimulating the diffusion of innovation. In the absence of a market mechanism, they sought another spur to the diffusion of information by popularizing S&T. Suttmeier has captured the essence of their system when he refers to its "supply-push" perspective, where the emphasis was placed on mechanisms for "launching" innovations, with an implicit trust that, once launched, such innovations would be readily accepted.[3]

As in any society, but especially centrally planned ones, the Chinese government had a variety of direct and indirect means at its disposal by which it could affect the rate and direction of innovative activity within the economy. Given the urgency of its needs and the absolute power it commanded, it is not surprising that the Chinese government favored such direct means of intervention as government funding of S&T and the reorganization of S&T resources into a command-type hierarchical system[4] that focused on central research laboratories both in CAS, for basic research, and in each of the ministries for directly relevant applied research. It developed a variety of science and technology plans designed to coordinate and focus the efforts of China's S&T resources.[5] Also reflective of supply-push inclinations was the establishment of mass organizational mechanisms for the popularization of S&T and innovation such as the All-China Association for the Advancement of S&T Knowledge.

Despite a few spectacular successes,[6] mostly in basic or theoretical science or in militarily related areas, the net result of China's investment in S&T has not been particularly noteworthy. This is especially true in industry, where outmoded products, obsolete technologies, and a reliance on reverse engineering (that is, the design or redesign of products based on imitation or careful review of others' innovations) as a primary source of new product and process ideas, remain the norm. China's weakness in this regard probably came from:

(1) the heavy bureaucratization of S&T activities, an outgrowth of the administrative structures established to organize, plan, and coordinate R&D resources; the relatively low level of national investment in civilian S&T between 1949 and 1979;[7]

(2) the destruction or dismantling of a significant portion of the nation's industrial capacity during and after World War II and the Civil War and the unwillingness of the Soviets to provide complete plans for, and complete training for technicians working in, the factories that they supplied to the Chinese during their period of close cooperation;

(3) inadequate linkages between the S&T performers and those organizations responsible for creating products and processes for Chinese industry and markets;[8]

(4) the occasional suspension of economic rationality in China, as

in the Great Leap Forward of the late 1950s and the Cultural Revolution which discriminated against the intellectuals employed in S&T activities;

(5) a neglect of the development of Chinese capabilities in engineering design, growing out of a reliance on reverse engineering as the primary source of innovative ideas;

(6) a general inadequacy in the planning mechanism, resulting from a lack of information necessary for effective planning and the employment within the planning bureaus of inappropriately trained administrators;

(7) and a low level of motivation in Chinese industrial enterprises to introduce, much less actively support, technological change.

In many ways, the state of Chinese industrial innovation in the period prior to economic reform could be viewed as suffering from many of the same problems observed in other centrally planned economies: the stifling constraints of a vertically segmented industrial structure[9] and the "subordination of efficiency and creativity to administrative needs and security concerns; risk aversion; a prestige bias toward theoretical rather than applied research; inadequate scientific and technical communications; poor S&T management; and an overall cultural milieu that emphasizes 'stability, security, and conservatism.' "[10]

MARKET INFLUENCES AND THE INTRODUCTION OF ECONOMIC REFORM

By the early 1980s, there were ample indications that the Chinese government was considering a basic change in the relationship between S&T and economic development. In July 1981, *People's Daily* carried a proposal that sought to strengthen direct linkages between the economic activities of enterprises and industrial innovation and emphasized the applicability to industry that was envisioned for S&T activities:[11]

(1) The plans for science and technology and those for the economy should be considered together, at the same decision-making forum.
(2) The (economic) planning authorities ". . .should set requirements on the departments of science and technology according to the needs of economic development."

(3) Major capital construction, technological innovation and technology import projects should be "checked and approved by specialists in economics and in science and technology."
(4) Attention must be given to the "connected sequence" from research and the import of technology to ". . .construction and the operation of production."
(5) The linkage between science and technology and production should be forged in the development of the technical innovation objective of the economic plans:

> "The economic development plan includes not only a production plan and a capital-construction plan, but also a technical-innovation plan, with the enterprises as its foundation and the development plan products and undertakings as its main content."

(6) Pilot and demonstration projects ("sample production devices") should get more attention as a means to facilitate the popularization of discoveries and inventions.

Although the emphasis on improving the ability of industrial innovation to serve the needs of an economic "market" is clear, the actual mechanism by which enterprises would be motivated to do this is not directly apparent. Under Mao Zedong, the Chinese had tried other non-administrative means to stimulate innovation. Suttemeier has observed that "one of the unique features of the Chinese development experience has been the attempts made in the past to abolish, or at least shrink, hierarchies without replacing them with markets."[12] The ineffectiveness of this approach in stimulating industrial innovation was expressed in interviews conducted by this author in 1980–1982. In too many cases, noted numerous Chinese managers, there was simply no need to innovate because either market demand so exceeded supply that it made innovation unnecessary or there was no direct financial linkage between market needs and enterprise benefits to make innovation worth the enterprise's while.

In December 1982, Premier Zhao Ziyang introduced the nation's 6th Five-Year-Plan "indicating that, in the future, as with other sectors of the economy, S&T would be increasingly governed by economic rather than administrative measures."[13] The introduction of increased managerial autonomy and the economic-responsibility system into industry also implicitly affected S&T activities, because the enterprise would presumably need to rely more on commercially attractive new technologies and products to succeed in the new, more competitive market environment. The implicit assumptions that market influences would translate into tangible industrial innovations as well as improved efficiency had been made somewhat explicit as early as 1979 in an article in the journal *Hongqi*:

> Competition between enterprises would certainly stimulate technology innovation, improve management, reduce costs, raise productivity, increase the variety, and enhance quality. So, competition will test an enterprise's efficiency and press an enterprise to satisfy the consumer's demand for better goods and greater variety, thus fostering the development of productive forces.[14]

By late 1982, changes were introduced in the management of the Chinese macroeconomy to create a variety of market-type influences which could affect industrial innovation. However, such a policy also shifted innovative energies toward product-design improvement, the upgrading of capabilities of existing equipment, and the improvement of the software aspects of commercialization, such as marketing, distribution, and quality rather than the creation of new technologies. In a technological sense, these results could be interpreted as more modest than what might have been expected from the changes called for in S&T.

POLICIES AFFECTING INDUSTRIAL S&T

There are a variety of direct and indirect means by which a government might affect industrial S&T. Among the most obvious are the direct funding of industrial R&D groups and the organization and coordination of R&D activities among organizations that would not otherwise coordinate their research. China has actively pursued both these approaches. The attractiveness of direct policies is the relatively straightforward cause-and-effect relationship that can be observed and measured. For the policymaker, such clarity also provides a high capability for focusing interventions so as to increase the probability of a desirable outcome.

Indirect policies to affect industrial innovation involve many different approaches, including mechanisms designed to stimulate market-type influences. These might include ways to make the pursuit of technological innovation directly profitable to the innovating organization. Central to such indirect methods have been recent Chinese government experimenting with increased market competition among industrial organizations and allowing managers limited autonomy in deciding how to invest their resources. In addition, reliance on market support for innovation has been encouraged by the gradual replacement of guaranteed government funding for industrial R&D by loans which must be repaid out of future earnings. National and regional forums for the identification and movement of innovations between originating organizations and those enterprises that need them (technology markets)[15] have also been developed to introduce market-type dynamics to the diffusion of innovation.

A major problem with indirect approaches, however, is the difficulty of actually relating the policy "lever" to the desired outcome. Indirect approaches tend to be diffuse; the linkage between cause and effect (a relationship that may often not be self-evident) can frequently be mediated by a number of other forces. Nonetheless, because they can fundamentally change the overall environment in which business decisions are made, indirect approaches may be considerably more significant than more limited direct approaches. In fact, recent S&T policy reviews in several different nations[16] suggest that the creation of an appropriate macroeconomic environment through the establishment of indirect inducements for innovation through market-type mechanisms is probably the single most effective policy intervention that can be undertaken to encourage and support industrial innovation.

The effectiveness of market forces heretofore on China's industrial innovation will be ascertained from interview and survey data collected by this author from 1980 to 1985. These data were obtained primarily at the Chinese National Center for Industrial Science and Technology Management Development at Dalian as well as through other interviews conducted as part of a variety of lectures and visits to China during this time period. Given the nature of the interview populations, these data are clearly not random but are thought to be indicative of what is going on at the leading edge of Chinese enterprises and should therefore provide insights into what the future might hold for Chinese innovation if the economic reforms continue.

CHANGES IN CHINESE ENTERPRISE INNOVATION AS A RESULT OF MARKET FORCES

The power of market influences as catalysts for innovation is visible almost everywhere in China in the form of new products and improved manufacturing processes. As the economic reforms of the early 1980s began to take hold in industrial enterprises, there was a noticeable improvement in product variety and design, attention to service, quality, and distribution, and a refocusing of China's S&T energies away from primary concentration on heavy industry toward light, consumer-oriented industries. Despite the almost undeniable evidence, however, the exact nature of the linkage between the market and innovation remains obscure.

The initiating mechanism for encouraging innovation in industry seems clear enough. Under the new economic reforms,[17] managers have increased autonomy in determining the means by which their enterprises will compete. They also have been granted considerable discretionary power over the disposition of investable resources by the enterprise in order to support their competitive decisions. The managers are also being held increasingly accountable for the generation of profits, the source not only of funds for capital investment and innovation but also of workers' bonuses and, ultimately, their quality of life. Along with these new responsibilities have come new "ground rules" for competition which would appear to facilitate innovative behavior. Although price determination remains beyond the Chinese manager's control in most industries, multiple product-design differentiations and innovative approaches to service and distribution are allowed him. The use of bank loans rather than government subsidies as a primary source of investment and working capital further emphasizes the desirability of achieving economic advantages through innovation and introduces a sense of accountability, which helps reduce complacency among China's managers.

The interviews indicate that, in those industries experiencing considerable managerial autonomy,[18] innovation substantially increased as a result of increased competition and the need to earn larger profits. An illustration of such behavior can also be found in the World Bank study of the Qingdao Forging Machinery plant.[19] In 1980, the directive plans and production quota for this enterprise were replaced by indicative plans supplemented by market regulation: "The enterprise's annual output, product variety, and specifications were to be determined by the market and users' needs." The introduction of market influences into the enterprise's planning environment led to the formation of user-service teams which visited more than 200 users to solicit (among other things) ideas about desirable product innovations. One of the lessons learned from such information gathering was that the design of the factory's products, which had not changed for thirty years, no longer met user needs; as a result, sales were stagnating. Futhermore, the fundamental readjustment in the Chinese economy from heavy to light industries created a whole new set of potential customers. The formation of R&D groups to serve these new product markets led to the development of 12 new products in 1981 alone, which accounted for 70 percent of all product varieties manufactured by the factory that year. Continued commitment to R&D and user service in this factory reflects a significant management change in innovation activities compared to what had existed prior to the economic reforms. Certainly, the linkage here between increased market influences and industrial innovation appears to be both strong and direct.

The Chinese press has printed a number of articles documenting the development of goods and services in a wide variety of industries that represent the effect of growing market influences on enterprise innovation. In electronics, for example, in the Shanghai components industry alone the Chinese have produced a 4K SRAM, a 4K DRAM, a 16K DRAM (not in production), an IC production line, a 25 Hz graphic alpha-numeric generator, and a LSI CAD mask-making system.[20] The recent development of the Yun-7 aircraft has been hailed by the *Beijing Review*, perhaps prematurely, as "ending the days of China's reliance on foreign aircraft imports for domestic air travel."[21] In the textile industries, a number of process innovations have been undertaken, including the recent development of a polyester staple-fiber production line designed and developed in China.[22] The list of innovations mentioned in the press is quite long; these few examples seemingly provide evidence of technological innovation and a sensitivity to market needs that demonstrate the role market influences have played in stimulating technological change.

Yet China continues to lag behind the international state of the art in all the fields noted; their innovations generally do not receive much attention in the developed world. Despite some obvious successes, interviews with Chinese managers reveal that, even with market influences, the level of industrial innovation is considerably less than they think possible. One important reason is the persistence of government involvement at a significant level in enterprise decisions not yet released to the enterprise's discretion. Despite the granting of limited autonomy in such management decisions as product design, marketing, distribution, and advertising, labor relations and pay, some capital investment, and capacity adjustment, the manager of an electronics enterprise interviewed in 1984, explained that the Ministry still retained control over:

(1) a major portion of the planning process for the enterprise

(2) the supplies of raw materials

(3) allocation of critical-component supplies

(4) control of investment, including large innovations and renovation projects

(5) funding of R&D projects

(6) appraisal of product quality

(7) control of imports and exports. (As of January 1986, only 130 Chinese state enterprises had the autonomy to deal directly with foreigners.)[23]

With control of such essential decisions remaining in the hands of the Ministry, a fact reconfirmed in interviews at the ministerial level in 1986, the risk-taking associated with innovation becomes uncertain perhaps to the point of being unattractive. Why take the risks associated with innovation under these conditions, when the Ministry affects so much of what happens? No wonder than an attitude of "better to do less than to do too much" still prevails among a majority of Chinese managers.

Another impediment is the perception that demand is so much larger than supply that enterprises have no incentive to undertake industrial innovations. It is felt by many Chinese managers that "virtually anything that is produced with acceptable product quality will sell." When told that the sales function in American firms suffers from high degrees of uncertainty, one Chinese manager replied: "How could the sales function have any uncertainty? After all, you simply produce as much as you can and sell them!"[24] In such markets, the prevailing strategy is to manufacture as effectively as possible with little or no thought given to either product or process innovation. Consider the case of a provincial transistor factory involved in making integrated circuits. Despite the pace of technical change in that industry, in the 70 percent of its markets characterized by many competitors but little real competition (because of the excess of demand over supply), it does no R&D, preferring to get the customer to buy the existing product with no changes at all.

Perhaps, because of this lack of innovation and/or managerial discretion over product pricing, quality serves as the primary competitive differentiator among products in China today. Even in industries as dynamic as computers, it appears that quality is the major concern of potential customers, with price second; product innovation is not even mentioned by a major Chinese computer manufacturer. Closely related to the emphasis on product quality is a continued preference for process innovation rather than product innovation. Part of this is probably attributable to the lower need for product innovation in markets where demand exceeds supply. A second reason may be the strong engineering and manufacturing background of Chinese industrial managers,[25] who are more used to pursuing cost reductions than new product development. Third, since process innovations qualify for preferential treatment in capital availability and political approval, given the current emphasis on "technological transformation," it is not surprising that it is relatively more attractive to the manager than the more risky product innovation. A fourth possible reason is the difficulty caused by product failures in a geographically large market with inadequate channels of repair and resupply. Such conditions make both innovator and consumer wary of new, untested, technologies.

An apparent tendency toward "technology-push" innovation behavior, cultivated by more than three decades of ignoring customer needs, probably persists because of a paucity of market information available to enterprises. In most industries enjoying substantial degrees of managerial autonomy, there are so many competitors entering the markets and so little market information that an enterprise has difficulty gathering and interpreting such information even should it wish to. This deficiency in market linkages frustrates the effect of market influences on industrial innovation.

While "hardware" innovations in product and process technology are by far the most obvious of the changes resulting from market influences, there have also been some significant alterations in the ways in which Chinese enterprises do business that can be recognized as innovations. If these "software" changes prove to be durable and widespread, they may have a greater impact on Chinese industrial innovation than the present accumulation of "hardware" innovations. Chief among these is a visible improvement in the servicing and distribution of Chinese products. Chinese enterprises are actively engaged in utilizing a variety of distribution channels previously not employed; emphasizing "customer convenience" in the displaying and servicing of products; and offering warranties as an inducement to the consumer. In addition, Chinese industrial organizations have become more active than previously in gathering market information, even arranging direct meetings with consumers to determine their response to new products. Recently, it has not been unusual for Chinese enterprises to shift their emphasis from sales and purchasing toward sales. The shift of engineers into the sales function, seen in many enterprises, may also serve to link the R&D function more directly to the market. The fledgling efforts at advertising one sees on billboards along Chinese streets, on television, and in magazines represent yet another departure from commercial practices of the past that suggests an increasing emphasis on addressing customer needs, although here it is the awakening of potential customers to previously unrecognized needs or enterprises serving those needs that is the primary objective.

The creation of new organizational forms, resulting from economic reforms, represents another type of "software" innovation reflecting the growing influence of market forces. An example is the establishment of the China Typical Sewing Machine Corporation, formed in 1981, combining 30 factories and 40 related businesses, linked through "mergers, joint ventures, co-management, and technical cooperation."[26] Another example is the well-known combination that produces the Jialing brand motorcycles. This particular venture, established in September 1980, was built around the Chongqing Jialing machine-building plant with 3 military and 5 civilian enterprises at its core, along with more than 100 other enterprises, to form a complete set of technical and production resources for the design and production of motorcycles.[27] Such ventures allow enterprises, which are relatively inflexible due to the existing constraints on enterprise diversification and capital availability, plus lack of personnel mobility, to overcome such barriers and to serve the market in a much more effective manner than would be possible otherwise. Their effectiveness comes from the advantages of vertical integration that provide economies of scale in manufacturing, capital investment, and R&D, and that make supplies and product quality much more controllable. It is doubtful that such organizational initiatives, however, would have been undertaken without market influences and the growing ability of Chinese enterprises to profit from such opportunities.

One of the most interesting "software" innovations developed in the early 1980s has been "technology markets." These meeting places for enterprises displaying innovations and enterprises desiring technological change emulate the market mechanism. Tianjin municipality was apparently the first to create a technology market in 1981, followed by Beijing, Wuhan, Shenyang, Chongqing, Dalian, Hangzhou, Xian, and Chengdu. These municipalities had accounted for 34 large-scale technology trade fairs up through mid-March 1985, "exhibiting 19,000 items of results and service, where 5,300 agreements to transfer technology and establish joint ventures were signed. The volume of transactions was worth 130 million RMB."[28] Illustrative of the type of transaction conducted at these technology fairs was the signing of contracts for the "H-pattern special energy-saving fluorescent light" developed at Fudan University and transferred to Shanghai's Lamp Bulb Plant No. 3 and other units.[29] The first national technology fair held in June 1985 in Beijing resulted in 1,500 contracts with a total value of 8.6 billion RMB (US $2.7 billion). A recent technology fair for transferring military technology to civilian industries resulted in 3,275 contracts with a total value of 1.12 billion RMB ($350 million).[30]

The interview evidence, therefore, indicates significant progress in a number of Chinese industries in producing a variety of new product, process, and commercial innovations. Clearly, however, this is not true across the board. Most Chinese enterprises still labor under considerably less than complete autonomy; the incompleteness of the economic reforms remains a significant impediment to the full achievement of innovative capabilities. Despite all this, there has been an undeniable increase in both hardware and software industrial innovation in China that must be associated with the increase of market influences. Among the most significant patterns developing are the following:

Hardware Innovations	Software Innovations
*Some new products emulating world market styles.	*Quite visible improvment in distribution and service.
*Heavy emphasis on product quality improvement.	*General recognition of the need for a marketing function.

*Much more attention paid to industrial design in light industry.

*Gradual shift from purchasing to sales emphasis.

*Creation of technology markets.

Government policies still continue to exhibit a much stronger supply-push initiative than need-pull. Most likely this reflects three decades of an administered economy and the engineering background of the new enterprise leaders. Since market information is extremely difficult to obtain, even the enterprise wishing to be more sensitive to market influences is left without that information needed to make innovations that are in tune with the market. Finally, the disparity between a much larger demand for products and their supply diminishes the perceived importance for market sensitivity among many Chinese managers.

CHINESE MANAGERS AND INNOVATION

The professional limitations of China's managers themselves represent a barrier to increased industrial innovation. Prior to the recent economic reforms, these managers most typically operated in an environment that required little in the way of real competitive decision making. Markets were made for them, prices standardized, product differentiation discouraged, distribution provided. Such an operating environment by its very nature not only discouraged innovation but placed a larger penalty on making a mistake than the reward for a successful gamble. Like any managers anywhere, China's managers adjusted to these conditions by behaving in a risk-averse fashion. This translated into paying considerably more attention to the maintenance of existing operations than to the creation of new ones. Since risk taking was not perceived to be rewarded, R&D activities were unattractive.[31] Moreover, prior to the post-Mao reforms, most directors of Chinese enterprises were chosen on the basis of their political suitability rather than for their technical or managerial expertise or commercial accomplishments. This also worked to suppress innovation.

MARKET INFLUENCES AND THE COMMITMENT TO R&D

Despite the very visible evidence of both "hard" and "soft" industrial innovation cited in the previous sections, it is not entirely clear how market influences affect the industrial R&D function within Chinese industry. The conclusions of a recent paper,[32] reporting on an empirical analysis of data collected in interview with 65 managers attending the Dalian training program[33] in 1984, suggest that, as of 1984, industrial R&D resources, assembled specifically to pursue innovation, were committed by a central planning mechanism that associated enterprise size and visibly dynamic industries with a priority need for R&D talent. Those enterprises that did not fall into either category apparently did not get the scarce human resources necessary for R&D. To the extent that such findings are representative of China at large, it appears that the ability to support an active R&D program was still very much the result of earlier centrally controlled resource allocations in which market influences or innovative performance played little or no role.

Even where market forces are introduced, in a country as big as China, where demand has traditionally exceeded supply and where, under central planning, there was little or no incentive for managers to manage anything other than factory operations, it should not be taken for granted that market forces will induce the desired result. In fact, statistical analysis to differentiate between those enterprises that perceive market influences as being important and those that do not suggests that it is the level of formal education of the manager of the enterprise that is the most important variable in explaining the perceived importance of market influences in the enterprise's planning. Introducing market forces into an economy, therefore, will be far less effective in stimulating innovation and industrial change if the managers in that economy are not predisposed, through education and incentives, to recognize the opportunities inherent in such market influences and respond to the new operating environment.

TECHNICAL DECISION MAKING AND MARKET INFLUENCES: SOME IMPLICATIONS

In many ways, the anecdotal impressions of Chinese innovation and the reports of empirical analysis suggest that market forces might be stimulating apparent change but that changes necessary to sustain and improve future industrial innovation may not be occurring. One way of checking this hypothesis is to look at technical decision making within Chinese enterprises to see if significant differences exist between those firms where the manager perceives market influences as being important to the enterprise's planning and those enterprises where this is not the case. Although preliminary in form and not yet analyzed, data collected on technical decision making in approximately 75 enterprises represented by managers at the Dalian center in 1983 and 1984[34] suggest that market influences are not particularly important in explaining differences in technical decision making. In fact, one is struck by the lack of weight given to considerations such as potential profit, the enterprise's reputation, or the enterprise's ability to bring a product to market among enterprises that characterize market influences as being important to their planning. There is little evidence, in fact, that market forces are actually influencing the process of technical decision making in Chinese enterprises.

It may be too early to evaluate the overall impact of market forces on industrial innovation in the Chinese economy. Something is happening, but just what is difficult to say. It seems clear that market forces are producing commercial "software" innovations and some improved industrial design of consumer goods. Although the effect of importation of foreign technologies has not been the focus here, it does appear that much adoption of these technologies has been the direct result of increased market forces. All these changes are important and have enlivened the Chinese economy. In many ways, however, they represent relatively "easy" changes.

Inside the Chinese enterprise, it is more difficult to make the case that market influences have changed the way China innovates. Interview data suggest that a substantial proportion of China's managers still are not concerned with market forces because they believe that "the market that they are serving is so large that they don't need to innovate."[35] Futhermore, it appears that an enterprise's commitment to innovation is frequently more the result of the historical centrally controlled allocation of scientific and technical talent to the enterprise than the presence of market influences. In short, it is not at all obvious that market forces have yet had any major or lasting impact on the process of Chinese industrial innovation. Perhaps this is merely a matter of the short time in which market influences have existed. Perhaps more serious impediments still exist which undermine the ability of market forces to stimulate innovation.

The one bright spot in this study lies in the recognition that education apparently leads managers to perceive market forces as an important influence on enterprise planning. This suggests that China's continuing efforts to upgrade its managerial cadre should lead to a more widespread and deeper appreciation of what a market means, which, in turn, could lead to improved industrial innovation. While any such linkages will undoubtedly take time to establish, they should, once established, prove to be more durable and dependable than changes instituted by government fiat. Thus, despite some of the short-term ambiguity described in this chapter, there is considerable evidence for optimism in the long term.

CHAPTER SIX

Organizational Reforms and Technology Change in the Electronics Industry: The Case of Shanghai

DETLEF REHN

During the last several years, the development aspects of regional high-technology economies have been a topic of great interest in the West.[1] Due not least of all to the success of California's Silicon Valley, the various economic, technological, and infrastructural conditions for establishing high-technology areas have been examined not only in the United States but also in Japan, France, England, the Federal Republic of Germany, and other countries. Even though each country and region has its own (and not always suitable) conditions for setting up high-tech centers, the establishment of Silicon Valley-type areas has seemed magically to promise not only a firm foothold in future-oriented industries, notably microelectronics, but also a safe means to overcome employment and structural problems.

Analyzing the conditions under which Boston's Route 128 was

transformed into a world center of high technology, Nancy Dorfman has called attention to a number of aspects for planners and policymakers to consider when they seek to duplicate the experience.[2] First, she suggests that a high-technology economy need not be planned, but may occur in a relatively spontaneous manner. Second, Route 128 has benefited from Massachusetts's excellent technological infrastructure, particularly academic institutions, which focus on research at the frontiers of high technology and electronics and thus push development once it had started. Third, advantages resulting from a spatial concentration of enterprises of certain branches in one area were a critical factor. And, fourth, development was propelled by a number of new firms founded after the onset of the electronics revolution, often started as spin-offs from other high-tech firms or from university laboratories.

Discussions in the West on the development conditions of high-technology areas have been followed closely in the PRC. When then Chinese Premier Zhao Ziyang, in October 1983, introduced the idea of a "new technological revolution" into the discourse on China's economic and scientific development, "Silicon Valley Fever"—to use the title of an American bestseller—became the rage throughout China.[3] Zhao Ziyang's speech triggered a response by China's leading social scientists in a series of nationally publicized lectures on the "new technological revolution."[4] Ma Hong, former President of the Academy of Social Sciences (CASS), for example, discussed how China should respond to economic, industrial, and technological changes in the West.[5] Other lectures analyzed the technological development of microelectronics and other new technologies (bioengineering, new materials, and so forth) and their future economic and social role in the twenty-first century.

As to electronics, and in particular microelectronics as the pioneer sector of the "new technological revolution," two of the aforementioned conditions for the establishment of regional high-tech economies are of special importance to Chinese leaders. They fully agree that, without spatial concentration, the establishment of a technologically advanced Chinese electronics industry would not be possible. Moreover, in order to reduce high investment costs and to reap high technological and economic benefits, this concentration is regarded as efficient only in areas that already possess an adequate technological infrastructure, an experienced and skilled manpower pool, a well-developed industrial sector, and suitable environmental conditions.[6]

In order to bring the regional high-technology areas into full effectiveness, the Chinese view as necessary the solution of several problems.[7] First, since the traditional industries (for example, steel, textiles) are the backbone of Chinese industry, the development of new technologies, and microelectronics in particular, must be directed toward the technical transformation and renovation of those industries. Second, to encourage a steady flow of modern equipment and facilities from the high-tech industries, modernization of basic manufacturing technologies has to be given high priority. In particular, machinery in use, frequently too old and outmoded to meet the demand for advanced equipment, must be upgraded. Third, since China's domestic technological and economic capabilities do not allow the establishment from scratch of regional high-tech economies, importation and quick absorption of foreign technologies will be a critical factor. And, fourth, short-term economic considerations have to be linked with long-term development perspectives, so that economic incentives as well as a solid future-oriented technological synergies can be realized.

China's emphasis on the "new technological revolution" and the reorientation of electronics development has had a particularly strong influence on the development strategy of Shanghai, historically, and in the contemporary period, one of China's principal economic centers. Shanghai leaders have conducted a lively, and at times contentious, debate over their appropriate response to what Alvin Toffler has labeled the "third wave" of the global industrial revolution. In many respects, Shanghai's reaction, or lack of reaction, is a microcosm of China's general response. It may, therefore, serve as an excellent example of the problems the Chinese will confront in their efforts to reform their industrial and economic system.

THE DEVELOPMENT STRATEGIES FOR CHINA'S ELECTRONICS INDUSTRY, 1978–1985

Since the inception of the modernization policy in late 1978, the contents of the development strategies for China's electronics industry have passed through two major stages. First, the proclamation of the Four Modernizations led to a reassessment of the position of the electronics industry within the Chinese economic and industrial system.

While, prior to 1978, the electronics industry mainly served military needs, the new post-Mao leadership labeled electronics the "hallmark of a country's level of modernization." Consequently, it set very ambitious goals: (1) to raise substantially the technological levels of China's electronics industry; (2) to let electronics play an outstanding role in providing other sectors with advanced technology; and (3) to close eventually the gaps with the Western industrialized countries.[8] Within this context, priorities were placed on the areas regarded as most critical for China's future electronics development, notably computers and sophisticated components, but also on consumer electronics so as to realize a more balanced product structure.

These ambitious objectives reflect the peculiarity of the Chinese situation at that time. The "open-door " policy, and the liberation of scientists, economists, and planners from the ideological constraints of the Maoist period, led to a new atmosphere of "starting fresh," in which everything and anything seemed possible. Although disillusionment soon followed and the targets of the modernization program had to be adjusted, the goals of 1978 are important for two main reasons. The definition of electronics as the yardstick of China's modernization level reflects the growing awareness by the Chinese leadership that electronic technology is not confined to national defense, but has an overall impact on the economy and the society. Second, identifying a number of priority areas indicates the leaders' recognition that the simultaneous development of *all* sectors of the electronics industry would not be a successful strategy, but would probably end up in the widening of the technological gaps with the industrialized countries. Moreover, it informed the various actors in the electronics area of the strategy regarded as best for China.

Second, the new "strategic goal" of the modernization policy promulgated at the 12th Congress of the Chinese Communist Party in 1982—that is, the quadrupling of the 1980 agricultural and industrial gross production value by the year 2000—was paralleled by the "discovery" of the productive character of electronics technology. Broad application of electronics in each economic sector promised large increases in labor productivity and financial revenues. In consequence, the focus of attention shifted to application as the ultimate goal of electronics technology. The result was a basic restructuring of the product mix of the electronics industry, notably by emphasizing the production of all sorts of civilian electronic goods. In particular, the computer sector was affected by the new strategy because its activities were directed toward user-friendly mini-and microcomputers, peripheral equipment, and especially software to support broad computerization.

As in other economic sectors, it was recognized that basic reforms of the industry's structure and the elimination of a number of bottlenecks would be required to realize the envisioned goals. Factors identified as major obstacles to the development of the electronics area were:

(1) the organizational structure, characterized by strong vertical barriers, for example, between research and production units, and between central and local institutions;

(2) the regional dispersion of China's electronics institutions, which does not correspond to the requirements of specialization, division of labor, and economic efficiency;

(3) the low technological level of existing enterprises and manufacturing technologies, which explains to a large extent poor innovation in the electronics industry and the low level of competitiveness of electronic products originating from China;

(4) the low knowledge level of workers and staff, lack of regular knowledge upgrading, and inefficient use of qualified personnel.

The decisions of the Party's Central Committee on the reform of China's overall economic and S&T system, adopted in 1984–1985, set the pattern for the electronics industry.[9] The general thrust of the reforms of the economic and S&T structure is to let the market forces play a larger role in stimulating technological development and enterprise performance. In response to these decisions, the Ministry of Electronics (MEI) in early 1985 elaborated a plan to institutionalize a system of *branch management* to overcome the vertical barriers between the many units involved in the management of the electronics activities.[10] One of the most far-reaching measures was the decision to remove (*xiafang*) all enterprises from under the direct control of the MEI and to put them under local supervision. In a second step, these enterprises were to be linked with local institutions to form branch or product-related corporations which would integrate research, production, sales, and service institutions.

In preparation for the 7th Five-Year Plan (1986–1990), a debate over the development strategy for China's electronics industry, which took place throughout 1984, culminated in the release of a document entitled "The Strategy for the Development of China's Electronics and Information Industries."[11] It was elaborated by the State Council Leading Group for the Revitalization of the Electronics Industry (Guowuyuan Dianzi Zhenxing Lingdao Xiao Zu). The leading group–formerly know as State Council Leading Group for the Computer and LSI Industry– was originally established in late 1982 as a government body to coordinate and control the various activities in the electronics field. In January 1985, the mandate of the leading group was expanded to include communications, a field previously outside its jurisdiction.[12] The new name reflects the expansion of responsibilities.

The strategic guidelines set for the electronics industry, as laid down in the decision of the Leading Group, can be summarized, as follows:[13]

(1) The overall goal of the electronic sector is to better serve the development of the national economy and society. Since microelectronics is a "multifaceted technology" that affects all economic and social sectors, its development will be given top priority.

(2) A specific task of the electronics industry is to support the modernization of the traditional industries by constantly providing modern equipment. To fulfill this task, the product mix of China's electronics sector has to be restructured. Emphasis will be placed on integrated circuits, computers, telecommunications, and software. In order to make Chinese computers more user-friendly, attention will be paid to the "Sinification" of computers, that is, the attainment of character-processing capabilities in computers.

(3) The selective import and rapid absorption of foreign manufacturing technologies, know-how, and equipment should help to realize more quickly envisioned goals as well as to enable China to "leapfrog" over a number of development stages.

While these guidelines define the long-term direction of China's electronics industry, the short-term goals, laid down in the 7th Five-Year Plan, indicate important policy changes in two major aspects.[14] The first concerns the relations between consumer electronics and industrial electronics; the second, regionalization as a development factor.

Since the inception of the modernization policy, the role of consumer electronics as a driving force in China's electronics industry has been interpreted in different ways. In 1981–1982, the pent-up demand of the Chinese people, together with the chances for the producers to realize high profits, quickly resulted in a massive and uncoordinated production boom in consumer electronics goods, which, however, were frequently of poor quality. Nonetheless, the Chinese leadership supported this shift toward consumer electronics, since it was in line with the prevailing strategy to study, and possibly follow, the successful example of other East Asian countries, like Japan, South Korea, or Taiwan, where consumer electronics has given the major impetus to overall electronics development.

By 1983, the Chinese leadership recognized that the physical limitations of the civilian electronics production capacity, which mainly manufactured consumer electronics goods, would not allow overall electronization, as implied in the "quadrupling goal." Therefore, during 1983–1984, the MEI repeatedly called for a development strategy focused on industrial electronics, and efforts were made to increase the output of industrial electronics goods.[15]

In the 7th Five-Year Plan, however, a mixed strategy will be carried out in the electronics sector. The *short-term* focus will revert to consumer electronics, because both the Leading Group and MEI recognize that consumer electronics constitutes the largest *existing* market for China's electronics industry. Although industrial electronics is also highly in demand (for example, CAD and other sophisticated equipment), due to the lack of an efficient manufacturing base, it is still only a *latent* market.[16] This, however, does not diminish its role as a major driving force in China's *long-term* electronics development.[17]

In addition to the realization of considerable profits, which will ease somewhat the burden of the state and local budgets, other positive effects of the focus on consumer electronics include production in large series, product quality, and a more efficiency-oriented production management as highest priority. Moreover, consumer electronics is one of the first sectors in China's electronics industry that is able to manufacture products of world-market standards.

The goal of setting up of a few regional high-technology centers, notably in the coastal areas, marks a definite departure from the old strategy of establishing comprehensive regional systems throughout the country. Given the importance of the electronics industry for China's future modernization, this linkage of a regional and sectoral policy means that enlarging regional disparities are accepted as necessary costs in the development process. The gains of such a policy are seen in the creation of a better structural environment to catch up with the industrialized countries. Moreover, the planners hope that the interregional gaps may also stimulate the economic and technological activities of the backward areas to compete with the advanced regions.[18]

STRATEGIC GUIDELINES FOR SHANGHAI'S ELECTRONICS INDUSTRY, 1978–1985

The quest for an appropriate strategy for Shanghai's electronics industry reflects the zigzags in the development strategy at the national level. Between 1978 and 1981, prior to the reformulation of China's regional policy and the debates over the most suitable site of China's first "Silicon Valley," the central government in Beijing looked on Shanghai mainly as the country's most important supplier of electronic as well as other industrial products. This attitude reflected the outstanding role Shanghai has traditionally played in China's overall economic development and was buttressed by the fact that Shanghai at the time still had a leading position in the production of many priority goods, notably in the consumer electronics sector. In 1981, for example, Shanghai accounted for 22.2 percent and 37 percent respectively of China's entire TV and tape-recorder production.[19] Consequently, in the long-term development program for the electronics industry, the center expected Shanghai to take advantage of its high concentration of R&D facilities and skilled manpower and to contribute especially in a number of technology and knowledge-intensive areas, such as computers and large-scale integrated circuits (LSI).

The central government apparently was not much concerned about decreasing growth rates of Shanghai's overall industrial performance in the late 1970s. But Shanghai's foremost social scientists, as early as late 1979, responded to the situation by starting a debate over the city's most appropriate development strategy. The debate was important because, for the first time, the advantages and weaknesses of Shanghai's industrial and economic system were broadly discussed and criticized. The majority of the discussants recommended that smokestack industries, responsible for many of Shanghai's pollution problems, be "evacuated" to China's hinterland areas. Moreover, they suggested that, because of Shanghai's high technological level and rich management experience, priority be given to the production of new high-quality, highly sophisticated, and high-precision products.[20] Shanghai's advantages would also facilitate the importation and the absorption of foreign technology, thus further supporting the restructuring of its existing industrial system.

The debate also clarified why electronics is advantageous for Shanghai's economic development. Electronic technology saves raw materials and energy; its level of pollution is comparatively low; and the electronics industry requires limited space for production. Moreover, electronics technology is both knowledge- and skill-intensive, areas in which Shanghai excels. Therefore, focus on electronics technology promised substantial gains in the long run for the city's overall economic and social development.

In the early stages of the debate, however, this view was not shared by the central government. One reason was that the vision of Shanghai's "vanguard role" in China's modernization drive had not yet developed. Furthermore, owing to limited local financial resources, Shanghai's economic and industrial restructuring would have meant a considerable monetary commitment for Beijing, because Shanghai's remittances to the central budget were calculated on the basis of its existing industrial structure.[21] Thus, any basic reform, notably the removal of certain industrial branches to other parts of China, would have forced the central government to review basically China's financial system as well as its industrial policy.

What is more, as mentioned above, with respect to electronics the central leadership in Beijing still lacked a clear conception of the links between regional and sectoral policies. Important high-technology developments were not necessarily supported in those areas where the existing technological and economic potential promised a quick and high return on investment. Rather, the affiliation with a certain administrative system or personal connections were more important factors in getting funds. This problem was illustrated in an article in the Shanghai journal *Social Sciences* of early 1982, which complained that, although China had imported seven production lines for manufacturing LSI in recent years, and although Shanghai had very favorable—probably the best—conditions to produce these components, not one of these production lines had been installed in Shanghai.[22] The decision to bypass Shanghai reflected the tendency of the MEI, which does not wholly control any production units in Shanghai, to support only its own units as opposed to those belonging to the various municipalities.

After 1983, the new policy to accept regional imbalances as a motor for development and to emphasize the establishment of economic zones and hub cities put Shanghai's role and the development of its electronics industry in a new perspective. Underlying this reassessment was the concern of the central leadership that Shanghai's continuing economic problems– a low growth rate of industrial production, the insufficient renewal speed of industrial products, and slow progress in science and technology–would sooner or later seriously affect the overall economy and thus have a negative influence on the whole modernization program.[23]

In response, the central authorities in Beijing began to participate directly in planning the new strategy for Shanghai's economic, industrial, and technological development. For instance, in February 1985, the State Council approved the "Draft Report on Shanghai's Economic Development Strategy," which was jointly elaborated by the central government and the Shanghai authorities.[24] Moreover, in November 1986, the central government in Beijing formally agreed with the "General Program for the Urban Development of the City of Shanghai" which set the guidelines for Shanghai's urban development until the end of this century.[25]

The new strategy gives the electronics industry a premier role in Shanghai's development. The actual objectives include:

(1) A substantial increase in the share of the electronics industry in Shanghai's agricultural and industrial gross production value; it is planned to increase this share from 4.4 percent in 1980 to 6.6 percent in 1990 and 13.3 percent in the year 2000.[26]

(2) A basic change of the product mix of Shanghai's electronics industry, particularly by increasing the share of the computer and microelectronics subsectors.

(3) Expansion of the supply and the application of modern electronics technology and equipment to improve the competitiveness of Shanghai's industry, especially in the world market.

SHANGHAI'S TENSIONS WITH BEIJING OVER THE ELECTRONICS INDUSTRY

The reorientation of the electronics industry to meet China's future technological and economic needs requires three major reforms:

(1) A basic change in the structure of decision making in order to overcome the vertical organizational barriers between the various institutions engaged in electronics. These barriers impede the concentration of the available resources to tackle and solve the major development tasks.

(2) Reform of the S&T system so that it can meet the increasing demand for advanced sophisticated electronics products.

(3) Reorientation of the electronics industry toward the market in order to make it economically efficient. Because Shanghai is one of the centers of China's electronics industry, it is to play a key role in carrying out these reforms.

The structural reform of China's electronics industry is a combination of measures to centralize as well as decentralize the organizational system. On the one hand, in order to coordinate the development of the electronics industry along clearly fixed strategies, it is necessary to have a powerful supra-organizational institution. Thus, the establishment of the State Council Leading Group for the Revitalization of the Electronics Industry in late 1982 gives the center control over an industry to which it attributes a crucial role in China's modernization. On the other hand, the central planners need to give more freedom to research and production units in order to facilitate advanced technological development. Organizational measures to attain this goal include the removal of enterprises under direct control of the MEI and the establishment of branch companies that act as economic entities.

Accordingly, Shanghai centralizes and decentralizes its electronics industry. Following the creation of the State Council leading group, the Shanghai Leading Group for Computers and LSI, headed by Vice-Mayor Liu Zhenyuan, was established in early 1984. Its original purpose was to coordinate and control overlapping activities, especially with respect to computer development and massive imports of microcomputers as China "discovered" the "new technological revolution."[27] With time, it assumed the broader responsibility of providing some coherence in the development of Shanghai's burgeoning high-tech industries.

While the role that the State Council Leading Group will play in China's electronics industry has become increasingly clear, the exact function of the local Leading Group within the decision-making structure of Shanghai's electronics industry is still uncertain. In principle, Shanghai's Leading Group is mandated to follow the guidelines set by the State Council Leading Group. However, unlike the local electronics industries of Beijing and Jiangsu province, which have strong personal links with the State Council Leading Group and the central ministries, Shanghai has no formal representatives in the central decision-making organs.[28] This has resulted in irritations and frustrations within the Shanghai leadership over Beijing's attitude toward the municipality's high-tech program. For example, in the face of Shanghai's efforts to establish a microelectronics center in its Caohejing district, Wuxi ascended to the position of a major Chinese electronics base.

Compared with Shanghai, Wuxi's role for the Chinese electronics industry was insignificant until the early 1980s. This changed when the MEI in 1980 decided to spend 250 million yuan for the importation of a production line for the manufacturing of linear integrated circuits (ICs) from Japan and to install it in the Wuxi Jiangnan Radio Equipment Factory, directly controlled by the Ministry. The factory's good economic performance and the advanced equipment quickly transformed Wuxi into one of China's leading centers for the production of ICs and into a major competitor for Shanghai. Wuxi's competitive edge was further strengthened by the establishment of the Wuxi Joint Corporation for Microelectronics Research and Production (Wuxi Wei Dianzi Keyan Shengchan Lianhe Gongsi) in June/July 1985.[29] It consists of the Jiangnan factory and the Wuxi branch of the Sichuan Research Institute for Solid-State ICs, one of China's most important semiconductor research institutes. The new company specializes in research and manufacturing of all sorts of ICs, notably for applications in computers, telecommunications, and consumer electronics. In January 1986, Vice-Premier Li Peng praised the new company as an outstanding example of development based on an efficient absorption of technology imports and expressed the hope that, within five years, the company would become a world-class microelectronics center.[30] In May 1986, the new corporation gave evidence of its performance by developing a 64K dynamic RAM.[31]

The effort to establish a microelectronics center in Shanghai's Caohejing district was originally a local project under the auspices of the municipal government, designed to pool scattered technological and manpower resources and to give momentum to Shanghai's IC industry. In line with its attitude to Shanghai's overall economic development, the central government in Beijing in the early planning stage of the project (1982–1983) was not very interested in giving any substantial support, since Caohejing primarily included local electronics institutions in Shanghai, and because the mere idea of setting up high-technology areas ran counter to the economic policy prevailing at that time.

The discussions about the establishment of a Chinese Silicon Valley changed the center's attitude toward Caohejing. In a lively debate, arguments in favor of both Shanghai and Wuxi as possible sites for a Silicon Valley-type microelectronics center were put forward.[32] In July 1986, a final decision was made at a work conference of the MEI in Shanghai—that further development of China's electronics industry would *generally* be centered around four areas which already had a certain technological and infrastructural basis: Beijing, Shanghai, Jiangsu, and Guangdong.[33] Division of labor among the different areas was to be based on the particular concentrations of research units, universities, key enterprises, and institutes. The Beijing area would focus on sophisticated electronic research and development; Shanghai would be a center for broad application of electronic technology especially aimed at transforming the "traditional" industries; Guangdong would be transformed into a Chinese electronics export area; Jiangsu would take advantage of its high concentration of electronics enterprises to realize economies of scale and set up large-scale corporations to produce world-standard products.[34]

Within this context, the central government, that is, the State Council Leading Group, has assumed responsibility for coordinating the management of the Caohejing project in line with China's overall microelectronics development. Today, Caohejing is a *national* project aimed at the creation of facilities for the development and production of LSIs. As a first step, the strategy calls for the import of a 64K-RAM production line (3-micron technology) to launch the effort. Initial investment will amount to approximately RMB Y 100 million; 80 percent is to be provided by the Shanghai municipal government and 20 percent by the central government. Key participants in the project include the Shanghai Institute of Metallurgy of CAS, the Shanghai Components Factory #5, and the Shanghai Radio Factory #14.

Caohejing's status as a national project most likely means that it will now have adequate funds. It also facilitates the gathering of a critical mass of semiconductor specialists in Shanghai, thereby possibly relaxing the constraints on manpower mobility which heretofore had limited development. It also may give Shanghai expanded access to advanced foreign technology–which largely has been unavailable because of foreign-exchange limitations. Moreover, due to the considerable financial commitment of both the central government and the Shanghai municipality, there may be stricter monitoring and control at each stage of implementation, thereby increasing the chances of overall success.

Within the strategy of China's IC development, the microelectronics centers of Wuxi and Shanghai very likely complement each other. Wuxi will primarily focus on linear ICs, although it also will manufacture digital ICs, such as microprocessors and memories; Shanghai, on the other hand, will concentrate on the production of memory chips.[35]

The concentration of resources to establish a high-technology center contrasts with the decentralization of the decision-making apparatus of Shanghai's electronics sector. At the national level, the municipality's electronics industry is characterized by the coexistence of different administrative systems. The Industrial Bureau for Electronics and Instrumentation oversees the planning, research, and production of the *local* electronics industry. Other important decision makers include the Ministry of Space Industry (MSI) which, via its Great Wall Corporation, substantially contributes to Shanghai's electronics research and production, and the MEI, which controls a number of research institutes located in Shanghai (for example, the East China Institute of Computer Technology). Cooperation among the institutions belonging to different systems was often prevented in the past by high vertical organizational barriers. Therefore, the reform of Shanghai's electronics industry stresses the dismantling of such barriers by setting up branch or product-related companies in all areas of the electronics industry to handle the full cycle of research, production, application, and service activities.[36]

The aforementioned MEI report on the reform of electronics management specifies that the Ministry represents the State Council in guiding the electronics sector. Accordingly, local industrial bureaus should represent provincial and municipal governments in planning and management of the electronics industry at the local level. Utilization of economic levers and indirect control by the state organs would gradually replace administrative measures and direct control of the business activities of the electronics firms. In Shanghai, these guidelines mean that the new electronics companies will be responsible for micro-level management, while the Industrial Bureau will primarily deal with macro policymaking.

It is difficult to assess the chances for the success of such a basic reorganization. In view of the many actors engaged in Shanghai's electronics industry, only a powerful organ that is able to harmonize the different interests can push through such a reorganization. As presently structured, the local industrial bureau seems too weak to implement the reforms, as indicated by the fact that in 1984 the coordination of Shanghai's local computer activities was placed under the control of the Shanghai Economic Commission. Therefore, in order to carry out the organizational reforms, either the position of the bureau has to be raised substantially or another institution, such as the local leading group which already acts as interorganizational body, must be placed in charge of restructuring Shanghai's electronics sector.

Market orientation lies at the heart of the S&T reforms as well as the economic system. The establishment of *technology markets* is regarded as a key element of S&T reforms. Because technology is now treated as a commodity, technology transfer from the research into the production sector is no longer free of charge, but is subject to supply and demand, which means competition and profit making. Technology markets aim primarily at improving innovation by directly linking suppliers and users of technology. They are also an important means of the reforms of the S&T funding system since government allocations of operating funds to research institutes are gradually cut, and the institutes have to open up new funding channels.

Since Shanghai's introduction of technology markets into its economic system in 1982, the volume of transactions handled through such markets has exploded. In 1985, the total value of business transactions in the technology markets amounted to RMB Y 580 million which is 25.2 percent of China's overall value.[37] Although detailed statistical information about the effect of technology markets on Shanghai's electronics industry is not available, the impact is apparently quite considerable. In 1983–1984, the Shanghai Components Factory #5, one of China's leading semiconductor producers, for example, received about 50 percent of its research and production tasks from the MEI, the State S&T Commission, and the Shanghai Bureau for Electronics and Instrumentation. But, in 1985, as a result of the S&T and economic reforms, about 90 percent of the factory's activities were self-initiated.[38]

This does not mean that a market orientation already pervades all levels of Shanghai's electronics industry. On the contrary, technological innovation continues to be primarily driven by technological opportunity and the desire for prestige rather than end-user demand. For example, as elsewhere in China, Shanghai's IC institutions are mainly interested in raising the integration density of the IC, but not in massproducing the newly developed chips, a very complicated process which requires a number of technological and managerial adjustments at the enterprise level. As a result, many newly developed chips remain prototypes which never reach the production stage.[39] On the other hand, Chinese ICs under production tend to be very expensive and of poor quality.

Market orientation, however, will undoubtedly influence the direction of Shanghai's electronics exports. In 1986, China's electronics exports within the China Electronics Import and Export Corporation (CEIEC) amounted to US$113 million of which the Shanghai branch of CEIEC accounted for around 7 percent (US$8 million).[40] At present, electronics goods exported from Shanghai are still relatively unsophisticated. For example, the 1984 major export products included 100,000 toasters (total value US$1 million), radios, and simple components. For 1990, China envisions an electronics export volume of around US$700 million. In order to realize this goal, the industrial structure has to be basically changed with emphasis on knowledge and intensive technology.[41] Consequently, fundamental changes may take place in the export structure of Shanghai's electronics industry, with the emphasis on sophisticated consumer electronics products (for example, color TVs and videorecorders), microcomputers, and especially computer software. These are areas where Shanghai has gained a reputation in international markets over the last few years. Contracts with Japan already exist, and the Nanyang International Technology Company, run by Shanghai's Jiaotong University, is developing software for the US market.[42]

As to the future development of Shanghai's software industry, Jiaotong University has also made a number of proposals. They aim at taking advantage of Shanghai's technology and manpower potential to establish a software-export development center.[43] The decision to set up a national software factory in the Caohejing district is an indication that Shanghai's software-related activities are also backed by the central government in Beijing.[44]

The success of a world-market-oriented electronics and computer industry as well as maintenance of Shanghai's leading position within the electronics area ultimately depend on raising the technological level of Shanghai's electronics industry. [45] Most of the equipment in use is of 1950s and 1960s vintage, which corresponds neither to export requirements nor to increasing demand for advanced electronic technology and equipment. Therefore, emphasis is placed on the acquisition of new equipment and the modernization of the existing research and production facilities. Both are important elements of the *technical transformation* of Shanghai's electronics industry. In 1984, investment in fixed assets of Shanghai's electronics industry totaled RMB Y175 million, the majority of which went to facilities manufacturing semiconductor components and parts rather than final products. Moreover, between 1983 (when technical transformation of Shanghai's electronics industry began) and 1985, the city was provided US $100 million for the purchase of foreign equipment, of which about US $50 million was spent for the IC (LSI) industry.

Yet, the measures introduced to improve the existing technological infrastructure of Shanghai's electronics industry so far have had modest results at best. Despite Shanghai's acquisition of a wealth of domestic and foreign equipment, its electronics research and production units have not achieved real technological breakthroughs, for many reasons. The efficiency of newly acquired equipment often tends to be low in Shanghai's electronics industry. To integrate the new equipment successfully with the existing equipment, a careful analysis of an enterprise's existing technological level should precede the acquisition. Such an analysis includes an assessment of the equipment in place, the educational level of the technical personnel, and the requirements of mass production compared with the usual prototype development. However, most often the introduction of new technology is not well coordinated, primarily because many enterprises are forced to undertake wild spending sprees in order to expend their foreign-exchange allocation in a certain period or lose it. Moreover, there is seldom regular manpower training to meet the technical requirements of the newly acquired equipment. All these factors have resulted in a wide variety of equipment from different sources in many of Shanghai's leading electronics research and production units. This may explain the difficulty of establishing a well-organized, efficient production system anywhere in China.

The foregoing discussion has shown that Shanghai's electronics industry is in the process of important reforms which are of crucial significance to China's overall electronics sector. The organizational reforms not only aim at the abolishment of vertical barriers and the installation of a flexible decision-making apparatus; they also reflect the efforts to create an institutional body that may be able to give a spur to and set the course for future development. Technological and economic reforms, on the other hand, are being introduced in order to make China's electronic products more competitive and to narrow the technological gap with the advanced industrialized countries.

In this process, the experience of other East Asian nations, such as Japan and South Korea, may serve as important illustrative material. The East Asian "electronics system" has the following characteristics:[46]

(1) It is highly integrated and diversified. By focusing on consumer electronics in the early stages of development, it has also influenced technological innovation in the semiconductor industry and is now moving toward industrial electronics.

(2) The formulation of national industrial policies set the scope of orientation, notably by making electronics the key area of the future industrial structure.

(3) Foreign technology has been used as a catalyst for development.

(4) Since most East Asian electronics manufacturers pursue global strategies, there is permanent competitive pressure to develop new products and introduce new manufacturing technologies. This requires considerable organizational effort as well as optimization of the various technological and economic factors, such as mastery of

the production technologies, development of managerial skills for mass production, and gradual creation of a marketing network.

China is attempting to move in this direction, as may be seen in the role the leadership attributes to foreign technology and to electronics becoming the pioneer sector for the country's economic and social development. Even more important, China is beginning to understand the dynamics of market orientation and economic efficiency. An indication is the lively debate among scientists and planners over the direction of China's IC industry which started in autumn 1986. The new element of this discussion has been the very critical attitude toward the technological and economic results that have been achieved in the sector since China's first IC was developed in 1963–1964. While it is acknowledged that some progress has been made, it is also undisputed that, without a basic reorientation toward the market of China's IC industry, no real breakthrough will be realized.

Against this background, China's decision to focus on regional high-tech economies and to establish science parks has a very real meaning. The concentration of scarce resources in a few areas may indeed be a means to achieve economies of scale and to move down the "learning curve" to attain a more efficient production and cost levels. Due to its rich technological and manpower resources, Shanghai has the chance to leave its mark on this movement. But, the municipality has to understand that it is integrated into an overall process, and that particularism is a hindrance to bringing its role into full effect.

CHAPTER SEVEN

Scientific Decision Making: The Organization of Expert Advice in Post-Mao China

NINA P. HALPERN

The post-Mao leadership's consensus that science and technology are crucial forces to promote economic modernization has led to the introduction of many reforms. Among these are reforms to promote "scientific decision making," the basing of government decisions upon scientifically derived knowledge rather than leadership whim. Concretely, this has meant adoption of measures to increase the role of experts—from both the natural and social sciences—in policymaking. Recruitment of educated individuals into leadership positions in both government and Party has been advocated; more important, organizational innovations in the form of new advisory bodies that increase and restructure the flow of expert advice available to the leaders have also been introduced.

The chapter examines the Chinese advisory system for S&T and economics, particularly the organizational changes introduced within the government since 1980. It asks what sources of expert advice the Deng leadership has established; how these new organizations operate, and how they relate to the rest of the government and to each other; and what their implications are for policymaking and politics in China.

BUILDING AN EXPERT ADVISORY SYSTEM

Post-Mao leaders, particularly Deng Xiaoping and his supporters, are committed to making better use of the nation's intellectual resources in order to achieve modernization. They have called for the Party and government to rely more on experts and expertise in formulating their policies. In May 1977, Deng argued: "We must create within the Party an atmosphere of knowledge and respect for trained personnel. The erroneous attitude of not respecting intellectuals must be opposed."[1] In July 1986, Vice-Premier Wan Li forcefully restated this need: "It is imperative to promote the practice of leaders' constantly exchanging ideas, communicating, and discussing questions on an equal and democratic basis with researchers and with those who have diversified knowledge and practical experience. Every leading department should have its own research group to rely upon in making policy decisions."[2]

As a result, many institutions capable of providing expert advice to the leadership have been established. Since 1980, a fairly extensive institutionalized expert advisory system has been created within the government, the political significance of which extends beyond its impact on the decision-making process. The new bodies were partly a response to increasing awareness of the inadequacies of bureaucratic advice. They also reflected a growing commitment to reformist policies that would restructure the economic and S&T systems in relatively fundamental ways; they were created partly to generate and legitimize such policies. They also place supporters of Deng Xiaoping and other reform-oriented leaders in high-level positions in the government and enhance the power of those leaders vis-à-vis potential challengers, such as Chen Yun. Finally, like Deng's simultaneous attempts to restructure the bureaucracy and alter the basis of cadre recruitment, they were also aimed at increasing the control of this reform group over the bureaucracy and ensuring implementation of his new policies. By placing experts in government advisory positions, the reforms alter the political balance of forces at the top. They also have the potential to alter the Party-dominated leadership's relationship to society by creating a powerful subgroup with an independent claim to legitimate political participation based on expertise rather than ideology.

Changing Perspectives on the Policymaking Problem

The new advisory apparatus reflects an evolving perspective on the problem of making good policy, quite different from the Maoist era and from the initial years after Mao's death. Mao made use of bureaucratic sources of advice, drawing in experts from outside the bureaucracy only to a limited extent. He was often dissatisfied with the quality of bureaucratic advice; among other problems, he complained frequently of bureaucratic conservatism and "departmentalism," or parochialism. His responses to these problems, although varied, centered on *motivational* rather than *structural* alternatives. To deal with problems of bureaucratic inefficiency, lack of control, and lack of coordination, "the most important device has been the establishment of political criteria for public office in addition to, or even in place of, technical and professional standards, and the development of programs of ideological indoctrination for public officials."[3] This response was premised upon the belief, both Confucian and Maoist, that properly motivated officials would not offer advice that reflected a narrow departmental perspective, but rather would demonstrate concern for the overall societal interest. Thus, the leadership would not be faced with conflicting advice from the bureaucracy and would have little trouble selecting the technically best policies.

When Deng Xiaoping regained power in late 1978, he sought to reverse many of the features of the Cultural Revolution period that he believed had produced "voluntaristic" decision making, including the "simplification" (that is, serious cutbacks) of the bureaucracy, the attacks on intellectuals and experts of all types, the extreme emphasis on political considerations, and the over-concentration of power in the hands of a small number of individuals. A better leadership, the reestablishment of bureaucratic structures and procedures, a new emphasis on intellectual activity and experts' advice, and a pragmatic attitude of "seeking truth from facts" would obviously lead to effective government policies that would permit rapid realization of modernization. Or so it appears Deng Xiaoping believed in 1978.

By mid-1980, however, the methods of achieving good policies did not seem quite so simple. At that point, Deng had achieved a number of impressive political victories and had established his preeminence within the Chinese political leadership. The bureaucracy had grown considerably, reestablishing and expanding its pre-CR data-gathering and research segments, such as the State Statistical Bureau and the many research institutes attached to the various economic ministries. Much had been done to improve the position and status of intellectuals. Certainly, the intellectual environment had loosened up considerably. Numerous bureaucrats and academic economists and scientists had been mobilized to do research on the economy and suggest new policies. Yet, decisions were still being made at various levels of the bureaucracy, and by the top leadership, that either created serious economic problems—such as the continued increase in capital construction—or failed to foresee and successfully address such serious problems as an inadequacy of energy supplies. At the end of 1980, the government announced a major economic retrenchment in order to cope with those problems.

These continued deficiencies in economic policy stimulated a rethinking of the procedures for successful policymaking. The Deng leadership did not waver in its belief than an effective decision-making process must depend on experts and must be largely bureaucratic rather than personalistic. But it no longer regards a strictly bureaucratically structured advisory system as adequate for government decision making. Like Mao, Deng and other leaders have expressed concern about certain negative features of bureaucratic behavior, such as the tendency to offer advice that suits the needs and interests of the particular agency without taking into account the broader picture, and the tendency to focus on the short run at the expense of long-run considerations. Unlike Mao's, however, Deng's responses to these problems since 1980 have been less motivational than structural. Instead of using campaigns to address these problems, Deng reorganized the advisory system by creating new bodies to supplement bureaucratic sources of advice and, more important, to structure that advice so as to overcome bureaucratic deficiencies.

In addition to differing from Maoist responses to bureaucratism, these bodies also differ from the mechanisms that the leadership used in 1978–1979 to obtain extra-bureaucratic expert advice and in part are a response to the shortcomings of those mechanisms. In 1978 and 1979, the leadership sought extra-bureaucratic expert advice on a largely ad hoc basis. In early 1979, a new Finance and Economics Commission, headed by Chen Yun, established four large groups made up of specialists drawn primarily from the bureaucracy and the Chinese Academy of Social Sciences (CASS) to provide suggestions on how to achieve "readjustment, restructuring, consolidation, and improvement" of the economy. These groups studied questions relating to the structure of the economy, the economic system, technology import, and economic theory and method, and submitted reports to the central decision makers.

These ad hoc efforts to incorporate experts into the decision-making process proved inadequate. They were no substitute for regular bureaucratic input; as one participant pointed out, they lacked "structure."[4] Groups without a permanent base in the government structure rarely have a sustained impact on decision making. One reason is that they are seldom given access to the full range of information available within the government. Moreover, these groups provided little help in reconciling the conflicting advice generated by various bureaucratic units, a problem the leadership was feeling most acutely. Finally, the groups focused on very general issues related to reform, but they did not provide a mechanism for the leadership to obtain an interdisciplinary perspective on concrete projects. For example, the proposal to build a huge dam on the Yangtzi River (the Three Gorges dam project) had been debated for decades, with various bureaucratic and local units providing sharply differing advice.[5] In deciding whether and how to move ahead with this project, the leadership needed expert advice to reconcile these conflicting views; it also needed to integrate various *types* of expert advice on different aspects of the project. In addition, the leadership needed to be able to enforce its priorities and control bureaucratic behavior.

The dissatisfaction with the existing decision-making process and mechanisms for obtaining expert advice led to establishment of a new advisory system to address these problems. Beginning in fall 1980, five advisory centers were set up directly under the State Council to provide advice on aspects of economic policy. In spring 1982, a System Reform Commission was established to organize studies and prepare proposals for the reform of the economic system. In late 1982, the SSTC created a National Research Center for S&T for Development (NRCSTD) to provide advice on science and technology policy. Numerous "leadership small groups," most important, one for science and technology, were established to promote bureaucratic coordination and provide advice on policy formulation and implementation in various areas of economics and S&T.[6]

Research Centers Under the State Council

The five permanent research centers created under the State Council between fall of 1980 and 1982 were: the Economics Research Center (ERC), headed by Xue Muqiao; the Technical Economics Research Center (TERC), headed by Ma Hong; the Price Research Center; the Rural Research Center, headed by Du Runsheng; and the Economic Legislation Center, headed by Gu Ming. In May 1985 ERC, TERC, and the Price Research center were combined to form an Economic, Technological, and Social Development Research Center (ETSRC), also run by Ma Hong. The centers bring together experts from both inside the government bureaucracy and outside it, from such organizations as CAS and CASS.

Each center has a particular purpose. The division of the ETSRC corresponding to the former ERC focuses on macroeconomic and long-term economic-policy problems, providing periodic reports on the state of the economy (much like the United States' President's Council of Economic Advisors). The division that was formerly TERC advises on technical-economic policies and feasibility of particular projects. The division corresponding to the Price Research Center (originally affiliated with the State Price Commission) works on issues related to price reform. The Economic Legislation Center, as its name suggests, advises on economic laws, reviewing and amending those drafted by different units under the State Council. The Rural Research Center advises on issues related to the rural economy. Although differing in focus, these bodies also have several common purposes, to:

(1) provide an alternative to bureaucratic sources of advice, geared more toward the needs of the central leadership than to a particular sector or institution;

(2) sort out conflicting views coming in from the different sectors of the bureaucracy and help the top leadership evaluate them;

(3) improve long-term planning; and

(4) channel extra-bureaucratic advice to the leadership on a regular or ad hoc basis.

TERC has been the most visible of the original five centers. Its relative importance is reflected in the fact that, when the three centers were combined, its director, Ma Hong, became the director of the new center while Xue Muqiao, the director of ERC, became vice-director. The following discussion will therefore focus on the operations of TERC and its corresponding division within the ETSRC ("TERC" will here refer to both of these units).

TERC: Structure and Functioning

TERC was set up in early 1981, under the official leadership of the State Council's Finance and Economics Leading Group (the group of officials most responsible for making economic policy). Nevertheless, the Leading Group was of small importance in supervising its work; the Center operated more as an advisor to the premier, or to the State Council more generally. Its purpose was to provide advice on questions of technical economics, that is, economic questions with a scientific-technical component.

In establishing TERC, the leadership sought not only to link scientific-technical and economic advice, but, more generally, to obtain a multidisciplinary perspective that could compensate for overspecialization and parochialism within the bureaucracy. Ma Hong, the director, explained that the center's purpose was to compensate for bureaucratic narrowness and biases and that it therefore must incorporate many types of experts:

> In this way we can avoid partiality, for example, the narrow viewpoint of a department. Everyone knows that a department regards everything it reports to the State Council for approval as reasonable; otherwise it wouldn't report it. But, if you seek opinions from related ministries and commissions, they'll put forward many opinions, even opposing ones. The reason is that each department, when reporting, frequently thinks only about departmental needs. But implementing these kinds of things affects many departments. Besides, sometimes some professional bias also exists. Because, if you follow a specialty for long, frequently you'll

overemphasize it and overlook related considerations. And there's another aspect, that some departments only emphasize short-term needs and lack a long-term perspective.[7]

Thus, because of its cross-departmental and interdisciplinary nature, TERC was expected to provide more objective advice than would be generated by bureaucratic agencies; likewise, it could address issues that cut across several departments, and thereby enhance policy coordination. In addition, as Ma Hong's statement suggests, it was expected to ensure utilization of a more long-term planning perspective.[8]

In addition to helping the leadership make better decisions, TERC and other centers were expected to provide leaders with the information needed to mediate more effectively inter-ministerial rivalries and turf battles.[9] Indeed, according to one of its officials, then-Premier Zhao Ziyang proposed TERC as a means of overcoming differences among bureaucratic agencies. Thus, it was also a way for the leadership to enhance its control over the bureaucracy and better enforce implementation of its chosen policies.

The makeup of TERC's staff also indicates the center's multiple purposes. It is composed of experts who differ in both training and likely policy commitment from the typical bureaucrat. It was originally run by a standing secretariat consisting of 10 officials drawn from research organs in the bureaucracy, as well as CAS and CASS.[10] When it became a division within the ETSRC, that leadership structure was replaced by 2 of the 6 commissioners who supervise the ETSRC's work. It has a research staff of about 30, including scientists, engineers, and economists. Many combine different types of expertise, such as a BS in science with a MA in economic management, and some are also trained in math or computers. Most have a higher degree, either a MA or PhD. Surprisingly, only a few have been abroad, and the quality of their training is reportedly of some concern to the officials of the center.[11] Although they have at least five years' work experience, most are under 40. Both their higher degrees and their relative youth suggest that they have not spent much time in any particular bureaucratic sector and are therefore likely to feel loyalty primarily to the leaders who appointed them. Reports by Westerners who have had some contact with these staff members suggest that they are bright and reform-oriented.

The smallness of the staff reflects the fact that the center was not intended to do much research itself, but rather to mobilize a wide range of experts from inside and outside the bureaucracy. The advantage of this structure, in the eyes of the leadership, was greater flexibility in bringing the appropriate type of expertise to bear on particular policy problems.[12] While the center sometimes turns to universities, professional associations, or even particularly qualified individuals to participate in a study, the desire for secrecy often limits the extent to which persons outside the bureaucracy can be drawn in. Accordingly, these centers draw educated individuals into the government, but open the decision-making process to nongovernmental experts only to a limited—albeit not unimportant—degree.

TERC's agenda is largely set by the offices of the State Council and premier, although it also independently initiates some studies it considers important. Its activities can be divided into two main categories. First, it provides advice on particular technical-economic policies or major construction items. This includes organizing feasibility studies on all major projects or technical-economic policies suggested to the State Council by a unit within the bureaucracy or by any other source, including a vice-premier and presumably even the premier himself. It also analyzes the results of policies already being implemented and suggests methods of improvement. More generally, it suggests policy options for resolving technical-economic problems that have been identified by the State Council or its own staff. Second, it organizes forecasting and planning studies for the economy as a whole or particular sectors or regions.

In evaluating and choosing among different capital construction projects, TERC appears to overlap in function with the State Planning Commission (SPC).[13] Undoubtedly, this reflects leadership's dissatisfaction with the Commission's performance.[14] Part of the motivation behind the establishment of TERC was to end fiascos like the Baoshan Steel Mill, on which the leadership had spent millions of dollars in 1978–1979 before realizing that the project was poorly planned, badly located, and entirely too expensive.[15] Such projects were the responsibility of the SPC. More generally, it was becoming clear that the SPC was either unable or unwilling to enforce the leadership's policy of cutting capital construction, perhaps partly because its ongoing bargaining relationship with the ministries made it difficult to make hard choices among their proposals. TERC, which has far more incentive to respond to the Premier than to the wishes of the ministries, was probably seen as a body that could assess the virtues of the latters' proposals more objectively. Since politics and policymaking are so often interrelated, Zhao Ziyang may also have wished to provide himself with an analytical capability to seize the initiative on the "readjustment" issue (that is, cutting capital construction and altering investment patterns), which had earlier rested with Chen Yun.[16]

Although one purpose of TERC is to help the State Council choose among competing departmental claims, neither its manner of operation nor its assigned role permits it to mediate those competing claims effectively. Because it must rely mainly upon the bureaucratic agencies for research, its primary functions are to ensure that important problems are studied by all relevant departments and to promote dialogue between the various actors that might resolve conflicts or illuminate important interrelationships. But conflicts inevitably remain, even after discussion, and TERC officials do not see their role as resolving these conflicts. As one researcher pointed out, TERC's relations with ministries and commissions are *pingxing*, meaning that in bureaucratic terms they are on equivalent levels. Thus, the Center does not have the authority to override departments in providing advice. Rather, it generally reports the full range of views to the State Council, and sometimes appends its own opinion and suggests several policy options.[17] According to one participant, TERC's final report to the State Council, laying out the different viewpoints, circulates through the bureaucracy, so that each unit has a chance to object to any misrepresentation of its position. The incentive for TERC to report accurately the views of different units is strengthened by the fact that each unit has the right to go directly to the State Council and plead its case if it feels it was unfairly treated (or, indeed, if it simply wants to tip the balance in its favor). Thus, TERC may promote better policy coordination and more informed leadership choices, but it clearly does not eliminate the competing claims on the State Council and its need to make choices among them.[18]

Early in 1986, the SPC moved to develop its own analytical capability. In February, *People's Daily* announced that the SPC and State Economic Commission (responsible for annual planning) would begin using a "China International Engineering Consulting Corporation" to evaluate and screen all applications for new large and medium-size construction projects. The Corporation, established in 1982, serves as an umbrella organization for a large number of technical-engineering expert consulting groups.[19] According to the *People's Daily* article: "Although in the past few years we have improved the decision-making procedures concerning construction projects, thus far we have not yet changed the situation in which decisions are made by administrative leaders alone. As a result, we are likely to pay sole attention to current needs, one-sidedly emphasize the interests of our own areas and departments, overlook scientific proof, and thus can hardly make comparisons between different proposals."[20] This statement suggests that TERC has failed to fulfill the role assigned to it. Equally possible, the SPC has enhanced its analytical capabilities for defensive purposes: to be able to point to the advice of its own experts in arguments with TERC or other advisory bodies. This suggestion is largely speculative, but the phenomenon of "using experts to fight experts" is widely practiced in the American bureaucracy. There seems no reason to believe that it would not be in China. Insofar as different bureaucratic units develop their own sources of expertise in order to fight bureaucratic battles, political conflict in China shifts to a new basis: expertise, rather than ideology. While still political, this "politics" operates in a new fashion and suggests that experts have acquired a more central position within the political system.

In addition to advising the State Council on specific projects and problems, TERC organizes long-term planning efforts. In 1983, Premier Zhao asked TERC to organize a large-scale study of "China to the Year 2000" and suggest appropriate policies.[21] About 400 experts over a two-year period produced a 13-volume report, examining such sectors as transportation, energy, population, and employment. These reports are said to contain concrete proposals, some of which were adopted in the 7th Five-Year Plan. The main, overall report was written by the TERC staff, but the 12 sectoral reports were done by other institutions; the S&T report, for example, was written by the SSTC Commission, while the transportation report was done by a research institute under the State Economic Commission in conjunction with some design institutes under the Ministry of Railways. TERC's procedures in this case also illustrate its use of extra-bureaucratic advice: Ma Hong asked the Chinese Association for S&T (CAST) to join in; the latter organized over 100 affiliated technical and professional societies to write sub-reports on such areas as electronics, manufacturing, and textiles, which were drawn on in writing the larger reports.[22] Without a body like TERC acting as intermediary, it seems highly unlikely that CAST's mandate to "play an advisory role for the government organs"[23] would ever be fulfilled, at least at the highest level of government.

TERC and the other centers, then, are a source of expert advice to counter the more narrow, shortsighted, and perhaps self-interested views of the bureaucracies, and an intermediary for transmitting views of a much broader group of experts to the leadership. They do not, however, challenge in any apparent way the leadership's control over policy development or ability—and need—to mediate bureaucratic conflicts. Nevertheless, their expertise and experts add a new dimension to such conflicts.

Other Sources of Expert Advice

The SSTC in October 1982 established a subordinate National Research Center for S&T for Development (NRCSTD) to improve policymaking in the area of science and technology.[24] Besides addressing real decision-making needs, this innovation may also represent a defensive reaction; the SSTC may have feared it was losing the initiative in the development of scientific and technical policymaking to TERC. The NRCSTD provides advice to the State Council through the SSTC and, on occasion, more directly. Its research topics are sometimes suggested by the State Council or the SSTC and sometimes internally generated. Like TERC, the NRCSTD relies primarily on experts within the bureaucracy and to a lesser extent outside it to perform much of its policy research. Unlike TERC, the NRCSTD does research on science policy and management of science, but in other respects its mission seems much like that of TERC.

One aspect of the NRCSTD's work involves long-term planning in the area of technology. For example, in 1983, the Center, in conjunction with the State Planning Commission and State Economic Commission, organized many different units to study 13 types of technical policy, such as energy, materials, and computers.[25] The NRCSTD first determined the general policy for the area (for example, the relative importance of each energy source); its members then chaired groups drawing in specialists from relevant ministries, CASS, and CAS, which studied more detailed questions, such as the exact number of mines and machines needed to produce a given amount of coal by the year 2000. Studies of this type produced 13 "Technology Blue Papers," which were adopted as policy by the State Council. These studies seem to cover much of the same ground as TERC's 13-volume study, although they may deal more with narrow technical problems.[26]

Also like TERC, the NRCSTD provides advice on specific projects. It too has been asked to help the State Council resolve the bureaucratic debate over whether and how to proceed with the Three Gorges dam project, and has commissioned studies from other institutions on different aspects of the project. Although one NRCSTD official implied that his organization had been given primary responsibility for resolving the debate over the dam project, recent leadership statements make clear that it is also looking elsewhere for expert advice: In April 1986, Vice-Premier Li Peng stated that a special committee would be established to perform the necessary feasibility research.[27] The State Council clearly has available multiple sources of expert advice on important policy issues.

An official of the NRCSTD, in explaining the difference between TERC and his institution, said that TERC was more economic in focus, while the NRCSTD concentrates more on scientific and technical questions. However, since TERC's staff contains more scientists and engineers than economists, one wonders how great this contrast really is. Moreover, the NRCSTD is apparently trying to give more of an economic focus to its work by recruiting more economists.[28] Thus, the training of the two organizations' staffs seems quite similar. In addition to providing similar kinds of advice, both institutions use many of the same agencies to conduct their studies; the Ministries of Coal and Petroleum, for example, presumably provided studies both for the energy volume of TERC's study, and for the NRCSTD's Technology Blue Paper on energy. It is quite likely there is some overlap, not only in the topics they address, but also in the sources of the policy analysis they provide to the leadership. The similarity in the types of tasks these two bodies perform increases the probability that "expertise" is being used at least partially as a tool of competition for policy influence.

This impression in reinforced by the apparent limited degree of interaction between them. There is some communication between the NRCSTD and the other centers; NRCSTD and TERC sometimes meet together. Still, the degree of coordination is limited, as illustrated by the fact that one SSTC member, who had worked on the energy study organized by the NRCSTD, knew little about the TERC's energy study done as part of its report on China to the Year 2000. Yet, this same individual reported that, during the course of the study, relevant price issues (such as the pricing of coal) were discussed with both TERC and the Price Research Center. Likewise, an official of the NRCSTD reported that those at his center who study agricultural issues have frequent contact with the Rural Research Center; closer relations, in fact, with this center than with the SSTC's own National Center for Rural Technology Development, because the former focuses on more general policy issues similar to the NRCSTD's own interests, while the latter deals with more technical issues.

Thus, the NRCSTD cooperates with the State Council centers occasionally, but also appears to be a competing source of advice, particularly on topics studied by TERC. Similar overlap appears to exist between the various centers and another advisory organ, the State Council's System Reform Commission.[29] The Commission was established in 1982 under Zhao Ziyang's leadership to map out and oversee the implementation of an overall plan for economic reform. It organizes economists and scientists from inside and outside the bureaucracy to perform studies related to reform. Since the State Council centers and the NRCSTD all study issues related to economic reform, the Commission clearly overlaps with them in its mission. The Commission also has an S&T group which probably performs studies similar to the NRCSTD's on reform of the S&T system (the NRCSTD helped draft the major decision on S&T reform adopted in March 1985). The Commission sometimes cooperates with the centers; in 1982, ERC worked with the Commission to organize 8 groups of specialists to study reform.[30] At the same time, the Commission also generates advice on its own; it has now established a subordinate Economic Reform Research Institute. In the words of one individual familiar with work done by the Commission and TERC, the research is both "overlapping and competitive," suggesting that, while some of the duplication is the result of poor coordination, some may be the product of inter-bureaucratic rivalry, perhaps deliberately created by the leadership to acquire alternative views on a problem.

Finally, additional sources of advice about which little is known but which may well be the most important are the many "leadership small groups" established under the State Council in the 1980s to help formulate policy on cross-cutting or priority issues. These small groups are composed of officials from the key bureaucratic sectors related to the particular area. Such leadership groups exist for electronics, nuclear power, Three Gorges construction, technical policy, technical market, environmental protection, water resources, and many other areas. The first three of these were headed as of 1985 by the Vice-Premier, Li Peng. The two most important in the area of economics and science policy are the Finance and Economics Leading Group and the Science and Technology Leading Group, both chaired by Zhao Ziyang. While some of the above-mentioned groups are ad hoc organs, designed to address priority issues, these two are permanent organs with sizeable staffs. The leading groups differ from the research centers in that they more clearly have the personal imprimatur of the premier or other important officials. Thus, they should be more authoritative and better able to resolve conflicts and enforce leadership priorities in policy implementation. But they also provide advice on policy formulation.[31]

The S&T group meets about once a month and is composed of officials from such key organs as the SSTC, CAS, the SPC, the State Economic Commission, the State Education Commission, the Ministry of Finance, the Ministry of Labor, MOFERT, and the NDSTC. It has an office, chaired by Song Jian, Director of the SSTC. The staff of about 20, drawn from the different ministries, carries out studies either requested by the Leading Group or at its own initiative. It also organizes exchanges of views among different units. The staff does not simply process the reports provided by other advisory organs; it clearly performs, or organizes, its own studies, thus serving as yet another source of competing advice.[32] This is also true of the Leading Group for the Invigoration of the Electronics Industry. Responsible for making policy recommendations, its office has a staff of about 20, but obtains advice from advisory subgroups composed of scientists from other institutions.[33] While very little is known about the staff of the Finance and Economics Leadership Group, it probably operates in a similar fashion. Despite the fact that TERC was nominally established as its subordinate unit, it is said to do little in directing the operations of TERC; it seems likely, therefore, that it uses its own staff to perform policy research on its behalf. Thus, the leading groups' staffs appear to play much the same role as the NRCSTD and TERC but, because of the greater authority of the leading groups, are probably more influential.

IMPACT OF THE ADVISORY SYSTEM

What impact have the organizational innovations described above had on the decision-making process and the political system? Although it would be foolish to pretend that we currently have enough information about the Chinese advisory system to answer this question definitively, the information presented above permits some tentative observations.

Since 1980, the advisory system has made available to the leadership much information, expert advice, and different viewpoints. TERC ensures that major policy proposals in the technical-economic area are vetted by other bureaucratic actors besides the one that proposes the policy, and contributes to the winnowing out of bad proposals. The system also enhances policy coordination by drawing in many of the bureaucratic actors relevant to a policy (and therefore likely to be affected by the policy's adoption and implementation). TERC not only helps ensure that poorly designed policies will be criticized, but also that trade-offs and linkages between different policy areas will be discussed. The multidisciplinary perspective of TERC and other bodies is also important in this regard. As we indicated, however, the system has by no means eliminated competition among different departments; the leadership continues to face the problems of reconciling competing departmental claims and attempting to enforce its priorities on the bureaucracy.[34] Although the quality of the planning remains open to question, the system has been more clearly successful in promoting a long-term planning perspective not generated by ordinary bureaucratic procedures.

The advisory system's contribution to improved decision making, however, has limitations. The *People's Daily* article announcing the new SPC procedures, quoted earlier, indicated that poor planning and project selection remain serious problems. Vice-Premier Wan Li, in a July 1986 speech, praised the improvements in the decision-making process of the last few years but complained:

> We have not yet established a rigorous system and procedure for making policy decisions; nor have we had a perfect support system, consultancy system, appraisal system, supervisory system, and feedback system for that purpose. There is [no] test to determine whether or not a policy decision is scientific, and it is hard to conduct effective and timely supervision to avert erroneous policy decisions. Even at present, the practice of relying on the leaders' experience in making policy decisions is still common and prevalent.[35]

A May 1986 article in the journal *Liaowang* also criticized the quality of the analysis performed in China: "At present, the studies on the feasibility of many of the PRC's projects have been extremely simple and crude. In reality, they are but expansions of the project's letter of proposal or recommendation and basically not a study of its feasibility."[36] This indictment may not apply to the work done by such organizations as TERC or the NRCSTD, but it is likely that their work is also not of extremely high quality. An indication that this may be so is the China International Engineering Consulting Corporation's plan to "recruit foreign specialists to join its consulting business in the hope of becoming a firm providing the state with reliable data needed for correct decision making."[37] "Experts" in China, at least those without foreign training, which includes the large majority, probably do not have the skills to do a sophisticated analysis of feasibility.

Finally, little is known about how advice generated by the various sources described here actually flows to the top leaders themselves. How is it decided which reports the premier and vice-premiers will actually read? Officials of some of the advisory bodies convey the strong impression that they have direct access to the premier. But, as indicated above, many voices continue to compete for the top leaders' ears; somehow priorities must be set and decisions made about whom they will see and what they will read. The State Council staff office and the premier's and vice-premiers' secretaries probably play a crucial role here.[38] If they do not act in an unbiased manner—if personal relationships or policy predilections determine the advice that they choose to pass on—the fairly good advisory system described above may actually do little to improve decision making. Thus, the interface between the top decision makers and the advisory bodies needs to be further explored before a judgment can be made about the influence of the advisory system on decision making. Finally, how the top leaders themselves process the advice that finally reaches them is still another issue. As Alexander George has pointed out, the quality of decisions probably depends less on the rationality of the decision process than on the "decisional premises" of those who actually make the final decisions.[39] A well-structured advisory system can make good policy decisions more probable; it certainly cannot guarantee them.

The institutionalization of expert participation in government has implications for the political system that go beyond its impact on the quality of decision making, narrowly defined. Experts, particularly those with foreign training, bring new ideas into the process, such as the notion of the "new technological revolution," which has been much in vogue in China in the 1980s and has been used to justify many reformist ideas. The availability of experts to flesh out the leaders' general reform ideas undoubtedly has helped maintain the innovative spirit of the Deng leadership. As in other socialist countries, the bringing of experts into the government has so far shown no signs of undermining Party rule.[40] Nevertheless, it has altered the nature of that rule. Experts in the mid-1980s appear to be well established within the political system, and the mushrooming of advisory bodies suggests that expertise has become an important political resource. Thus, the Party still rules in China, and politics continues to pervade the policy process, but political competition appears to be conducted at least partially through competing use of experts and expertise. This is a very different kind of politics than before.

CHAPTER EIGHT

The Impact of Returning Scholars on Chinese Science and Technology

O. SCHNEPP

The exchange of students and scholars between the United States and China was originally sanctioned by the Understanding of the Exchange of Students and Scholars, concluded in October 1978 between the US Information Agency and the Chinese Ministry of Education (MOE). The Understanding was replaced by a new Protocol on the Exchange of Students and Scholars, signed in Washington in 1985 with China's State Education Commission, the successor agency to the Ministry.[1] USIA sources have reported[2] that "some 28,000 Chinese scholars, students and lecturers have come to the United States since 1978." According to 1984 reports attributed to Chinese MOE sources, between 1978 and April 1984 a total of 18,500 students and scholars had been sent abroad under government programs, of which 8,900 were sent to the United States. Of this total,

81 percent or about 15,000 were said to be scholars and 12 percent were graduate students.[3] By August 1985, more than 29,000 state-supported students and scholars had reportedly been sent abroad, and a further 7,800 had gone abroad to study at private expense, according to a Chinese report.[4] It was also noted in this report that more than 15,000 students and scholars had already returned after completing their studies. The breakdown according to broad discipline categories, as quoted in the report, was: 28 percent in science, 39.6 percent in engineering, 7.7 percent in agriculture, and 13.1 percent in liberal arts.

A comprehensive study of the United States–China educational exchanges during the years 1978–1984 has been carried out by the Committee on Scholarly Communications with the PRC.[5] According to this study, a total of about 19,000 students and scholars (including self-supporting or "self-paying") came to the United States during the period 1979–1983. In the academic year 1983–1984, there were about 12,000 PRC students and scholars in the United States, about 4,000 of them scholars. MOE sources[6] reported 3,300 visiting scholars on the official program in the United States as of March 1984, which is not inconsistent with the 4,000 figure, particularly since the dates are not identical. Orleans[7] has discussed what he calls "the numbers game" in some detail. Clearly, statistical information concerning the visiting scholars and students is confusing and sometimes inconsistent. Nevertheless, it is clear that appreciable numbers have been involved in studies abroad approaching the order of 10 percent of the total of science and engineering R&D personnel in China, which are estimated to be in excess of 300,000.[8]

During the beginning years of the exchange program, the great majority of the Chinese exchange scholars were men and women in their forties who had been deprived of the opportunity for graduate studies because of the Cultural Revolution, 1966–1976. Many of them, however, had a number of years of practical experience in some field of research or technology, or they had been active in university instruction. In the United States, they primarily did research at universities. Only a very few changed their status to graduate student and completed PhD requirements before returning to China. Since the great majority of returnees so far have belonged to this category, it is this group of visiting scholars who have to date made the initial impact on China science and technology. In recent years, however, the fraction of younger students enrolled in graduate degree programs at American universities has steadily increased.[9] MOE and CAS, in the long term, aim to balance the numbers of graduate students and scholars[10] sent abroad. However, only small numbers of doctoral graduates have returned to China to date.[11]

Chinese sources have identified three major objectives for the scholars and students study-abroad program.[12] First, the quality of teaching in universities was to be improved. Second, the knowledge of natural and social science was to be furthered, with particular emphasis on applications. Third, more attention was to be paid to science. It was generally recognized in China that the extended isolation of the country had resulted in a science and engineering work force that had not kept up with more recent developments in the world. The prevailing level of knowledge in universities and research institutions was badly in need of updating, and the content of teaching as well as research programs had to be overhauled.

To study the impact on science and technology in China of returning scholars who had participated in the exchange program, 134 returned scholars and 44 senior scientists, administrators, and officials were interviewed in China in May–June 1984.[13] The sample was selective, in that visits and interviews were limited to a number of high-status ("key") universities and research institutes of CAS, listed in Table 1. The profile of the interviewed scholars by institutional affiliation and by current academic rank was as follows. A total of 55 (41 percent) were faculty in universities under the direct administration of MOE and 79 (59 percent) worked in institutions administered by CAS—63 were on the staff of research institutes and 16 were on the faculty of the University of Science and Technology of China (USTC) in Hefei, Anhui province (see Table 1). Among the interviewees were 3 (2 percent) professors, 49 (37 percent) associate professors and 82 (61 percent) below professorial rank, that is, lecturers, assistant researchers, and engineers.

The indicators for which data were obtained are academic stature as characterized by academic rank held and promotions in academic rank, appointment to positions of administrative responsibility, knowledge and skills acquired during the stay in the United States, research contributions made after return to China as indicated by research activity and publications, relationship between research pursued after return to China and work done in the United States, facilities for research available to the returned scientists, impact of US experience on teaching ability in advanced courses, and interaction of the returned scholar with industry and other institutions outside his or her own. The results of this survey will be discussed in the context of numerous reports which have appeared in the Chinese media describing and assessing the conditions surrounding the returned scholars and the level of efficiency at which the country is making use of their knowledge and experience. CAST conducted a countrywide survey[14] covering 4,000 scholars in 15 provinces and municipalities (that is, Beijing, Tianjin, and Shanghai), collecting data through group discussions, written surveys, and home visits.[15] Another survey was carried out in Shenyang[16] (Liaoning province in China's industrial northeast) and included 116 of the 321 scholars sent abroad by 22 universities and 14 scientific research and design institutes of the city. There have been additional surveys of returned scholars in Shanghai[17] and in Beijing.[18] A thoughtful analysis of information available in the public domain can be found in Orlean's article.[19]

Before assessing the impact of the returned scholars, it may be useful to discuss the potential for impact as indicated in the number of scholars sent abroad relative to the total number of eligible academic staff at individual institutions and relative to the total R&D manpower potential of China. These data were obtained during the survey conducted in China. The second criterion considered for the potential for impact on return to China was the professional level of the scholars by world standards. This was determined on the basis of the judgments of the American faculty hosts of those Chinese visitors who worked in their research groups for periods of one to two years. Altogether 7 universities were visited and 52 faculty interviewed in person by this author to obtain information during the American phase of the survey[20] on 131 visiting scholars. The professional level was assessed according to the contributions made by the visitor to the research project he was associated with and his/her potential for independent original research. In addition to the overall evaluation, the faculty host was asked at what academic level he would be willing to accept or recommend the Chinese scientist in comparison to the personnel available in the United States.

POTENTIAL FOR IMPACT

Many of China's leading institutions had, by May 1984, sent abroad over 10 percent of their eligible faculty (21 out of a total of 30 institutions visited), as shown[21] in Table 1; 13 of the institutions had sent over 15 percent. Eligible faculty are defined as professional faculty holding ranks of lecturer, associate professor, or professor at universities and their equivalent ranks at CAS research institutes. Actually, only 4 out of the 134 returned scholars interviewed held ranks below lecturer or its equivalent prior to going abroad. The institutions listed in Table 1 are all located on or near the east coast of China, but represent locations that supplied fully 64 percent of all holders of J-1 visas issued by US consular offices in the PRC during 1983; this category of visa was mostly issued to students and scholars on the official program.[22] Given the relatively large number of leading institutions that have sent an appreciable fraction of their eligible faculty abroad, the potential for significant impact of the exchange program, at least on the scholars' home institutions, was considerable.

Probably the factor most relevant for assessing the potential impact of the returned scholars on China's science is the scientific and intellectual stature of the returnees. The scholars included in this survey with very few exceptions returned to their original places of work in China after stays in the United States ranging from one to two years. Table 2 summarizes the results of this phase of the survey. The institutional profile of the scholars is evident from lines 1–3 of the table.[23] All but one of the scholars included in this sample, who were sent abroad by institutions of higher learning, came from major universities under the direct administration of MOE.

Based on the survey of faculty sponsors in the United States, there is reason to believe that a sizeable fraction of the Chinese visiting scholars made major contributions to their research projects. The results indicate that over two-thirds (71.9 percent—line 11 of Table 2) were in that category, while the other third (27.2 percent—line 12) made only technical and lesser contributions. The criterion used was a concrete contribution to at least one research paper published in a refereed journal (a serious effort was made to ascertain that the scholar's name was not added to the paper merely as a courtesy). Altogether, over two-thirds (71.0 percent—lines 5–9) of the sample would be welcomed back if they wished to return to their host research groups, while 28 percent (line 10) were not acceptable in competition with available research staff. There is good agreement between the fraction of the sample that made major contributions to their research projects (71.9 percent) and the fraction that would be welcomed back by their former faculty hosts (71.0 percent). Since these two groups may be expected to be correlated to a high degree, the above result is satisfying. About a fourth of the sample (28.0 percent—line 13) were judged to be of a caliber to break new ground and to achieve world stature, given adequate conditions for development, according to the testimony of their American faculty hosts; an additional third were believed to be capable of making solid contributions to scientific research with the same working conditions (36.3 percent—line 14).

The question naturally arises as to how the fraction of the sample that would not be accepted back (28 percent) compares with the overall population of people in the same category. Since the Chinese visiting scholars fall somewhere between graduate students (research assistants) and postdoctoral fellows, they are a unique group in American faculty experience, making valid comparisons difficult. Faculty experience with postdoctoral fellows seems most relevant. An informal poll of faculty colleagues indicated a disappointing performance by 30 percent of postdoctorals to be the norm. Therefore the Chinese visiting scholars, in this respect, are in line with the overall performance at many American universities.

We conclude that there is evidence for promising scientific level and intellectual potential among returned Chinese scholars; and that their impact on Chinese science and science education, if effectively used, could be considerable. The fact that a relatively large fraction of eligible staff were sent abroad by major institutions (10–25 percent by May 1984) reinforces this conclusion.

ASSESSMENT OF IMPACT

There was virtual unanimity among the senior Chinese scientists and officials interviewed in their enthusiasm for the program as a whole and their satisfaction with its results. While most agreed that the program already had had significant impact on all aspects of Chinese science and higher education in science and engineering, particularly at the home institutions of the participating scholars, they believed that the returnees' full impact would be felt only after some time had passed and their acquired skills had become fully evident. The exposure of a large number of Chinese scientists to world-level research, modern equipment, and up-to-date methods was considered a major contribution to Chinese scientific research and graduate education. Also, the connections made with the United States and international scientific communities were felt to represent a major, long-term benefit.

According to the interviewed senior scientists, the scholars have acquired a large body of knowledge and information about state-of-the-art scientific methods and experimental techniques. But, even more important, they have learned just where the frontiers are in their respective fields. Furthermore, they have been introduced to new research directions in basic and applied science and have brought home new ideas for promising research projects. The influence of their initiatives has often extended to high-priority state "key projects," which were thereby significantly furthered. The returnees were also said to have brought back with them familiarity with the use and applications of computers, and they were credited with making a significant contribution to the introduction of automated and computerized procedures. Of the 134 returned scholars, 111 believed that their knowledge, orientation, and research ability in their scientific fields had been advanced by giant steps as a result of their stay in the United States.

On the other hand, Chinese sources have warned that the results of the exchanges were not as good as they could have been. *GMRB* succinctly summarized the results of the CAST survey:

> The results show that many among the returned scientific and technical personnel have been active in making contributions in their fields, but a significant number have not been able to develop their potential.[24]

Of a sample of 1,059 returned scholars from which replies to the survey were received, 21.8 percent stated[25] that they made full use of their specialized knowledge, 62.9 percent said that they were no worse off than they were before their sojourn abroad but their skills were not fully used, and 15.3 percent responded that their working conditions were inferior to what they had been before they went abroad. The Shenyang survey report[26] expressed a similar "mixed review."

As a rule, these scholars returned to their former work units.[27] It has been the Chinese government's policy, in principle, to support returning scholars and to encourage them to develop their potential to the fullest extent possible. Many articles have been published in China in the 1980s advocating "making full use of science and technology personnel returning from abroad," training those among them who have administrative ability, and eventually appointing those suitably equipped to leadership positions.[28] In practice, however, Chinese senior professors said that there was no special treatment for returning exchange scholars in CAS institutes or at universities; the returning scholar was generally put on an equal footing with other research staff or faculty members and competed for the relatively scarce resources on equal terms. Some senior scientists said that the chances for obtaining research support and promotion were good, but others said that only few could obtain special support, owing to budget constraints. At Peking University, a professor said that the returning scholar typically was provided with laboratory space and two assistants to help him get started in his research program. At the universities, some equipment had been purchased during 1982–1983 with World Bank loan funds.

Promotions in rank for returned scholars have by no means been guaranteed but have been carefully considered and based on concrete individual accomplishments. The scholars reported that there was quite a backlog of pending promotions; returning scholars were said to have been recommended by their departments and institutions but the actual promotions had been held in abeyance due to a countrywide freeze on promotions to professorial ranks, in effect between 1983 and 1985.[29]

There is, nevertheless, some significant evidence for the successful integration of at least some of the returned scholars. Many were reportedly promoted in academic rank soon after their return to China and others were entrusted with administrative responsibility, both being symbols of official recognition as well as signs of increased influence. Of the scholars interviewed by this author, a total of 32 were promoted either during their stay in the United States or shortly after their return; 19 others had been promoted within the two years prior to their leaving the country. In addition, 7 lecturers and assistant research professors, among those interviewed, said they had been recommended for promotion to associate professor by their institutions, but the promotions had not been officially confirmed because of the freeze referred to above. An additional 11 returned scholars below professorial rank (lecturers, assistant research scientists and engineers) referred to the freeze during their interviews by way of explanation for not having been promoted; it could not be confirmed whether the scientists' institutions had completed evaluation procedures and had actually recommended promotion.

Nine associate professors and one lecturer in the above groups who have been promoted in academic rank were also given new administrative responsibilities as deputy department heads or division directors and deputy institute directors at universities and research institutes. In addition, 23 below professorial rank were entrusted with similar new administrative duties without being promoted in academic rank. At Zhongshan University in Guangzhou, 20 returned scholars (one-third of the returnees) were said to have been promoted to associate professor. At Peking University, 25 returnees were reported to have been appointed heads or deputy heads of departments and another 25 had been named leaders of teaching and research groups. Many of the returned scholars at the University of Science and Technology of China in Hefei were said to have become group leaders. Similar observations came from senior personnel at the Organic Chemistry Institute and the High Energy Physics Institute of CAS. However, a total of 48 (or 36 percent) of the returned scholars in the sample, all below professorial rank, experienced no change in official status, either in academic rank or in administrative responsibility as of May/June 1984.

Chinese media sources have corroborated the above findings. A Xinhua News Agency release[30] dated June 1985 reported that over 90 of 180 returnees at Fudan University in Shanghai had been promoted to positions of heads of department or leaders of research and teaching groups, and the returned scholars had had beneficial effects on the city's 45 universities and 700 research institutes. Another article, referring to the Shanghai survey, commented:

> Actually, many returning scholars have assumed key positions as a result of the concern of leaders and the support of the older generation of scientists.[31]

Another news item reported that about 26 of the 63 returned scholars at Shanghai's Jiaotong University had been appointed to positions of academic administrative responsibility, one of these being "vice-president of the graduate school."[32]

Exemplary placement of returned scholars was described in a news item[33] from Shanghai, which praised the CAS Organic Chemistry Institute for appointing returned scholars to positions of leadership and the older generation of scientists for voluntarily relinquishing their positions in order to give the younger returnees more influence. Three deputy directors of the Institute were said to be among the faculty who had recently returned from studies abroad as well as half of the 24 heads and deputy heads of research divisions. According to the article, the older personnel had "retired to the second front."

A number of reports, however, have complained that the scientists' effectiveness was seriously limited. The seniority system prevented many of the returnees under age 50 from being given adequate authority to do their work effectively. But, once appointed to positions of administrative authority, they spent 70–80 percent of their time on such duties and their academic productivity was reduced to very low levels. Other sources have also reported complaints by returned scholars[34] that they spend too much time doing administrative work and not enough time doing their professional work. The participation by returned scholars in administrative functions is likely to make beneficial contributions to building up China's technology. There certainly is some loss to professional work, but it is widely acknowledged that technically educated personnel make essential contributions to administration and are mobilized to perform such functions under most, if not all, systems of accepted organization and management. The contributions made by technical personnel in administrative functions as opposed to their productivity in research laboratories is the subject of continuing debate. In the United States, the question is usually decided by individual preference at different stages of a professional career and by the incentives offered. It is not clear if individual scientists are free to make these choices in China, but, given their limited job mobility, it is probable that they are more constrained to accept functions assigned to them than their American counterparts. It also appears that promotion and salary advantages make administrative positions in China attractive,[35] as they do in the United States.

Raising the level of instruction at universities was one of the primary objectives of the policy to encourage academic personnel to go abroad. There is good evidence that this aim has been largely fulfilled. Altogether, 49 of the returned scholars (72 percent of the 68 interviewed) who are on university teaching faculties reported that they had gained knowledge in the United States that significantly furthered their ability to teach advanced courses; 14 reported that their teaching ability was improved to a lesser degree, and 5 claimed their teaching ability had benefited only slightly. Of the 66 research scientists interviewed who did not primarily have teaching duties, 27 taught graduate courses at their institutes or at the CAS Graduate School in Beijing. Of this group, 23 reported notable gains in graduate teaching competence, while 4 said that they had gained only slightly. All the interviewees, both scholars and senior professors, had no hesitation in asserting that many new advanced courses had been successfully introduced by the returning scholars. Another source has confirmed this conclusion by reporting that "returned students at Qinghua University have started 46 new courses."[36]

Research was not considered essential in the mission of universities during the Mao era; following the Russian model, they were primarily regarded as teaching institutions, while research was to be pursued at specialized institutes. However, during the early 1980s, scholars returned from abroad to their universities with a strong motivation to continue carrying on research projects. The returned scholars had accepted the concept prevalent in the United States and other industrialized countries that research was important for the improvement of the faculty's teaching effectiveness, primarily at the graduate level but also in undergraduate instruction. In most instances, the senior scientists interviewed credited the returned scholars with significant contributions to the improvement of the general academic climate at their home institutions, primarily by their own activity and by engaging their colleagues who had not had the advantage of experience abroad in their research projects. In the majority of cases, these programs were continuations and extensions of research from abroad. It has not been possible to determine the level of results obtained relative to world standards but the benefit to students is believed to be significant. At CAS institutes where the research tradition was stronger, the impact of the returned scholars was also said to have been critical to updating the general level of science and promoting awareness of the frontiers of their specialties.

In concrete terms, it was found that a sizeable fraction of the returned scholars had made significant contributions to research in China as evidenced by publications based on completed projects, all of which were begun subsequent to their return. The sample of returned scholars interviewed included 94 (70 percent) who had built up laboratories and/or had launched research groups in meaningful research programs. An additional 25 returned scholars (19 percent) had begun research work but had not yet carried any project to the point of publication. A group of 15 interviewees (11 percent) presented no evidence of research activity or accomplishment. Of the last group, 3 had returned to China only as recently as 1984 and 9 had returned in 1983. A majority of the papers resulting from the research of returned scholars were said to have been published in Chinese journals; a minority had published outside China–9 who said they had already published in foreign journals and 3 who planned to submit manuscripts for publication abroad.

According to the survey conducted in China by this author, only 25 percent of the returned scholars interviewed said that they had encountered no problems doing research. Of this group, 9 had only modest requirements by virtue of their research fields (mathematics and theory), 14 said that they had good equipment, 6 required only computer facilities and had those on an adequate level. On the other hand, one third of the sample (43 of the interviewees) cited inadequate equipment as their major difficulty in doing research and almost one quarter (31 scholars) said their work suffered from difficulties obtaining materials; 14 of these two groups cited shortages in both equipment and materials. A lack of adequate computer facilities was cited by a smaller group (18 or 13 percent) of the interviewees, including problems with computer access, software, and availability of sufficient machine time. A lack of support funds was explicitly cited by only a few scholars, although at least part of the equipment shortage must reflect deficiencies in funding. Science Fund grants[37] have been the source of research support for some returned scholars, according to the information given by the interviewed sample. Some said that overall efficiency of facilities and supply procedures was low. Office space and shortages of books and journals in their libraries were also said to pose problems. Many scholars confirmed that reprinted foreign journals arrived at their libraries only a full year after their original publication.

Since 1980, there has been substantial improvement in equipment at CAS institutes and, more recently, also at universities. Most professors interviewed said that, in spite of significant progress in recent years, research facilities, even in major Chinese universities and CAS research institutes, still lagged significantly behind those at good laboratories in the United States. Many senior professors held that there also was a definite shortage of materials and reagents for research as compared to their availability in industrialized countries; this confirmed the opinion of many returned scholars. Since the facilities and supply channels were considerably less efficient than in the United States, the time required to complete a given project was significantly longer, even though more manpower was available in China for auxiliary work, such as the preparation or purification of materials. One scholar quoted in the CAST survey report estimated[38] that her productivity in China, relative to abroad, was lower by as much as a factor of 7. A *People's Daily* article described the difficulties encountered by returnees and their near-heroic efforts to overcome them:

> Faced with conditions for experimentation that are inferior to those in some countries abroad, primitive equipment, insufficient materials, and a shortage of funding, they have worked creatively to eliminate difficulties and have created a new situation in scientific work.[39]

Many of the leading Chinese scientists interviewed dwelled on the benefits of the scholars' exposure to the organization of research groups and general administrative systems in other countries' academic and research institutions. The returned scholars not only supported reforms in progress but often came forward with suggestions for administrative changes aimed at making the larger unit more efficient in its performance. Without exception, the interviewees stressed the importance of communication between scientists in their immediate environment as well as with those in other departments and, beyond, in other institutes and universities. According to the senior scientists' reports, the returning scholars had organized a large number of seminars within their research groups and also convened seminars in which scientists interested in a given subject met regularly, including some from different institutions.

Only 10 of the returned scholars interviewed reported that they were consulting for industry or had industrial research support. These scientists contributed their newly acquired knowledge outside their home institutions for the benefit of the country's industrial development and modernization. Seemingly, only a fraction of the sample (7.5 percent) had made direct impact beyond their home institutions. However, many of the first rank universities in China have been given important roles[40] in the execution of priority "key projects" and contributions to these projects would have significance on the national level. It is therefore probable that the returned scholars have had wider impact than is apparent from the above results.

The various surveys reported from Chinese sources as well as the survey conducted by this author all have indicated that 20–30 percent of the returning scholars felt their abilities were being used at or near optimal levels. On the other hand, a not insignificant number, ranging from 8 percent[41] to 15.3 percent[42] reportedly said that their working conditions and professional effectiveness were inferior after their return from abroad relative to before their departure. The remaining 60–70 percent thought their skills were not being fully used. Several articles decried this situation and stressed the need for corrective measures. For example, a news item concerning a study in Shanghai reported:

> Not an insignificant portion of the returnees have not been able to make full use of their know-how. Denied opportunities to put their expertise to good use, some researchers have given up their specialties, changed careers, struck out on their own, or even have gone overseas. Clearly this is a waste of talent, a loss that should not occur in scientific research.[43]

There is some evidence of efforts to improve the placement of returnees. A *GMRB* article detailed the conscientious efforts of the Cadre Department of the National Defense Science, Technology, and Industry Commission (NDSTIC) to rectify cases of unsuitable employment of returned scholars.[44]

The CAST survey and the Shenyang study investigated the economic and social conditions of technical personnel who had returned from abroad. Not surprisingly, their incomes were found to be no better than those of the other scientists; only 20 percent of the Shenyang sample earned more than 100 yuan per month and 45 percent earned less than 80 yuan. By comparison, CAS statistics of about the same period showed[45] that 80 percent of scientists in the age group of 40–50 earned 78–89 yuan, which, "when you consider subsidies and total bonuses, is lower than the national per capita monthly income of 47 yuan based on population averages." Presumably this means that CAS science personnel do not obtain the bonuses added to basic salaries in industry. In addition, most of the scholars lived in crowded accommodations that were not considered conducive to a productive life, according to the Shenyang survey report. This condition was also reported[46] to be widespread among the general science community. Several articles have, in recent years, complained of the economic hardships that have beset the "young and middle-aged technical personnel" (ages up to 35 and 36–55, respectively)[47] who, at the same time, bear the heaviest weight of responsibility.[48] The group of middle-aged scientists includes the bulk of the returned scholars. It has also been reported in these articles that this group's health is inferior to that of the older scientists, a difference generally explained on the basis of the harsher treatment suffered by the younger individuals during the CR years.

PROSPECTS FOR THE FUTURE

The survey conducted by this author and published Chinese reports agree that returned scholars have profited immensely from their sojourns abroad. For almost all, the experience advanced their scientific and research capability by a giant step; they brought back with them new knowledge and new skills. Most important, however, the returned scholars acquired a good grasp of the frontiers of knowledge and generated ideas on which research projects could be based after their return to China. On the other hand, all sources also agree that only 25 percent of the scholars interviewed said that they had not encountered any problems in doing research since their return, and the remainder reported difficulties of various kinds. It should, however, be noted that the returned Chinese scientists compared their productivity at their home institution with that in the laboratory abroad where they had no teaching or administrative responsibilities. This difference probably accounts for a factor of at least 2 for most faculty, although support by collaborators and technicians would be expected to make up for a part of this loss. Furthermore, research productivity is lower by a factor of 2 to 3 in countries at levels of industrialization[49] lower than the leading industrialized countries, primarily because many materials and spare parts are unavailable locally and must be purchased from abroad. The import process is often complicated by shortages of foreign currency and customs formalities of various sorts. As a result, it is reasonable to expect the returned scholar to find his research productivity reduced relative to his experience abroad. In addition, the Chinese scholar is likely to experience a lower level of stimulation in his environment at the home institution, further affecting his productivity adversely. Advances in availability of research equipment have been considerable during the past five years, although the gap between China and the United States is still large. Although shortages can be remedied by purchases from abroad, materials used routinely in laboratories cover a very wide range, and shortages in this category may be expected to impede progress for many years. Most scientists anywhere, and particularly the most innovative, find that they are short of means to achieve their research aims; it is therefore the rare researcher who will say that he does not lack resources. The returning scholars' replies must be considered in the light of this general situation.

The promotion and appointment to positions of administrative responsibility of returned scholars is evidence for their acceptance and integration as well as for their increased influence. This status permits them to contribute to the improvement of the administrative atmosphere and procedures. The complaints by returned scholars that they spend too much time doing administrative work and not enough time doing their professional work are difficult to evaluate. They are not unlike the complaints of scientific administrators elsewhere.

The model for future development of the R&D sector proposed here assumes that, in many instances, the visiting scholars had not experienced innovative environments before their recent visits abroad and they were then already at an age when it is not expected that many would undergo a basic change and become creative. Therefore, only a minority of this group may be expected to become leaders in their fields and break new ground in their scientific work. Greater contributions more likely will come from the younger group of graduate students who will return after a significant period spent in innovative environments at an early stage of their careers. This younger group's contribution would be very much advanced by the groundwork which, it is hoped, the older group of scholars will prepare. Although the returned scholars are not used as effectively as many would like, there nevertheless is already evidence that their influence is making itself felt. It is not clear, however, if they have already raised the academic standards sufficiently to ensure a uniformly good level of graduate education, particularly in the research component. The authorities concerned are meanwhile moving forward cautiously with the authorization of graduate programs, but the development of world level PhD programs is likely to prove difficult for some time to come.

As in other sectors, China has made enormous progress in science and technology, but there is a long way to go to reach the level of industrialized countries. During 1978–1986, the future seemed bright, and technically educated intellectuals worked with enthusiasm as they sensed the dawn of a new era—an era of diminishing political and administrative restrictions and controls. Optimism for the success of the reforms and the new policies was in evidence everywhere. Everybody knew, of course, that there were conservatives who opposed the policies of reform and of loosening the reins on the population which were instituted to encourage personal initiative and creativity. However, nobody knew how much power these opponents of reform still held. It was widely assumed in China and abroad that Deng Xiaoping could hold the opposition in check. The events of November 1986 to February 1987 showed that this was not so. Hu Yaobang, Secretary General of the Communist Party of China and acknowledged leader of the reform movement, was forced to resign in January 1987.

Closer to home for the scientific leadership, Lu Jiaxi and Yan Dongsheng, the President and Senior Vice-President of CAS, both elected by their peers in 1981, were retired, and a new leadership of the Academy was appointed, apparently without consultation with the science community. These events followed the dismissal of the President and Vice-President of the USTC (administered by CAS) on apparently political grounds.

These events doubtlessly sent a chill through the scientific intellectual community. What we know for certain is that they chilled the community of Chinese students and scholars in the United States. Between 500 and 1,000 students went so far as to risk their future in China by signing a letter of protest, which was sent to Chinese authorities. According to persistent stories circulating among the community, the immediate reaction of some students was to delay their return to China, if it was imminent; there can be little doubt that the enthusiasm of this group for returning and helping to build China has been appreciably dampened. Since then, the results of the Communist Party Congress held in Beijing in October 1987 have reassured the world that China will continue its policy of opening up to international cooperation and of pushing key reforms. The rate at which changes are to be instituted and the depth to which reforms will be allowed are still not clear.

The returning students, after obtaining their doctorates abroad, would provide the real reservoir of scientists with potential for innovation and creativity. The impact of, say, 5000 well-trained and vigorous young PhD scientists and engineers, if well used, on the country's R&D capability would be far-reaching. The country has at present fewer than 3,000 active holders of formal doctoral degrees,[50] most in the age range 50–70. Unfortunately, the incentives for the return of students and for their vigorous support of the country's modernization are still weak. Much remains to be done to upgrade the working environment and living conditions of scientific and other intellectuals, according to Chinese sources as well as foreign observers. However, even if the integration of the returning students were to be successful, again, the change would be most significant, but still a great gap would continue to separate China from industrialized countries in the domain of high-level technical manpower for a long time. The United States in 1983 had 260,000 PhD natural and life scientists and engineers.[51]

TABLE 1 Statistical Data, by Institution, on Visiting Scholars Sent Abroad

Institution	*Eligible Staff or Faculty A*	*Visiting Scholars Total B*	*Visiting Scholars to U.S. C*	*% Total Visiting Scholars B/A*	*% Scholars to U.S. of Total C/B*
Physics Institute, Beijing, (CAS)	700	140	100	20.0	71.4
Mathematics Inst., Beijing, (CAS)	87	20	12	23.0	60.0
Systems Sciences Inst., Beijing, (CAS)	110	25	12	22.7	48.0
Photographic Chem. Inst., Beijing, (CAS)	205	32	15	15.6	46.9
Electronics Inst., Beijing, (CAS)	620	55	31	8.9	56.4
Atmosph. Physics. Inst., Beijing, (CAS)	350	40	30	11.4	75.0
Geophysics Inst., Beijing, (CAS)	150	8	5	5.3	62.5
Biophysics Inst., Beijing, (CAS)	400	79	60	19.8	75.9
Acoustics Inst., Beijing, (CAS)	400	20	7	5.0	35.0
Chemistry Inst., Beijing, (CAS)	480	80	53	16.7	66.3
Botany Inst., Beijing, (CAS)	400	40	20	10.0	50.0
High En. Phys. Inst., Beijing, (CAS)	1100	125	35	11.4	28.0
Metallurgy Inst., Shanghai, (CAS)	700	25	15	3.6	60.0
Opt. and Fine Mech. Inst., Shanghai, (CAS)	510	32	12	6.3	37.5
Organic Chem. Inst., Shanghai, (CAS)	600	105	33	17.5	31.4
Biochemistry Inst., Shanghai, (CAS)	380	65	35	17.1	53.8

TABLE 1 *Continued*

Institution	*Eligible Staff or Faculty* A	*Visiting Scholars Total* B	*Visiting Scholars to U.S.* C	*% Total Visiting Scholars* B/A	*% Scholars to U.S. of Total* C/B
Geol. and Paleant. Inst., Nanjing, (CAS)	160	30	10	18.8	33.3
Purple Mt. Astron. Obs., Nanjing, (CAS)	200	16	6	8.0	37.5
University of S&T of China, Hefei, (CAS)	1120	230	115	20.5	50.0
Peking University, Beijing, (MOE)	2300	416	292	18.1	70.2
Quinghua University, Beijing, (MOE)	2700	250	118	9.3	47.2
Beijing Normal Univ., Beijing, (MOE)	1035	143	72	13.8	50.3
Fudan University, Shanghai, (MOE)	1500	300	110	20.0	36.7
East China Normal Univ., Shanghai, (MOE)	1500	113	54	7.5	47.8
Jiaotong University, Shanghai, (MOE)	1700	225	102	13.2	45.3
Nanjing University, Nanjin, (MOE)	2000	250	125	12.5	50.0
Zhongshan University, Guangzhou, (MOE)	1600	160	80	10.0	50.0
All CAS Research Institutes Visited, Total	7552	937	491	12.4	52.4
All Universities Visited, Total	15455	2087	1068	13.5	51.2
CAS, Total	30000	2400	1550	8.0	64.6

TABLE 2 Summary of Results of the Survey of U.S. Faculty Sponsors

Category	*Number*	*%*
1. Number of vis. sci. from Min Education	46	34.8
2. Number of vis. sci. from CAS Inst.	32	24.2
3. Number of vis. sci. from Industr. Mins	20	15.4
4. Number completed PhD degree	4	3.0
5. Number acceptable as faculty	17	12.8
6. Number acceptable as post-docs	41	31.0
7. Number acceptable as PhD students	18	13.6
8. Number acceptable as MS students	4	3.0
9. Number acceptable as res. assistants	14	10.6
10. Number of vis. sci. not acceptable	37	28.0
11. Number contributing to research	95	71.9
12. Number not contributing significantly	36	27.2
13. Number expected to be leaders	37	28.0
14. Number expected to contribute	48	36.3

PART THREE

Application of the Science and Technology Reforms

CHAPTER NINE

Issues in the Modernization of Medicine in China

GAIL HENDERSON

The Swiss medical historian Henry Sigerist once observed that "medicine is the mirror of society." Education, law, or art may also reflect the values of a society. But, because medicine deals with the fundamental issues of life and death, it is an especially poignant barometer of the way a society resolves conflicts between basic needs and limited resources. In the United States, much has been written about the current medical-care crisis brought about by a combination of a rise in chronic illness, increasingly expensive technology, and (until recently) an insurance system that paid for whatever the doctor ordered. Although our medical care system is a result of particular historical circumstances,[1] this crisis highlights the difficulty we have in developing a society-wide consensus on how medical care resources should be allocated. Our fragmented, individualistic medical-care system has relied upon the market to ration medical resources. We conduct ethical debates in the language of medical economics,[2] rarely confronting the fundamental conflict between social responsibility for medical care and the value we place upon individual rights.[3]

The post-1949 Chinese medical care system has evolved with a very different set of characteristics and ethical dilemmas, based upon the health care needs of its population, its cultural and political values, and its economic system. Under the Communist regime, decision making was centralized and the distribution and development of scarce resources highly controlled. Not only was health care for the common person declared a priority in the PRC, but it was also defined in terms of the value to the whole society. Disease was to be conquered as an "enemy of the people."[4] The public health program developed during China's first three decades attracted considerable attention in the West, particularly from American physicians searching for alternative models for the delivery of care to those most neglected under our system.[5] In 1979, however, in a frequently cited article in the *New England Journal of Medicine*, Dr. Robert Blendon concluded that the innovative characteristics of China's medical care system could not be separated from the social and economic system that produced them.[6]

Since the late 1970s, medicine in China has taken a new turn. Deng Xiaoping's reform program has swept medicine up in its technological revolution. The system has been transformed from one that focuses on primary care and public health/community medicine, relying upon mass campaigns and political mobilization, into a system that increasingly emphasizes hospital-based care, depending upon physician services and imported technology inputs. This shift is not only a product of the change in policy under Deng. It is also the result of an increase in demand for a different kind of care. The Chinese population has passed through the epidemiologic transition from infectious disease to chronic illness as the leading cause of morbidity and mortality. In 1957, the major causes of death were respiratory disease (16.86 percent), acute infectious disease (7.93 percent), tuberculosis (7.51 percent), and digestive disorders (7.31 percent). In 1984, the leading causes of death were cardiovascular disease (22.65 percent), stroke (21.13 percent), and cancer (21.11 percent).[7] This transition, in combination with the aging of the population, means that the demand for treatment of expensive, chronic illness can only increase in the future.

The technological revolution in medicine illustrates the issues, dilemmas, and even contradictions characteristic of the current period. Like other fields of scientific and technological development, medicine illustrates the range of issues that contemporary Chinese society faces as it modernizes. These issues include managing foreign-technology importation, adjusting to the ambiguities of the transition to a decentralized authority structure, and incorporating the specific reforms in personnel, financing, and management. But, for medicine, there is another dimension that sets it apart from the other sectors of the economy, and makes it a particularly useful focus of inquiry. Medicine is a public good, not just an economic product. The way in which it is distributed reflects cultural values of social responsibility, and the power relations within that society. In the starkest terms, variations in health indicators, such as infant mortality rate and life expectancy, can be viewed as one measure of the distribution of resources within a society.

In this chapter, I describe the impact of the reforms upon the medical-care system, first, by outlining the characteristics of the Maoist model of medical care and, second, by exploring the major areas of change under the reforms. These include (1) hospital reforms, (2) re-professionalization of medicine, (3) investment in high-technology medical care, (4) reliance upon returned scholars for transfer of information and leadership skills, and (5) the substantial increase in regional and sectoral variation in the financing and distribution of care.

Another way to describe this impact is through the social, organizational, and ethical dilemmas emerging from these structural changes. The reforms in medicine have raised four major areas of conflict:

(1) The conflict between an administrative style dominated by the new, expert professionals and that of the old work unit (*danwei*) system.

(2) The conflict arising from the increased dominance of expensive, high-technology medicine over the public-health-oriented, Maoist model of medical care.

(3) The conflict raised by an increasing need to plan for the regional development of modern medicine, but a decreasing ability on

the part of the state to direct this process under the decentralization reforms.

(4) Conflict between the socialist value, which views health care as a government welfare benefit, and the reformist trend, which views medicine as an economic product and a privilege, increasingly distributed by the market.

The modernization of medicine in China involves a broad range of issues, including the production of both scientific research and clinical care, and the structural and ethical issues involved in who is consuming this product and how they are paying for it. Each of these aspects is essential to an understanding of the contradictions produced during this transition era. Data presented here are derived mainly from an ongoing study of technology transfer in medicine. They are based upon structured, open-ended interviews with returned scholars, hospital administrators, physicians, and nurses at four research sites in Beijing, Guangzhou, Harbin, and Wuhan.

STRENGTHS AND WEAKNESSES OF MEDICINE UNDER MAO

The Maoist model of health care in the countryside was influenced by the rural health programs in Europe and in China during the 1920s and 1930s.[8] Its success has made China the focus of the World Health Organization's primary-care movement.[9] This model was guided by both pragmatic and political concerns. Faced with a limited number of physicians and medical facilities, and a political commitment to providing care to its citizens, the post-1949 Chinese government created a system that emphasized prevention, the utilization of traditional practitioners and non-MD personnel, and reliance upon local, collective resources for the financing of the majority of care. Developed during the late 1950s and early 1960s, its basic feature was the coordination of three levels of care, from the county hospital, to the commune clinic, to the barefoot-doctor station at the brigade level. Integral to this system was a referral network from the brigade to the county (and in the cities, from the neighborhood clinics to middle-level and tertiary-care facilities). Reimbursement for care at higher levels depended upon adherence to the referral regulations. By the late 1960s, the central figure in the primary care network was the barefoot doctor, who was recruited from the local brigade, continued to work part time as a farmer, and was paid by the collective income. While not a traditional-medicine practitioner, the barefoot doctor dispensed a combination of traditional and Western-style medical therapies.

The government limited the development of technology-based tertiary care, expensive biomedical research, and medical education, which have been a major drain of resources in many other developing countries. Instead, they developed networks of disease-prevention stations (*fangyi zhan*) and maternal- and child-health stations (*fuyou zhan*), which provided an infrastructure of personnel and information to address the health problems of a rural population. Although variable in coverage, most rural areas also provided some collective insurance programs (*hezuo yiliao*). Many health improvements can be directly related to the public health measures taken on as the responsibility of the collective rather than the individual. The results of China's low-cost, public-health model of medical care included a decline in infant mortality from approximately 200 deaths per 1,000 births before 1949, to 34.68 per 1,000 reported in the 1981 Chinese census.[10] Without a comparable rise in per capita income, China transformed its health profile and emerged as one of the few developing nations with a list of major causes of death that closely resembled those of the industrialized countries.[11]

These results were not achieved without some cost. The weaknesses of the Maoist model as it pursued quantity and wide distribution of services lay, as one would expect, in the quality of care and the research and educational support for such a system. The waste and misuse of talent during the Cultural Revolution (1966–1976) are well known. Medical teachers and scientific researchers were cut off from the international scientific community; few investments were made in the human or physical capital that supports the advancement of medical science. The quality of care at all levels was very uneven. In many instances, poor medical care was provided by inexperienced paramedics, thrust by the politics of the era into positions of authority. Furthermore, despite the desire for equitable distribution of resources, certain disparities in the type of care available continued to exist. These were the result of differences in investment in rural and urban areas, and the different insurance programs available to farmers, compared to workers and government employees. The insurance for state-enterprise workers and government employees covered essentially all medical care. Insurance for farmers (and for workers employed in collective enterprises), constituting over 80 percent of the population, was directly dependent upon the productivity of the collective and rarely equaled that of the state workers and employees.

In summary, the Maoist model of medical care demonstrated strengths in the areas of primary care and public health, and was well suited to a country with limited resources and a lack of highly skilled technical manpower. A system such as this can effectively respond to a population whose major health problems stem from infectious disease. The administration of immunizations, well-baby care, and sanitation measures require an organized health-care system, but not a medical-school education. However, as the proportion of people who suffer chronic, debilitating disease increases, the Chinese confront a challenge equal to that which they faced in 1949. In the West, chronic disease is the source of increasingly hospital-based, technologically sophisticated, and costly medical care. It is precisely this type of care that is now being developed in China's hospitals, and competing effectively for China's limited medical-care resources.

REFORMS IN MEDICINE

Overall, more money is being spent on medical care in China in both the urban and rural areas.[12] Between 1949 and 1984, the percent of the state budget devoted to medical care rose from 1.08 to 3.23.[13] Half this increase took place between 1978 and 1984, following the implementation of the reforms (see Table 1).

Although there continues to be real growth in funding for rural health care, the modernization of medicine is first and foremost occurring in China's cities. The proportion going to the urban (versus the rural) areas has risen. There has been a shift in hospital beds, manpower, and Ministry of Health funding from the countryside to the urban areas. The percentage of hospital beds in the cities between 1975 and 1984 rose from 39.9 percent to 42.4 percent, while in rural areas it has fallen from 60.1 percent to 57.6 percent. Likewise, the percentage of medical personnel has risen in the cities, from 41.9 in 1975 to 50.2 in 1984, with a corresponding drop in the countryside.[14] A study of government health expenditures between 1976 and 1983 reported that the proportion of state funding for provincial hospitals

Table 1 National Health Expenditures between 1949 and 1984

Time Period	*Total Health Expenditures (million yuan)*	*Health Expenditures as a Percent of Total State Expenditures*
1949–1952	0.559	1.52
1953–1957	1.455	1.08
1958–1962	2.345	1.02
1963–1965	1.884	1.56
1966–1970	3.835	1.52
1971–1975	6.562	1.67
1976–1980	11.365	2.17
1976	1.696	2.10
1978	2.242	2.02
1980	3.016	2.49
1981–1986	N.A.	N.A.
1981	3.274	2.94
1982	3.766	3.27
1983	4.195	3.25
1984	4.816	3.25

Source: Zhongguo weisheng nianjian (China health yearbook); Beijing: The People's Medical Publishing House, 1985), p. 59.

almost tripled, while the proportion devoted to the commune hospitals and disease-prevention stations declined by approximately one-third.[15] Since that time, the Chinese government has attempted to reverse this trend, and recent articles report increased funding for disease-prevention stations.[16]

In the countryside, the de-collectivization of agriculture after 1978 resulted in a breakdown of many of the collectively supported village clinics and insurance programs. Many barefoot doctors were unwilling to continue to provide medical care when they realized there were greater profits to be made in agriculture or township enterprises. Between 1975 and 1984, the number of barefoot doctors dropped from 1,559,214 to 1,251,204.[17] Because of the difficulty in agreeing upon a universal definition of collective insurance, there are varying estimates of the number of farmers who were left without insurance. Hsiao[18] cites Ministry of Health figures that the percentage of farmers enjoying some coverage dropped from about 85 to only 40–45 after the reforms.

During the 1980s, the reconstitution of health-care services in the countryside has included both collective and private ownership of village health stations.[19] The type of ownership and level of management of these stations varies considerably in different regions across China. This will be discussed in greater detail below. Approximately half the barefoot doctors have passed a licensing examination to qualify to be village doctors (*xiangcun yisheng*). Under the newly organized village health stations, they receive higher salaries, either paid directly from the collective, or as a proportion of the health station's income.[20] In fact, in some very wealthy rural areas, the village doctors may earn two or three times as much as their better-trained, urban counterparts.[21] Insurance, however, remains a serious problem, particularly in the poorest areas where farmers can least afford to pay for medical care.[22]

REFORMING THE MANAGEMENT AND FINANCING OF HOSPITALS

Under the reforms, hospitals are focusing on updating techniques, and eradicating the inefficiencies and the low productivity created by the system where all staff "ate from one big pot." The July 1985 wage-reform policies introduced more occupational differentials and rewards for seniority.[23] In addition, under the responsibility system, bonuses are given to reward attendance and hard work. Doctors are encouraged to work privately in off hours. Hospitals have increased their incomes through processing more patients, raising registration fees, charging more for patient visits to senior physicians, increasing the charges for medicines and for new diagnostic treatment and technology. The state- and enterprise-sponsored insurance programs have responded to increased costs by introducing their own reforms. Deductibles, prepaid-benefit experiments, and other incentives to reduce the waste that characterizes cost-reimbursement insurance have been introduced in some locations. The state has attempted to promote greater efficiency by standardizing subsidies to hospitals and limiting reimbursement for hospital deficits. The entire system is constrained by excess demand for high-level hospital beds on the one hand, and state-set prices for medical care on the other.

The source of the increase in hospital income varies. During the 1980s, there have been a small number of government-regulated price increases for procedures, diagnostic examinations, and medicines, and, in many places, this is the main factor in increased patient income. For example, in Heilongjiang, Jiamosi Municipal Hospital, the proportion of income from patients for medicines rose from 38 percent to 59 percent of the total patient income between 1980 and 1985.[24] In the hospital affiliated with Hubei Traditional Medicine College, the charges to patients for medicines have doubled in the past five years. In 1981, the average outpatient charge was 3.09 yuan per person; in 1985, it was 6.54 yuan.[25] At a teaching hospital affiliated with Hubei Medical College, the main source of rising income is from increases in diagnostic examinations.[26] The yearly income from a new ultrasound machine in the department of obstetrics, for example, has been over 100,000 yuan ($27,027).

In the coastal cities of Guangzhou, Shanghai, Beijing, and Tianjin, increased income is also derived from the implementation of a 1981 State Council directive to introduce a two-price system for insured (*gongfei*) and uninsured (*zifei*) patients.[27] Reports from Shanghai, where it was first introduced, indicate that the additional income from the higher charges for insured patients has boosted the income of all Shanghai hospitals by one-third of their total yearly allocation from the Ministry of Health.[28] Less progressive regions, such as Hubei, have not attempted to implement this system, although uninsured patients are charged less for very expensive diagnostic examinations.

Other hospitals assert that better patient management is responsible for their increased income. At one hospital, contracts with radiology-department staff provide bonuses for increased patient examinations. Throughout China, "home beds" (*jiating bingchuang*) have been introduced to move convalescing patients out of hospitals and increase patient turnover, thus raising income. In 1984, there were almost 500,000 home beds in China.[29] At Peking Union Medical College (PUMC), a model for its 1983 reform program, hospital income rose from 5 million yuan ($1.35 million) in 1982 to 8.4 million yuan ($2.27 million) in 1984. This dramatic increase is attributed to more efficient patient management, including an increase in inpatients, outpatients, occupancy rate, and number of procedures performed, and a decrease in patient length of stay.[30]

The increased cost of care is inevitably affecting the insurance system and the individual self-paying patient. The two-price system, in particular, places a greater strain upon the government and state-enterprise work units which must bear the additional cost of increased patient charges. The state and worker insurance programs have always generated waste. Although many of the reforms are now attempting to address this problem, insured patients still have overly long hospital stays, over-use of medications, and, in general, lack incentives to reduce medical costs.

In contrast, the cost of medical care is carefully monitored by physicians for the uninsured or partially insured patients. Many of the new technologies and procedures are beyond the budget of most rural residents. For example, treatment with peritoneal dialysis for acute renal failure averages over 1,000 yuan ($270), much more than the annual income of many farmers in China. Chronic kidney dialysis costs between 12,000–15,000 yuan ($3,243–4,054) per year. One CT scan is approximately 200 yuan ($54). Expensive hospitalization costs are not limited to the Western-style hospitals. In one traditional-medicine hospital, the average total cost per inpatient in 1985 was 367 yuan ($99), generated primarily by the longer average length of stay, 28 days, compared to less than 20 days at a comparable Western-style hospital. The average cost per patient per day at both hospitals was 12–13 yuan ($3.25–$3.50).[31]

At the same time that hospitals are reforming the management and financing of medical care, the government has attempted to standardize its regulations for hospital subsidies. In the early years of the Communist regime, the government decided to subsidize hospital care by setting prices below cost and providing funds for personnel salaries, maintenance, and small-equipment purchase, varying the amount according to the size and quality of the hospital. Thus, for example, at a provincial-level hospital in Wuhan, the income from patient charges in 1985 was almost 3 million yuan ($0.8 million), the state subsidy was over 1 million yuan ($270,270), and 700,000 yuan ($189,189) was allocated by the provincial government for large-equipment purchases.[32] However, deficits beyond the income generated by patient care and government subsidies continued to accumulate under this system. In 1984, the average deficit for a provincial-level hospital such as the one in Wuhan was 693,000 yuan ($187,297). The total annual deficit for medical care was 950 million yuan ($257 million).[33]

In an effort to eliminate these deficits, the state has adopted a more stringent policy toward hospital financing. This, in addition to the other hospital reforms, is producing a transformation of medical care from a public-supported, welfare benefit into a more self-sufficient, service-sector product (*disan chanye*).[34] Beginning in Shanghai in 1984, and now in many parts of the country, the government has set stricter standards for the distribution of state subsidies and limited its contribution to hospital deficits.[35] This has placed a great deal of pressure upon hospitals who are still constrained by the state-regulated pricing system. In many cases, particularly for surgical procedures, the charges for patient care are still substantially below the actual cost.[36] At present, the way in which the state and the market should combine to plan and distribute medical-care services is a complex and controversial issue.

This issue is more complicated in medicine than in other sectors. High-technology medicine has the capacity to generate substantial profits, but several factors operate in medical care which undermine its ability to function as an ideal market. These have to do with the difficulty the consumer has making an informed choice for medical care, and the ability of the supplier (physician) to influence and even generate demand for the services supplied.[37] It is further complicated by the belief, on the part of many, that it is unlikely that a public good can be justly regulated by the market.

In China, the transformation of medical care from welfare benefit to service-sector product is occurring at a time when the system is under additional pressure from a substantial unmet demand for more medical-care services. A recent Ministry of Health Report[38] identified the demand for hospital beds as one of its major problems. In 1985, there were 2.14 hospital beds per 1,000 persons. While this is not a remarkably low number, the average length of time patients stay in Chinese hospitals is two or three times as long as that in the United States.[39] The Ministry of Health report stated that, out of a yearly potential inpatient population of 50 million persons, 20 million are unable to obtain needed hospital care. Although the supply of beds in the cities is three times that in the countryside (4.53 per 1,000 persons compared to 1.53 per 1,000), the report cited statistics on patients in Beijing, showing that, in 1985, 60,000 patients were unable to find hospital beds.

To remedy this excess demand for beds, particularly in higher-level hospitals, a new medical-care network has been introduced in some areas. While similar to the earlier referral system, which linked different level hospitals in the same region,[40] this system bases cooperation between hospitals upon financial incentives. Physicians from tertiary-care hospitals are given bonuses for temporary work at neighborhood or district hospitals, paid from the resulting increase in patient fees. The goal of the cooperative network is to provide better care at lower-level hospitals, thus inducing patients to stay at those facilities with available beds.[41]

In medicine, as in other sectors of the economy, the reforms are intended to unleash productive forces. The basic administrative structure remains, but the system of finance and management of medical institutions is gradually being decentralized and made more economically efficient. Although the reforms are intended to make hospitals more sensitive to economic factors, this does not mean that market forces have been unleashed in China. Prices are not determined by supply and demand; rather, the state continues to set prices to satisfy both social and economic needs. In rural areas, where some clinics are privately owned or managed by countryside doctors, and many patients have no insurance coverage, the market is more important. In the more than 67,000 hospitals at the country level and above, however, the reformers are more cautious.

RE-PROFESSIONALIZATION OF MEDICINE

The conflict between medical professionals and Communist Party administrators in China has been longstanding.[42] In social-science terms, professionals are members of a service-oriented occupation which has established control over the information and skills used by its practitioners, and over the qualifications needed to be a member.[43] According to this definition, Chinese physicians were never completely in control of their profession, but this was particularly true during the more egalitarian phases of the Maoist era. In the CR, the Soviet-style, joint rule of party leaders and technical experts in organizations was replaced by the single rule of the Communist Party secretary; medical care, education, and research were under the control of these secretaries; and physicians were often sent to the countryside to be reeducated according to Party- rather than physician-defined standards of medical practice.[44]

Under the reforms, the anti-professional, anti-technology ideology of the CR has been replaced by the values of efficiency, educational achievement, and technical competence. In all organizations, the Party has recruited experts to join its ranks, and medical professionals have been promoted to positions of authority for fixed terms. Medical-college presidents are noted scientists. Hospital and department directors are experts. Hospital committees to evaluate new technology (*yuanwu weiyuanhui*) and advisors to purchasing departments are more technically knowledgeable. Since state policy now requires a diploma for promotion, administrators with no formal training are attending two-year health-management training schools. Medical education and technical training are gradually being strengthened, and key medical schools receive additional funds from international agencies and philanthropic organizations.

It is difficult to evaluate the extent of the decline in Party influence. Medical experts are now directing the updating of medical knowledge and skills through the transfer of technology and information from abroad.[45] As one professor of internal medicine noted, "They must listen now. They know I am right."[46] On the other hand, the Party is still involved in the major decisions made by medical institutions. And the extent of professional influence depends upon a range of other factors, such as age, length of employment, and connections with influential leaders in the organization.

Furthermore, medical professionals still face the inertia of the work-unit (*danwei*) system, thwarting efforts to modernize science and create more efficient medical care. The basic element of work in urban China, a work unit, can be a hospital, a factory, a school, or an office. It combines the functions of production, politics, and, often, residence for its members, much like an urban village.[47] Unit members are assigned by the government, and, until recently, it was almost impossible to transfer out. In the past, unit leaders were responsible for all aspects of life and work. This leadership style appears increasingly inefficient as work units face the demands of performing as complex bureaucracies. For example, as work units are implementing reforms aimed at increasing efficiency, it is clear that the addition of clerical secretaries would significantly increase the productivity of unit leaders. Many medical personnel cited the need for leaders to be able to delegate authority, so that "they are not constantly arranging for the repair of a light bulb," and the need for the duties of different kinds of personnel to be clarified.[48] One young scientist, promoted to vice-director of his institute upon his return from abroad, railed against the work-unit personnel system, "People can't move and can't be fired. I can't make my technician work for me. I can't say I will fire you if you don't work. I must do everything for everyone . . . babies, housing, retirement, relatives, people wanting more money, grant application forms, everything! The top leaders can't control the middle leaders and must involve themselves in all the details of the work unit (*danwei*)."[49]

Yet, the values upon which this socialist system was founded are not easily put aside. Thus, when asked if the system of contracts and ability to fire nonproductive personnel would aid modernization, one medical college president cautioned, "But if we just fire such a person, he will have nothing to eat, and we will have created a social problem."[50] This fundamental principle of social responsibility remains an incontrovertible force in the modernization process.

INVESTMENT IN NEW MEDICAL TECHNOLOGY

During the 1980s, large urban, tertiary-care hospitals, as well as county and district hospitals, have invested a considerable amount in new medical technology. The Ministry of Health annual budget for imported medical equipment alone is 720 million yuan ($200 million).[51] The main focus has been to replace or upgrade older technology, and to purchase new diagnostic and monitoring equipment. Large hospitals now typically own advanced x-ray machines, ultrasound machines, blood-gas and biochemical analyzers, heart monitors, fetal monitors, respirators, and, by 1985, there were several hundred CT scanners in China. The highest level hospitals in China have intensive-care units. Many are beginning to develop critical care beds to care for cardiology or cardiac-surgery patients. Basic critical care for infants is common at county-level hospitals. Hospitals owned by wealthy enterprises, the army, and those located in areas receiving large overseas Chinese contributions have made extensive investments in new medical technology. for example, the first NMR (nuclear magnetic resonance-imaging machine) was purchased by the premier military hospital in Guangzhou.

These changes in expensive technology are not matched by improvements in the support system and ancillary services needed for daily patient care. The minimal laboratory facilities, hand-made swabs, reusable rubber tubing, and other small equipment remain essentially unchanged. Furthermore, as described below, there is inadequate technical support for the acquisition of so much new equipment, in terms of informed purchase, coordination for efficient utilization, and maintenance.[52] Equally important, there have been very few efforts to assess the acquisition of so much new technology, or to plan for the systematic, coordinated distribution of such expensive medical resources.[53]

Funding for the purchase of new medical equipment which costs above a set amount (previously 10,000 yuan [$2,702] now 20,000 yuan [$5,404]) is provided by a separate equipment budget from the state. For most hospitals, this equipment allocation process has not changed under the reforms. Application for new equipment is considered by the hospital technology committee, then the hospital finance department, and finally, the upper levels of the health bureaucracy. Hospitals that generate additional income from patient charges may use a portion of these funds for equipment purchases. Thus hospitals that are making more money from patients are able to upgrade their facilities, and presumably attract more patients. This is not an issue for the highest-level hospitals with their 100+ percent occupancy rates. For those hospitals with lower occupancy rates, those serving uninsured patients who are no longer constrained by the old referral regulations, and for those who are trying to attract foreign and overseas Chinese patients, this is an important concern.

When the machine to be purchased is manufactured abroad, the hospital must make a special application to the Ministry of Health for foreign currency. Ministry control of foreign currency allocation for expensive medical equipment is intended to be a planning tool. This would undoubtedly be more effective had there not been a simultaneous explosion of new "channels" for funding. These channels include grants from foreign foundations, international agencies, private companies, and from overseas Chinese and Hong Kong Chinese who wish to return to China to receive medical care (paying in foreign currency). Perhaps most detrimental to the state's ability to control technology acquisition has been the trend, in provinces with access to foreign currency through trade or tourism, to independently fund lower-level hospitals under their direct jurisdiction, particularly county-level hospitals. Finally, even domestic channels are proliferating, with some work units contracting hospitals to provide medical services in exchange for imported equipment, buildings, or other items. There are substantial regional variations in these trends; coastal areas and those provinces rich in foreign currency have much greater access to these channels.

One of the most salient issues in considering the acquisition of so much new technology is the way in which the Chinese are able to plan and control this investment. Much has been learned in the past decade, and much more informed purchasing is currently carried out. The media and academic journals have featured articles about the problems of waste, poor planning in technology acquisition, and "blind purchase" of unnecessary machines. For example, in a recent issue of *Jiankang bao* (Health daily), the Chinese Academy of Medical Sciences was criticized for waste in the purchase of new machines.[54] Several cost-benefit studies evaluating new technology, such as a study comparing the CT scanner and traditional x-ray machine, have appeared in technical journals.[55] Despite these encouraging trends, however, there are major difficulties in developing a sensible planning capability for the acquisition of new medical technology.

There are two main problems. First, there is no coordinated information system to advise hospitals in purchasing new equipment, and to provide the technical information needed to develop a planning capacity. Second, particularly in the coastal areas and wealthy provinces, decentralization of authority has resulted in much less state control over the purchase of new technology. Guangzhou, for example, has more CT scanners per capita than London;[56] and hemodialysis, initially supported by foreign currency from Hong Kong, is expanding rapidly in south China.[57] With the rising cost of medical care and the insistent desire to purchase the most up-to-date equipment, the ability of the Chinese health planners to control the allocation of resources in this arena is clearly a central concern.

THE ROLE OF RETURNED SCHOLARS

The technical information and skills needed for medical modernization have been brought back to China by clinicians and biomedical scholars studying abroad. There has been a conscious strategy of "importing human capital" (*rencai yinjin*),[58] involving both the use of foreign experts within China and Chinese scholars who have returned from abroad. Between 1978 and 1986, as many as 3,000 Chinese students and scholars and 5,000 short-term visitors from biomedical fields studied abroad.[59] Western-trained returned scholars have been the main source of knowledge and planning for the way in which each of the medical disciplines will modernize. They are actively supporting the development of academic societies in China, fostering important horizontal, discipline-based linkages in a predominantly vertically integrated society. They are also, in many cases, the people promoted to leadership positions within their organizations who are making the technology acquisition decisions. Since most of the scholars are from the main coastal cities, however, this process has also heightened the uneven development of medical centers throughout China.

Although sending Chinese abroad to learn the latest skills and information is part of the overall strategy of the modernization effort, returned scholars have encountered a number of difficulties once back in China. Interviews with medical university officials in several locations, as well as reports in the Chinese press, confirm that a large portion of scholars often return to their original jobs, do not make full use of the technologies they have learned, and find it very difficult to move to a more desirable work unit. In addition, the lack of central coordination of study abroad, and the fact that half those who go abroad are self-financed (and free to choose their own subject of study), further undermines effective integration of returned scholars. Interviews with scholars in four cities revealed universal problems with productivity, caused by lack of funds, space, technical personnel, and equipment. This is complicated by the need to resume teaching and clinical work, as well as by the common request for the scholar to assume additional administrative duties. For some, the major stumbling block is the inability of their work-unit leaders to provide a supportive environment. For those returning to units where few have gone abroad, the readjustment back to the conditions in China is even more difficult without a critical mass of Chinese colleagues with whom to share technical and clinical information.

Recognizing these problems, the Chinese government has attempted to facilitate job transfer for returned scholars. It has established central laboratories with facilities similar to those abroad, and, in some cases, given funding priorities to scholars to build their laboratory research capabilities. Through the media and other avenues, the state has emphasized the importance of the role of the intellectual in the modernization drive. Despite these efforts, it is often impossible for the returned scholar effectively or completely to transfer the new technology to China. Instead, the contribution of the scholars may be better viewed as a bridge between Maoist China and the world of modern science. Their functions are to communicate information and organizational skills, and build networks with Western scientists and clinicians for further exchange of technology and personnel. It is the next generation of scholars, educated abroad for longer periods and at an earlier point in their training, who will carry out the modernization work begun by their elders.

The focus upon studying abroad as the most desirable route to modernization has resulted in an elevation of the people involved and a legitimization of the information and perspectives that are brought back to China. Consciously and unconsciously, the Chinese are influenced by the Western model of clinical medicine. In contrast to the earlier public-health model, which promoted prevention, mass participation, and a combination of traditional and Western-style therapies, the Western biomedical model depends upon scientific, hospital-based medical care. The doctor is removed from the patient by diagnostic and treatment technology, and non-biomedical factors are often ignored in the technical process of treatment decision making.[60] Although Chinese physicians are anxious to modernize their medical equipment, they are also critical of this approach to the treatment of patients. In fact, that philosophical perspective in the West that advocates a more holistic consideration of psycho-social and environmental factors, as well as the strictly biomedical, has found proponents among returned scholars, particularly in schools of public health.

Modernizing medicine through sending scholars abroad to transfer technology has historical as well as philosophical implications for China. The question of the appropriate Chinese uses for foreign technology was a subject of intense debate before 1949. The current embrace of high-technology medicine in post-Mao China raises this debate once again. Yet the way in which Chinese and Western perspectives will ultimately be combined remains unresolved.

REGIONAL VARIATION: THE URBAN-RURAL GAP

It is assumed that the reforms will increase differentiation and diversity in China. Reacting to earlier norms of implementing policy by "one slice of the knife" (*yi dao qie*), current policies support "an uneven but coordinated development in different areas."[61] The resulting variation is exemplified most keenly in the comparisons between urban and rural health care.

In general, medical care in the countryside is improving. The countryside doctors are now required to pass a standardized course of study, and the state is funding a program to update county-level hospitals throughout China.[62] Farmers are demanding better care, and their rising incomes have fueled their ability to pay for it. On the other hand, the reforms have introduced greater variation in the type of health-care services, the financing for those services, and the range of services covered. In its 1984 report on China's health care, the World Bank identified inequities in access to rural health care as one of China's most pressing problems.

According to the *Chinese Health Yearbook*, in 1984, half China's 707,168 brigade health stations were collectively owned, 10 percent were owned by the village but managed by the village doctors, and almost one-third were privately owned[63] (see Table 2). A study of medical care in southwest China in 1986 reported that 64 percent of the brigade health stations were collectively owned, 8 percent were collectively owned but run by village doctors, 18 percent were privately owned, and 19 percent were either owned by the township (former commune) or staffed by traveling doctors.[64]

Although many collective insurance programs disappeared as the responsibility system was implemented,[65] there are currently a number of different programs in operation or in the planning stages at the township (*xiang*) level. (Township has replaced the term *commune* and is the level of organization above the village. Brigades, which consist of 5–10 villages, have lost most of their previous administrative functions.) Reflecting the philosophy of decentralization, there is a wide diversity in programs and in approaches to resolving the dilemma of financing medical care in the countryside.

The township can obtain funding for medical-care services from various sources. These include the district (*qu*, the level above the township and below the county), the county, the local township-run

TABLE 2 The Situation of Health Organizations in the Brigade Health Stations in China, in 1984

	Number	*Percent*
Total number of brigade health stations:[a]	707,168	100.0
Total number of brigade stations run by the collective	360,079	50.9
Total number of brigade stations jointly run by the collective and village doctors	71,305	10.1
Total number of brigade stations run by the township hospital	23,967	3.4
Total number of brigade stations privately owned	222,771	31.5
Other	29,046	4.1

Source: Zhongguo weisheng nianjian, p. 45.

Note: a. Out of the total of 715,265 brigades in China in 1984, 623,665 (or 87.2%) had health stations. The fact that there were 707,168 brigade health stations is explained by the fact that some larger, wealthier brigades have two health stations.

enterprises, and contributions from local village residents. This funding can be used to support services of countryside doctors and health aids, primary-care services, and prevention services such as immunizations, maternal and child health, and anti-epidemic work. The prevention services are most often subsidized by higher levels due to their intrinsic value to the entire community.

Wealthy townships are often able to provide other forms of health insurance to their residents, covering a percentage of inpatient and outpatient charges. The variation in coverage is related to the income of the township, the demand of the farmers, and the historical experience with health insurance in each location. Examples of the different financing packages include: insuring medical care but not medicines (*heyi buheyao*); various combinations of management and finance at the village and township levels (*cunban cunguan, xiangban xiangguan,* and *cunban xiangguan*); insurance coverage for only maternal and child health and anti-epidemic work; and a local contracting system (*chengbao*) in which the countryside doctors jointly run a clinic for which the township or village residents make yearly contributions. Experiments with prepaid medical services through giving each participating farmer a set amount for medical-care costs, are also being considered.

The breakdown of rural cooperative insurance has produced access problems for farmers in very poor regions. It has undercut the functioning of the primary-care referral network, which based reimbursement for health-care costs, in part, upon the referral from lower-level medical personnel. Those who can afford to pay for their medical care or who still enjoy insurance coverage, however, are increasing utilization of county and higher-level facilities. In fact, at one mid-level-income county in Guangdong province, the health director stated that the major problem he faced was meeting patient demand for better medical care, citing statistics on occupancy rates for facilities below the county hospital as 60 percent or less, compared to 99.9 percent and even over 100 percent in county and higher-level hospitals.[66]

The most important question regarding these changes in the rural health-care system is the impact upon the health status of farmers in China. On the positive side, with the rise in rural incomes, the ability to pay for better health care has also risen, perhaps offsetting the loss in collective insurance protection. The price of care continues to be subsidized by the government. The training of rural practitioners has improved and become more standardized. In most places, community prevention work is collectively funded. Since the implementation of the one-child policy, pre- and post-natal care is particularly emphasized. Furthermore, partly due to the perception of a decline in rural health conditions, medical care in the countryside has received attention and financial support from several international agencies, including the World Bank,[67] the World Health Organization, and UNICEF. Lastly, probably the greatest impact of the reforms upon health has been the substantial improvement in diet and nutritional status due to increased income and availability of foods.

Factors on the negative side are mainly related to an increase in regional differences in access to health care. The differential in the amount spent on medical care per capita seems to have increased considerably in the countryside. In poorer areas, official figures cite about 3 yuan per person per year,[68] but in wealthy regions it can be as high as 50 yuan.[69] The loss of insurance coverage is reported in the poorest areas. Decline in insurance coverage is associated with decreased utilization of health services, which has negative consequences for health status. These are also the areas with the poorest nutritional status and greatest difficulty in funding collective prevention efforts, such as childhood immunization programs.

Health-status indicators that would measure the impact of these changes in the countryside are difficult to obtain. According to Chinese maternal- and child-health leaders, and central government statistics,[70] the infant mortality rate has continuously declined. Other studies[71] have shown that there has been an increase in the infant mortality rate in some regions since the implementation of the reforms. If these figures are correct, then this basic measure of equality points to the conclusion that increased regional variation under the reforms has produced greater inequality in the distribution of the society's resources.

During the past decade, there has been a substantial professionalization of medical care; a significant increase in Western influence in the training and funding of modern medicine; a decentralization of authority in all Chinese organizations, creating more choices for spending health funding and more avenues for obtaining that funding; a shift in emphasis from rural to urban medical care; and a decollectivization of basic-level, rural medical services. The result has been a much more hospital-based, physician-directed, and curative (versus preventive) model of medical care. The quality of medical care, education, and research have all risen substantially.

As more patients in China are paying for medical care themselves, the market is beginning to play a greater role in the rationing of care. Furthermore, as always when a valuable commodity is in short supply, the use of connections to gain admission to high-level facilities becomes increasingly important.[72] These factors have not suddenly appeared with the onset of the reforms in post-Mao China. But the contradiction between the previous philosophy of medical-care distribution as a state-supported welfare benefit and the current focus upon efficiency and distribution based upon ability to pay presents new and difficult dilemmas for those in the health-care arena.

It is easy to exaggerate the extent of the change that has taken place, and to conclude that China has once again reversed its "line," replacing Maoist innovations with Western science and technology. Yet, this interpretation is inaccurate. The basic infrastructure of the health-care system has not been rejected. Rather, technical and organizational changes have been introduced, building upon the foundation of primary and hospital-based care. Professionalization, decentralization, and more efficient management of medical care are responses to policy shifts from above. But they are also a reflection of shifting demands on the part of the Chinese population. The changing disease patterns, the increase in income, and the greater sophistication of medical consumers have all brought pressure to bear upon this system.

For the future, several important questions will need to be resolved. First, can China afford to pay for the much more expensive care of chronic illness which is increasingly the focus of clinical medicine and research? Second, given the greater regional disparity in access to services, to what extent can the market be permitted to distribute medical services? Third, how can China reconstruct a planning capability that will permit a more systematic regional development of expensive medical technology? Last, what will be the long term impact of foreign study and foreign influence in the course of medical modernization?

These questions are not unique to medicine. The reforms in medicine are part of a broader initiative in technology and other sectors of the economy. Many of the issues discussed here are similar to those faced in other areas of science and technology. Questions of state spending priorities, the role of the market, the difficulty in managing new technology, the new role of the expert in work organizations, and the impact of the returned scholars are essentially the same. What is special about medicine is the extent to which it highlights the ethical dilemmas of a society in transition. The contradictions described here reflect the issues of the whole society as it recreates ideologies and organizations more consonant with its new economic system.

CHAPTER TEN

Science and Technology in the Chinese Countryside

ATHAR HUSSAIN

Whatever the course of modernization in the Chinese economy, technical progress in the rural areas, especially in farming technology, will be a central component. In 1984, agriculture still furnished 44 percent of national income and employed 68 percent of the labor force. Agriculture in China is a diverse sector covering not only the usual farming, aquaculture, and animal husbandry but also the increasingly important sideline production. Ranked according to shares of gross agricultural output, farming heads the list with a figure of 58 percent;[1] next, with a share of 22 percent is sideline production, including village-run industries.[2]

Farming technology is conditioned by the endowments of agricultural resources and climate. With only 1 ha of arable land for over 10 mouths, the ratio of arable land to population in China is very low.

Although the ratios for Japan and Korea are even lower, the proportion of the labor force employed in agriculture is higher in China, where the number of agricultural laborers per cultivable ha is probably the highest in the world.[3] In all, two features single China out. The low ratio of arable land to population is a constraint within which the longstanding aim for near self-sufficiency in grain and other crops has to be realized; second is the exceptionally high expenditure of labor per unit cultivated area. How might these features change by the year 2000?

The population will keep on increasing well into the twenty-first century. Given the official forecast of a population of 1,200 million for the year 2000, by then each arable ha will have to support an additional two mouths.[4] Consequently, the demand for agricultural goods will keep on increasing, and further progress in production will be necessary to keep pace with demand. Although the proportion of the labor force in agriculture has been decreasing since 1952, the absolute numbers have been going up, with a corresponding decrease in the ratio of arable land to the agricultural labor force.[5] Since this trend is likely to persist into the 1990s, agricultural technology will not be called upon in the near future to accommodate a decrease in the number of hands.[6]

Infrastructure and improvements to the land, intermediate inputs and machines and implements, and climate also influence the course of farming technology. China is well endowed with hydraulic infrastructure, a heavy user of fertilizers, yet still at a low level of agricultural mechanization. Furthermore, China has fewer frost-free days and a shorter growing season than elsewhere in the Eurasian continent with the same latitude and altitude.[7]

Thus, the environment for the operation of China's agricultural technology is marked by an extreme shortage of land, abundance of labor disposed toward hard work, painstaking and meticulous husbandry, extensive water management, a heavy use of fertilizers, and a low level of mechanization.

TARGETS AND COMPARISONS

The Ministry of Agriculture targets up to the year 2000 for agricultural output (excluding sideline products) are an annual growth rate of 4.4 percent, and a radical change in the output composition. The

TABLE 1 Growth Rates for the Volume and Yield per ha (% per annum)

	Volume		*Yield*	
Crop	*1978–84*	*1985–2000*	*1978–84*	*1985–2000*
Rice	4.5	1.8	5.14	1.8
Wheat	8.5	2.6	8.25	2.6
Corn	4.6	2.6	5.3	2.8
Cotton	19.3		12.51	
Non-grain crops		6.8		6.6
Oil seeds	14.7		8.5	

Sources: Target rates are from *China:* "Agriculture to the Year 2000," Appendix 2 to *China: Long-Term Development, Issues and Options,* p. 58, and the rates for 1978–1984 are from *Statistical Yearbook of China,* 1985, pp. 253–255.

share of the livestock sector is to increase from 15 to 25 percent, implying an annual rate of growth of 7.4 percent for the livestock sector and 3.5 percent for crops. The former is regarded as ambitious, the latter feasible. The composition of crops is to undergo a change as well: Grain is to grow at 2.4 percent and the non-grain crops at 6.8 percent per year.

More relevant for present purposes are the targets for specific crops. Table 1 displays a sample of them, covering 67 percent of the cropped area in 1984.

From a historical perspective, these targets err on the side of modesty, but extrapolating from past historical trends can be misleading. The important point is that the targets are to be achieved by an increase in yield per cultivated ha. Even after due allowance for the expected increase in the agricultural labor force, this will require technical progress. A distinction employed by economists posits an increase in agricultural yield consisting of two components: first, "catching up" with current best practices (fuller exploitation of existing technical possibilities); and, second, improvements in best practices themselves. In the case of "catching up," S&T's role is to adapt best practices to the habitat and widen their use; in the second case, S&T have the additional task of pushing forward the margin of technology. As modern farming technology is centered on high-yield-variety seeds (HYVs), these tasks, for the most part, amount to introducing and diffusing new varieties, and determining inputs and

TABLE 2 Comparison of Crop Yields (tons per ha)

Crop	*(1) China 1984*	*(2) Best Comparator Yield*	*(3) China 2000*	*(4) 2/1(%)*	*(5) 3/2(%)*
Rice	5.37	6.50	7.26	121	112
Wheat	2.97	4.22	4.47	142	106
Corn	3.96	6.14	5.97	155	97
Seed Cotton	2.64	2.88	No Target	109	

Source: Except for cotton, the Chinese figures are from *Statistical Yearbook of China* 1985, pp. 253–255; the rest are from *FAO Production Yearbook,* 1984. In order to allow for fluctuations in agricultural yields, the figures in Columns (1) and (2) are the highest over the period 1982–1984.

services complementary to them. Thus, "catching up" may amount to no more than more fertilizer and more irrigation, and depend therefore entirely on investment in agriculture and the fertilizer industry.

To obtain quantitative estimates of what percentages of crop targets for the year 2000 could be achieved by catching up with the best practice, China's targets can be compared with the maxima of yields from Japan, South Korea, North Korea, and Egypt, which, like China, have a low ratio of arable land to population. Since the three East Asian economies do not produce cotton, the comparator sample for that consists of Egypt, Turkey, and Mexico.

As we move from rice to corn, the gap between the current Chinese and the best comparator yield (Column 4) widens. This ranking also holds for research performance. As Column 5 shows, to achieve the targets for rice and wheat, China has to improve upon the best East Asian yield (among the highest in the world) by 12 percent in rice and 6 percent in wheat. In contrast, the target for corn can be achieved merely by closing the gap.[8] After the spectacular increase since 1978, Chinese cotton yields are now among the highest in the world, marginally short of the best of recent Mexican yields. These estimates are rough. While China is vast and diverse, the comparator East Asian countries are homogeneous and small, more akin to Chinese provinces. Interprovincial differences can be large in China.[9] Furthermore, the gap between the current Chinese and the best comparator yields may be smaller than indicated above, since the index of multiple-cropping is higher in China, and more crops usually imply a smaller yield per crop.[10]

This comparison yields one general conclusion. The targets for the year 2000 are more ambitious than they appear from a projection of the spectacular rates since 1978, most of which could be attributed to a catching up with best practices. With a further closing of the gap, the Chinese rates of growth will decelerate to the rates of improvement in best practices. Whatever may be its source, growth would involve extra investment in services and inputs, particularly in S&T in development and diffusion of HYVs, water management, and fertilizers.

Until the recent shift from collective to family cultivation, apart from the government, the commune, the brigade, and the team were the sources of agricultural investment, which, in Chinese statistics, covers infrastructure, machines, and implements, but not fertilizer production. The three tiers of the commune together accounted for around 60 percent of the total.[11] But, since 1980, at the same time that agricultural production was booming, direct investment in agriculture both by the government and the commune declined.[12] the government's decline was due to budgetary problems; the commune's to the end of collective cultivation, which drastically reduced the resources at the disposal of the commune (the production team in particular) and severely curtailed its ability to mobilize labor for capital projects. Government investment in agriculture is expected to pick up, but only 30 percent of it will be directed toward crops and over 50 percent toward dairy production, livestock, and forestry.[13] Now that households are allowed to own agricultural implements and machines, however, they are an important alternative source of investment in agriculture, and have the means to finance it. Between 1978–1984, the savings deposits of commune members rose almost 8-fold.[14] However, because a substantial part of communal investment went for hydraulic structures and other construction projects, households can cover only a part of it. Ultimately, there seems to be no alternative but for the government (including a new tier of local government) to assume responsibility for infrastructure and large-scale services.

ORGANIZATION OF S&T ACTIVITIES

While the organization of research activities has remained fairly stable, S&T extension and diffusion have followed the transformations in the rural economy. The institutional core of the present-day research network dates from 1957, when the Chinese Academy of Agricultural Sciences (CAAS) was established. In structure, it follows the Soviet rather than the American land-grant-college model. Research activities are separated from postgraduate education and dispersed over a multiplicity of highly specialized institutes. Schematically, the network is shown in Table 3:

TABLE 3 CAAS Research Network

Level	*Coordinating Bodies and Institutes*
National	CAAS and 30 National Institutes
Provincial	Provincial CAAS and Provincal Institutes
Prefectural	Prefectural Institute

Source: WB, *"China: Agriculture to the Year 2000,"* pp. 50–51; FAO, China: *Agricultural Training System,* 1980.

Its multi-tiered structure mirrors the federal structure of the government. The tier indicates the principal source of funding—the national institutes are funded by the central government and so on—and the domain of research interests. The national institutes concentrate on issues deemed nationally important, such as the breeding of high-yield varieties of rice; the focus narrows with each step down the ladder. Thus, prefectural institutes are confined to adapting technology to their area. In fact, this division of labor is almost the same as in Japan.[15] In terms of numbers, the network is dense, consisting of 390 institutes at the national and provincial levels and 200 prefectural institutes (almost 1 per prefecture). Moreover, there are institutes outside the CAAS network that also conduct research bearing on agriculture, those of the Chinese Academy of Sciences, for example.[16] In the past, CAAS was completely given over to the applied and adaptive research, but it has started to branch into related pure research. A recent breakdown of its activities is as follows: (1) 11 percent basic research, (2) 71 percent applied research, and (3) 18 percent development.[17] The orientation of the research network still remains overwhelmingly applied and adaptive, due not so much to choice as to the acute shortage of qualified researchers.

Although the network seems elaborate and well-structured, it is overlaid with a tight administrative hierarchy but a loose coordination of research activities. The channels of information among the institutes are rudimentary and informal. As a result, there is unnecessary duplication of research activities and a wide variation in the performance of institutes. Keenly aware of these problems, the political leadership has sought to grant the institutes greater power to determine their activities and to change their methods of functioning.[18] There has also been an important change in the methods of funding: Institutes are expected to raise funds by contract research and by marketing research products.[19]

The loose coordination of research activities can be attributed to a variety of factors. Institutes were treated as if they were departments of the government. The channels of communication have been predominantly political and administrative rather than professional and scholarly. During the CR period and until 1978, professional publications ceased altogether. While contacts between researchers and the peasantry were encouraged, those among researchers were suspect. Although professional publications and societies are now allowed to flourish, China still lacks the network of communications needed to create a community of researchers and scholars. Moreover, the ideology of self-reliance, which in multiple forms has permeated China, engenders local autonomy within bureaucratically delineated boundaries. The slogan of self-reliance has dropped out of circulation, but many of its institutional effects still remain.[20]

In the past, weak internal linkages went hand in hand with national autarky. Until 1978, barring occasional visits and sporadic exchange of plant varieties, the Chinese research network was cut off from the outside world. But things have radically changed. Now China participates in various international arrangements for cooperation in agricultural research—in the work of International Rice Research Institute (IRRI) in the Philippines and International Maize and Wheat Improvement Center (CIMMYT) in Mexico, for example. It has gone as far as actively seeking aid from foreign governments and international organizations. One consequence is aid from the Rockefeller Foundation, the UN, and the World Bank toward the construction of the new China Rice Research Institute in Hangzhou. Prima facie, the "open-door" policy has increased the effectiveness of the S&T network. It has provided China with easy access to foreign varieties, research results, and the possibilities of training abroad for researchers. But it has also highlighted the glaring gaps in research activities.

Research was, in the past, confined to problems with an immediate bearing on production and processing and was highly selective, focused above all on breeding new varieties of rice and then wheat. Important fields of applied research, such as fertilizer and irrigation, and other grain and non-grain crops, were neglected.[21] Soybean, which originated in China and is an important ingredient of the Chinese diet, was also a neglected crop. But, since the late 1970s, the range of crops covered is much wider. Research into cotton has borne fruit in the form of new varieties. Increasing attention is being devoted to coarse grain and oil-bearing seeds. Furthermore, China has now a cooperative agreement with the United States for research into soybean.[22] Research activities which once concentrated exclusively on yields now pursue a diversity of aims.[23]

Generally, the performance of a research network depends on its endowments of trained personnel, equipment, facilities, and style of work. The personnel in Chinese research institutes are short on postgraduate and research training, but long on experience in the field. Given some notable successes in the few areas in which Chinese researchers had been active, the qualifications of the personnel were probably adequate for the highly selective and practical research strategy followed in the past. With the broadening of research activities and China's participation in the international agricultural research network, the poverty of educational qualifications of research and extension personnel has become a major weakness. As for equipment and facilities, during the CR, many of the institutes that escaped closure were summarily dispatched to the countryside. The institutes visited by foreign scientists are reported to have facilities adequate for the task at hand.[24] But the sample visited are likely to be among the best equipped. There are reports of the lack of such basic facilities as experimental fields and greenhouses, especially in the institutes in outlying areas. Moreover, whereas all visiting scientists in the 1970s were impressed with the devotion and enthusiasm of the research staff, they remarked upon the informality of research methods and a relative absence of controlled and replicable experiments.[25] With the participation of Chinese researchers in international activities, it would seem that at least the best Chinese researchers are attuned to established research methodologies. In sum, the agricultural research network in China is not as academically sophisticated as the Indian or the Brazilian networks, which, apart from the Chinese, are the two most outstanding among developing economies.[26] However, the Chinese network is more practical and probably in closer touch with the tillers of the soil. With the responsibility system, and a shift in emphasis from "being a red to being an expert," this may no longer be as true as it was before 1978.

"If you could hybridize the Indian and the Chinese agricultural research and extension systems, you would have a very good system," remarked an American scientist after a visit to China in the 1970s.[27] While Chinese agricultural research has been marked by occasional bright sparks in the midst of the unexceptional, the performance in diffusion and extension of agricultural technology has been outstanding. In the past, it was organized along the four tiers of the county, the commune, the brigade, and the team—hence the appellation "4-level network." It was a massive undertaking, involving 14 million persons spread over half of China's counties.[28] Notwithstanding their deleterious effects on research, "open-door research" and "learning from the masses" did despatch scientists into the field, prompt them to develop contacts with the local communes, and bring peasants into research institutes. These practices were one way of tackling the universally difficult problem of instituting a feedback loop from the experimental field to the farm and vice versa. To give an example of the effectiveness of the 4-level system, within 13 years of the introduction of the high-yield rice variety in 1964, it covered 80 percent of the rice-sown area, as rapidly as any recorded in Asia.[29]

In the past, the operation of the 4-level network depended heavily on collective cultivation. The shift to family cultivation has radically transformed the task of extending and diffusing agricultural technology. In magnitude, the task has become greater. It has to cover 200 million agricultural households compared to the previous 6 million production teams. Moreover, its nature has changed. Before, diffusion consisted in getting cadres to introduce the technical change, which could be virtually accomplished by an administrative fiat. But, with family cultivation, the task is to persuade an economically motivated peasantry.

Like so many other rural institutions, the network for diffusing and extending agricultural technology is in the midst of reorganization. In 1982, the government embarked upon a 10-year plan to establish agro-technical centers (ATEC), 1 per 2,000-odd counties. These centers are to regroup and consolidate within one unit hitherto disparate local agricultural services, including seed production (a weak link in Chinese agricultural technology), fertilizer distribution, and soil analysis. They are to play an educational role as well, providing short-term training courses. By 1985, around 300 ATECs had been established, mostly in developed agricultural areas. A third of them were financed by the MOA and the rest by the province and tiers of the government below.[30] It remains to be seen whether they can match the success of the 4-level network in popularizing agricultural technology. In sum, the research and extension network is in a state of flux, and has yet to settle down to a definite modus operandi.

EDUCATION AND S&T ACTIVITIES

Further developments in S&T activities in China are severely hampered by an acute shortage of researchers with advanced training and of technicians. For example, of the total research staff engaged in agricultural research, less than 3 percent hold postgraduate degrees, and more than 35 percent lack tertiary education. A large proportion of leading figures in agricultural research are over 60. The shortage of trained technicians is even greater than that of researchers. Moreover, of the personnel in extension and popularization of agricultural research, 65 percent are farmers with experience but no formal education, and almost half need further training.[31]

The government estimate is that, with the present level of postgraduate enrollment, the demand for researchers will exceed supply up to 1990, and even beyond that for some specialties. The implication is that China is at least a decade away from accumulating enough manpower for a network spanning main areas of agricultural and related research. As well as putting a constraint on the extent of research activities, the shortages also limit the capacity of the educational system, which is further aggravated by the separation of research from education.

In terms of absolute numbers, the educational system is large. There are some 85 institutions of higher education in agriculture and related disciplines, accounting for 5 percent of university graduates in 1984.[32] As elsewhere, agriculture is not the first choice of university candidates. There is the additional problem of retaining trained personnel in agriculture. The adverse selection of students in agricultural sciences and a high depletion rate of trained personnel are not problems unique to China, and they defy an easy solution. The CR period set out to tackle the problem by exiling students to the countryside and deliberately undereducating them so that they would not regard it as below their station to work in the country. That had some beneficial effect on the diffusion of farming technology, but at a heavy cost in terms of formal education and training. Several cohorts of youth missed out on post-secondary and tertiary education. Since revolutionary coercion is no longer feasible, the only option left is to improve the economic conditions and career prospects of the personnel in the S&T network for agriculture.[33]

Educational qualifications of those engaged in agricultural S&T are dismal, and the rate at which they can be improved is limited. Indeed, post-Mao China lacks human capital sufficient to support a comprehensive and sophisticated research system. But is such a system necessary to support the rates of growth envisaged up to the year 2000? Formal educational qualifications are imprecise indices of research competence. To a degree, "training on the job" can compensate for formal education, especially in the case of adaptive research, diffusion, and extension. Furthermore, if competence is assessed by performance, then the Chinese record would compare favorably with records of economies far better endowed with educated manpower, such as Brazil and India. For crops other than rice and wheat, we have seen that China can achieve its targets by catching up with the current best yields. Moreover, rice is one crop in which the Chinese researchers are on the frontiers of technology; and they are well-advanced in wheat research. Once a high-yield variety is available, increasing yields may require no more than extension, a higher dosage of fertilizers, and a better water management, which do not make a heavy demand on research competence. The Chinese economy is well geared to their provision.

IRRIGATION

Heavily reliant as Chinese agriculture is on high-yield variety seeds and multiple-cropping, water management has always been a central component of agricultural technology. The geographical and temporal distribution of precipitation and river discharge is highly skewed.[34] Among the main agricultural regions, that south of the Yangtzi River has ample water, but the north suffers from a shortage, especially in the fall and winter, and the vast northwest is perennially arid. The intensive exploitation of aquifers under the North China plain has, to a degree, compensated for the low average and high monthly and yearly variations of surface water. This regional pattern holds particular implications for different crops. Because water is plentiful in the main rice-growing areas, rice yields are not constrained by water availability. In contrast, water shortage heavily affects the winter crops of the north; since wheat is the most important of them, it constrains increases in the wheat yields.

Generally, water management in China has consisted in: (1) prevention of floods, (2) storage of water for use in dry months and years, (3) spatial redistribution of river waters, (4) exploitation of aquifers, (5) economy in water use, (6) drainage, and (7) counteracting the adverse side effects of irrigation. Since 1949, there has been a heavy investment in the first four due to mobilization of labor in the slack season. Although labor mobilization for hydraulic construction has a long history in China, it assumed staggering proportions with the advent of collective agriculture. China's colossal hydraulic system is a monument to collectivization and bears the hallmarks of bursts of Herculean but ill-planned activity with vast gaps in the system.[35]

Nevertheless, China is well endowed with flood-prevention structures and irrigation facilities. Of the total cultivated area, around 45 percent is irrigated; the proportion in India is about 25 percent.[36] In well-irrigated areas, the principal natural constraint on yields is the annual temperature range. Thus, there is an extensive water-management system to support an agriculture based on multiple-cropping, hybrid seeds, and massive doses of fertilizers. The target to increase the irrigated area by a further 10 percent by the year 2000 would generally take the form of building structures complementary to the existing system.[37] In addition, the government is committed to realizing the historical ambition to divert northwards the headwater of the Yangtzi River–*Nan shui bei diao* (northwards transfer of southern waters).[38]

Since the further extensions of the irrigation system are increasingly costly, the main problem may instead be efficient utilization of an already vast capacity. Although higher than in other developing economies, the efficiency in water use still falls short of, for example, the level in California. Sprinkler irrigation is still rare and costly.[39] Even after all the economically feasible extensions have been made, the irrigated area would still fall well short of the total cultivated area. This points to the need for more resources in the hitherto neglected research into irrigation methods, the response of plants to water, and increasing yields in rain-fed and arid regions.

As with many rural institutions, the shift from collective to family farming has undermined the foundations of the previous mechanism for the construction and maintenance of the water-management systems. Added to that, government investment in water management declined by almost 50 percent during 1979–1982.[40] The neglect of water management systems is borne out by the statistics. Starting from 1979, there has been a decrease in the irrigated area: The total irrigated area in 1984 was around half a million ha less than the level in 1979–a decline of over 1 percent.[41] Until the government succeeds in building an alternative water management organization, it may be difficult to maintain constant the irrigated area, let alone to extend it.

SEEDS AND CROPS

With the coining of the term *green revolution*, the technical modernization of agriculture in developing economies has become almost synonymous with the introduction of "engineered" HYVs, in cereals, especially. As elsewhere, much of China's S&T activities in agriculture have been concentrated in developing HYVs of cereals. As suggested by its name, "seed engineering" consists in embodying in a new variety a menu of characteristics picked from a population of extant varieties left behind by human and natural selection. The central feature of the HYV seed is that it is but one component in an interdependent network of inputs (such as fertilizer and water) and farm practices (such as weeding and controlling diseases and pests). To use the Chinese idiom, it is appropriate to describe the technological combination with the slogan "Three-in-one combination of HYV seed, water, and fertilizer." Because water conservation and fertilization have a long history in China, and were encouraged by the Communist leadership even before HYVs came in vogue, Chinese agriculture has been a fertile ground for HYV seeds.

The reception of technology depends on the organization of agriculture, and technology, in turn, has wide ramifications for agriculture as an economic activity. It affects seed production, institutes new linkages between agriculture and industry, and ties agriculture permanently to S&T activities.

With the introduction of HYVs, seeds are no longer a mere by-product but the principal product of a specialized activity. Not only is this true of the original stock but also of seeds for planting. To maintain an unblemished stock requires a separation of seed production from cultivation. Since this separation is tantamount to establishing an entirely new activity, it can only gradually be achieved. In the interim, properly processed seeds remain in short supply. This is true in China as in other developing economies.[42] China's National Seed Corporation (NSC) supervises the work of 2,545 local seed companies and stations. Besides these, the so-called specialized households are also entrusted with seed production. But the seed produced under NSC supervision constitutes a mere 12.5 percent of the total requirement. Most farmers, therefore, still rely on retained seed of uncertain quality and dubious progeny. To ameliorate the situation, the government plans to quadruple by the year 2000 the volume of seed produced under NSC supervision and establish a network of centers for testing the seed produced by specialized households.[43]

Imports of HYV seeds are an important source of primary stock for crossbreeding. After a period of sporadic and circuitous imports of foreign varieties, China resorted to international imports on a more consistent basis. In the early 1970s, it imported substantial quantities of HYVs of wheat developed by CIMMYT for crossbreeding and planting.[44] By the late 1970s, it had a program of exchange of rice varieties with IRRI. As a full participant in the international network for agricultural research in the 1980s, China has access to the world stock of germ plasm and the techniques of crossbreeding. But the differences in habitat and cropping pattern limit the role of imported seed for planting. For example, while CIMMYT varieties are a spring crop, wheat is mostly a winter crop in China.[45] Besides, reliance on imported seed for planting is limited by the fact that China is among the largest producers of all major crops, which makes her requirements large relative to the international supply.

Based on fine tuning of the biochemical environment, HYV technology increases the dependence of agriculture on extra-agricultural inputs and services. This has interlinked financial and quantitative implications: financial, because extra-agricultural inputs, unlike by-products of cultivation, have to be purchased; quantitative, because the supply of extra-agricultural inputs is limited by the capacity of the industries to produce them. Thus, a shift from traditional to HYV seeds leads to an extension of the financial constraint on the acquisition of inputs and services by farmers. Therefore, the provision of rural credit is a crucial determinant of the speed of adoption of HYV seeds. Moreover, with a shift from the traditional to HYV seeds, there is an increase in money cost per unit of revenue, which happened in China in the early 1970s.[46] This should not be regarded as incontrovertible evidence of the mulcting of agriculture by the government, as some China specialists have suggested.[47]

Correlated with the diffusion of HYVs is a surge in the demand for chemical fertilizers. Despite the legendary thoroughness with which Chinese peasants recycle waste into agriculture, organic fertilizers can neither match the scale nor the composition of nutrients needed to realize the full economic potential of HYVs. This entails a massive investment in the chemical fertilizer industry, and often in irrigation. Since investment takes time, the consequence is a shortage of chemical fertilizer and irrigation facilities. China has suffered from both since the 1970s, as have other developing economies.

Given the constraints on the rate of investment in industries and the infrastructure supplying inputs and services to agriculture, the economic potential of HYV seeds is realized not all at once but gradually. This renders intelligible two commonly observed phenomena: first, an initial disappointment that actual yields of HYVs fall significantly short of expectations, as documented by the first crop of the literature on the green revolution; second, once HYVs have struck roots, yield per ha may keep on increasing under the momentum of an increased application of inputs, made possible by a technology introduced in the past. The latter is exactly what has happened in China from 1978 on. As we saw in Table 1, the rates of growth of total and per ha yields have been spectacular without any contemporaneous technical innovations major enough to account for them.

It must be remembered that Chinese agriculture began introducing HYVs of rice from the early 1960s, of wheat from the early 1970s, and of some other crops such as cotton before 1978. Furthermore, the consumption of chemical fertilizer between 1978–1984 increased by as much as 97 percent.[48] Consequently, a large credit for the rates of growth since 1978 must go to technical change, interpreted broadly to include increased application of inputs made possible by the introduction of HYVs in the past. The shift over to family farming has indeed led to a more efficient utilization of technical possibilities, but it was the introduction of HYVs and the massive investment in irrigation that created those possibilities in the first place. Since the possibilities offered by current HYV technology are not yet exhausted in China, an increased supply of chemical fertilizers and a more efficient use of irrigation facilities may on their own lead to a further increase in the yields of at least some major crops.

The third feature that marks out the HYV technology is that it requires a continual support of S&T activities, not merely for extension but for research as well. Because HYV seeds have an inbuilt obsolescence, new stock has to be continually developed to maintain resistance against diseases.[49] Unlike traditional varieties, which are more robust with respect to variations in habitat, HYV seeds need to be fitted in more closely with the local biochemical environment. In fact, the HYV technology is basically a set of breeding techniques that can be used to develop a wide range of varieties, each with a different combination of characteristics. Ideally the spatial diffusion of HYV calls for its adaptation to the location and its slot in the local seasonal cycle. In the past, the political pressure for immediate results led to a neglect of local differences, with disastrous results in a few cases. Although its S&T network has become more discriminating, China has still some way to go before it has a full range of HYVs covering different crops and tailored to different locales. Finally, since the optimal environment for HYVs is different from traditional varieties, it is at the outset an unknown to be discovered by research.

In the past, seed engineering in China focused first and foremost on a few HYVs of rice and then wheat. In the 1980s, the range has broadened to include coarse grains, cotton, and soybean, facilitated by China's growing international research contacts and access to the world stock of germ plasm. Although much criticized now, the earlier narrow focus on rice and wheat had ample justification. First of all, technically trained manpower was and still is too limited for an effective pursuit of a broad research strategy. Second, rice and wheat were and still are the principal food-crops, constituting over 60 percent of grain production in recent years.[50] Despite a decrease since 1978, they still covered around 43 percent of the cropped area in 1984. Grain has the first claim on the cultivable land, and the land available for other crops is a residual. This continues to be true, with the difference that, because of the increase in grain yields, the area now needed for grain is less and the residual higher. Hence, it may be argued that the narrowly focused strategy was not misconceived but has been rendered obsolete by its own success.

As elsewhere, the Chinese went in for high yields (that is, extending the range of positive response to fertilizer and water); but, in contrast, they have placed a much higher priority on early maturity. While in the United States the trinity of desirable characteristics in HYVs is yield, resistance to diseases, and quality, in China it is yield, resistance, and early maturity.[51] Home-bred HYVs of rice, wheat, and cotton are marked by a shorter maturity than their international counterparts.[52] Early maturity is an obvious route to increasing yield per cultivated ha, but, beyond a limit, it has a cost in terms of quality and yield per crop. Early maturity and multiple-cropping are no longer emphasized to the exclusion of attributes such as resistance to diseases, a special problem for wheat.[53] Nevertheless, they still are of a central importance for China because of its limited arable area and a relatively short growing season.

Rice-seed engineering in China has been the center of China's agricultural research efforts. China grows two rice varieties—the temperate Japonica (*geng*) and the semi-tropical Indica (*xian*). In the past, the latter yielded most of the crops, but, in the 1980s, they are about equal in importance.[54] While high-yielding Japonica varieties were developed by simply adapting the ones from Japan, the high-yielding Indica varieties were bred from scratch.

Chinese breeders first released a short-statured HYV of Indica in 1964, two years before the release of a similar variant developed in parallel by IRRI (IR8). Aside from a sporadic exchange of varieties, they had no contact with IRRI until 1978. Since then, they have exchanged personnel, varieties, and research results. Although both Chinese and IRRI breeders went for short stature, they diverged in the choice of other characteristics. Whereas the Chinese accorded primacy to early maturity, the IRRI breeders emphasized resistance to diseases and pests.[55] In economic terms, while early maturity serves to increase annual yield per cultivated ha, resistance to disease decreases the dispersion of yield per crop. Thus, the Chinese pursued a riskier course of forgoing a lower dispersion for a higher average yield per cultivated ha.

Chinese rice breeders scored their next success in 1977, when they released the first ever hybrid rice. Hybrid varieties are now a commonplace for corn, but they have been especially difficult to develop for rice and wheat. The hybrid rice developed in China has up to 20 percent higher yield than the usual HYVs, but as yet it is profitable only in some areas.[56] It is not only to play a central role in achieving the target for rice production for the years 2000 but also has been partially responsible for the rapid increase in production witnessed in recent years. Prompted by political pressure, its rate of diffusion has been rapid. In 1984, eight years after its general release, it covered 23 percent of the area sown with rice.[57] However, the hybrid rice remains a technology yet to be fully developed. Its period of maturity is too long to fit in easily with a double-cropping schedule, its range of variants is still limited, and its seed-cost high.[58]

Mirroring the ranking of yield per ha in an international context, research performance in rice has been impressive but becomes progressively patchy and low as it moves from fine grains to coarse grains and other crops. As the production of fine grains became adequate in the 1980s, the diversification of the diet away from grain toward meat and vegetables has displaced "providing enough to eat" as the central consideration. This points to a need for broadening research activities to coarse grains and oil-bearing crops, where performance still falls short of that in rice. Besides, research has to respond to regional considerations. The emphasis on rice and wheat has favored South China and the North China plain, which have also benefited from a preferential allocation of inputs. As elsewhere, China's "green revolution" has bypassed areas not readily suitable for HYVs.[59] As regional inequalities have come to the fore as a major problem in the Chinese economy, there is a pressing need for redirecting research to crops suited to backward regions.

FERTILIZERS

The rate of application of chemical fertilizers in China is already quite high, as Table 4 demonstrates:

TABLE 4 Consumption of Chemical Fertilizers (kgs per arable ha)

India	39.4	W. Europe	224.5
United States	104.5	S. Korea	331.0
China	120.6	Japan	437.0

Source: The Chinese figure is calculated from *Statistical Yearbook of China, 1985,* the rest are from FAO, *Fertilizer Yearbook,* 1985.

Since the multiple-cropping index is higher in China, a comparison in terms of kgs per cropped ha would put China below the United States level. Yet, the Chinese level would still be at least 2.5 times that of India. And, since organic fertilizers supply almost as much nutrients as chemical fertilizers in China, the Chinese rate of fertilization probably approaches the West European level.[60]

Currently, China produces around 85 percent of its fertilizer consumption and spends about a billion dollars a year on fertilizer imports. The margin of self-sufficiency is higher for nitrogenous than phosphate and potassium fertilizers. The fertilizer industry consists of a mixture of a few large-scale efficient plants and numerous small-scale rural plants, still supplying around 70 percent of the output. Many of the former were imported in the mid-1970s, but, since then, China has acquired technical capacity to design and produce large-scale nitrogenous fertilizer plants. Most of the small rural plants date from the 1960s and are inefficient in the use of energy and raw materials. Besides, rather than urea (the usual nitrogen carrier), they produce ammonium bicarbonate and liquid ammonia, which are bulky and volatile, losing a fair proportion of nutrients during transport.[61]

After the closure of ineffective small plants in the early 1980s, the Chinese fertilizer industry has been making a profit, but its level of efficiency still falls well short of its East Asian counterparts. This has important economic implications for a future increase in fertilizer application. For the rate of fertilizer application in China to approach rates in East Asia, the Chinese fertilizer industry has to narrow the efficiency gap with East Asian fertilizer industries. The target is to increase the chemical-fertilizer consumption form 17.4 million tons in 1984 to 30 million tons in the year 2000, that is 240 kgs per arable ha, still short of the Korean and Japanese levels.[62] But, given the efficiency of Japanese industry and the government subsidy for fertilizers, the Japanese level is most likely beyond the range of economic justification for China. Even to approach the Korean level would require a considerable increase in the efficiency of utilization of energy and raw materials in the Chinese fertilizer industry. That would depend crucially on the speed of replacement of older plants. Moreover, the target probably underestimates the increase in demand, for it is based on the optimistic assumption that the share of nutrients from organic fertilizers would slightly increase. If Japan's past and China's own recent experience is a guide to the future, then an increase in rural incomes will lead to a steady substitution of chemical for organic fertilizers, which are cumbersome to use.

A central feature of fertilizer application in China is the marked imbalance in the composition of nutrients: nitrogen (N), phosphate (P) and potassium (K). See Table 5 below. The low dosage of phosphates and even lower dosage of potassium are due to a number of factors. Since some varieties of nitrogenous fertilizers can be easily produced from widely available raw materials, they have been especially suited for the rural fertilizer industry. Apart from technical simplicity, there is a good economic reason for initially concentrating on nitrogenous fertilizers. For, when the rate of fertilizer application is low, the marginal yield response to nitrogenous fertilizers on its own is high. The composition assumes a special importance only when the rate of application of nitrogenous fertilizers is high and the marginal yield response to them has started to level off. Furthermore, organic fertilizers, which are applied in abundance in China, are rich in phosphates and potassium, and thus redress the balance. Yet, despite all these considerations, the single-minded concentration on nitrogenous fertilizers until the 1980s was due also to neglect of research in fertilizers. None of the scientific delegations that visited China in the 1970s report examples of fertilizer research. Contrary to the earlier presumption in China, recent work by Chinese and foreign researchers points to the need for a high dosage of phosphate and potassium fertilizers on intensively cultivated land.[63]

As for the distribution of fertilizer across crops and regions, 70 per-

TABLE 5 Composition of Nutrients

	N	*P*	*K*
China	100	27	8
India	100	29	13
N. America	100	46	50
S. Korea	100	47	49
Japan	100	109	90

Source: FAO *Fertilizer Yearbook 1985.*

cent of the fertilizer is used on grain and 30 percent on other crops.[64] These proportions are roughly in line with the current allocation of cropped area to grain and other crops. But the rate of fertilization varies greatly across the provinces, spanning 61 kgs per cropped ha in Tibet to 210 kgs in Fujian.[65] In general, the rate is high in agriculturally developed areas. This imbalance partly results from the routing of a portion of fertilizer supplies through the State Marketing Network in return for future purchases, a procedure that automatically favors developed regions with a surplus to sell. This has implications for equity, because it amounts to a preferential allocation for high-income areas; and for efficiency because, as CAAS studies show, the marginal-yield response is lower in the areas with a high rate of application.

Problems concerning fertilizers have gained in importance with the rise in the rate of fertilization, but their solution is within the range of competence of the Chinese economy and its S&T network. The problem of composition of the nutrients can be solved by imports or investment. But to improve the efficiency of the Chinese fertilizer industry would require not only a massive investment in replacing small rural plants with large-scale plants but also a change in work practices and a radical improvement in China's inefficient transport network.

FARM PRACTICES

Multiple-cropping has had wide ramifications for all aspects of Chinese agricultural technology: irrigation, seed breeding, fertilization, and mechanization. It is an old practice in East Asia and has always been one of the principal methods for increasing the intensity of cultivation. In China multiple-cropping includes:

(1) sequential cultivation of 2 or even 3 crops on the same land within the same year;

(2) relay planting of the next crop between the rows of the preceding crop planted earlier;

(3) intercropping of two or more crops at the same time.

Both sequential and relay cropping are geared to increasing the margin of utilization of the plant-growing season, which in most regions of China is considerably shorter than 365 days.[66] All three increase annual yield per ha and widen the assortment of crops grown within a geographical area.

By opening up the possibility of shortening the crop period, seed engineering fits in well with the traditional practice, which has also been encouraged by the Communist leadership. Starting from the 1960s, there was both a steady geographical extension of multiple-cropping and an increase in its intensity in those areas where it already existed. Schematically, the vicissitudes of multiple-cropping are displayed in Table 6.

Propelled by political enthusiasm, multiple-cropping has, in the past, been carried much further than warranted by purely economic considerations.[67] However, in the 1980s, increasing the multiple-cropping index is no longer considered a panacea but simply one of a number of cropping strategies for increasing yields.[68]

Yet, the singular emphasis until the 1980s on stretching the sequence of crops has cast a long shadow on plant breeding and on all aspects of crop management in China. In plant breeding, it put an extra premium on early maturity, which permeated plant research in China and limited the scope for transplanting foreign HYVs, even when suitable. Furthermore, multiple-cropping affected farming operations by imposing a taut time schedule. By putting sole emphasis on the timeliness of labor, it restricted the possibilities for smoothing seasonal peaks in labor requirements and thus often created a seasonal labor shortage, even in densely populated regions. It restricted the assortment of crops and varieties, because some had a period of maturity too long to fit in with the tight time schedule. Finally, multiple-cropping required a much higher rate of application of inputs. Because it is a highly fertilizer-intensive system, nutrients

TABLE 6 Changes in the Pattern of Sequential Cropping

Area	*Historical Pattern*	*End of the 1970s*	*Reversal to Historical Pattern, 1980s*
Huang he Valley	2 crops in 3 years	2 per annum	partial reversal
Chang Jiang Basin & Delta	2 per annum	3 per annum	almost complete reversal
Guangdong & the Far South	2 per annum	3 per annum	partial reversal

Sources: F. Leeming, *Rural China Today;* FAO, *China: Multiple-Cropping and Related Crop Production Technology;* He Kang, "Rice Production in China," in IRRI, *Impact of Science on Rice;* John L. Scherer, *China: Facts and Figures Annual 1985;* T. B. Wiens, "The Limits to Agricultural Intensification," in R. Barker and B. Rose, eds., *Agricultural and Rural Development in China Today.*

have to be provided not only for the additional crop but also to compensate for the lost fallow period. Moreover, by furnishing a fertile environment for the proliferation of pests and diseases, it necessitates additional measures for protecting crops. And, the greater the number of crops, the heavier the demand on irrigation facilities.[69]

MECHANIZATION

The general economic presumption is that agricultural activities in China have to be labor-intensive, and will remain so well into the twenty-first century. As compared to developed economies, the level of mechanization in China is still low; for example, mechanized plowing covers only 24 percent of the sown area, and the figures are much lower for other activities such as harvesting and rice transplanting.[70] However, since the mid-1960s, the level of mechanization has increased significantly. Furthermore, despite having three times as many agricultural laborers per cultivated ha, China's agriculture is significantly more mechanized than India's.[71]

In order to analyze the incidence and the pace of mechanization in Chinese agriculture up to now, and its likely pattern in the future, we may start with the activities that constitute modern field agriculture: (1) ground preparation and plowing, (2) sowing, (3) fertilization, (4) crop protection, (5) irrigation and drainage, (6) harvesting, (7) processing, and (8) on-and-off farm transport.

Although these activities are diverse, they all involve control and rely on at least one of the three sources of motive power: humans, animals, or internal combustion or electric motors. Chinese agriculture deploys all three in particular combinations across the spectrum of activities. Generally, the degree of mechanization, which can range from complete to negligible, is determined by technical possibilities and socioeconomic calculations. For some activities there may not be a feasible technical alternative to mechanization, such as irrigation from a deep aquifer. But most agricultural activities leave open a margin of socioeconomic choice between human, animal, and inanimate power.

The hypothesis is that the increase in the intensity of agriculture, that is, the scale of activities per arable ha and their range has been and continues to be the central influence on the process of mechanization in China. The main factors responsible for the increase in intensity are HYVs, farm practices complementary to them, and an increase in multiple-cropping. Table 7 contains a stylized list of their effects on the scale and composition of major agricultural activities.

Furthermore, HYVs require a higher level of weeding and crop protection; and harvesting and transporting increase pro rata with the yield. Clearly, a shift over to agricultural technology centered on HYVs implies an increase in the range and the intensity of farming activities, which requires more labor per cultivated ha. Therefore, with the introduction of HYVs, mechanization need not lead to a displacement of labor or draught animals; on the contrary, it may be coupled with an increased employment of both, as has happened in China. Since the early 1960s, the number of agricultural machines has risen sharply, but so too has the number of agricultural laborers. Although the number of draught animals does not show any sustained trend up to 1978, it has risen steadily with the shift to family farming.[72]

In China, the increase in the multiple-cropping index has been the most important factor responsible for increasing the intensity of cultivation, because multiple-cropping not only increases the repetition of agricultural activities but also the amplitude of seasonal peaks in the level of activity. In most areas growing more than one crop a year, the harvest of the preceding crop and the sowing of the next crop have to take place within a very short time period. As a result, there is an overlap of activities that are highly labor-intensive. This may create a seasonal shortage of labor even in densely populated areas.[73] On particular days of the year, the paddy fields of the Nanjing-Wuxi

TABLE 7 Effects of HYVs on Farming

Activities	*Effect on the Activity*	*Reason*
Plowing	Deeper plowing	Deepens the roots, hence increases the fertilizer response
Sowing	Controlled seeding and sowing in a line	Ensures uniformity in application of fertilizer and the date of maturity
Fertilization	Much higher level	Yield is directly dependent on the level of fertilization
Irrigation & Drainage	Higher level of water management	Complementary to fertilizer use

area undergo three changes of color: yellow in the morning before the crop is harvested, brown in the afternoon when the field has been plowed and harrowed, and then, green in the evening after the next crop has been transplanted.[74] Moreover, in order to squeeze in a sequence of crops within the limited growing period, Chinese farmers resort to the highly labor-intensive activity of transplanting from nursery beds a wide range of crops that are sown directly elsewhere.[75]

Obviously, the change in the scale and the composition of these activities did not take place in one step but was spread out over a period in time; its speed has varied from region to region. The mechanization that has taken place in China has indeed displaced labor but not so much away from the agricultural sector as from one agricultural activity to another—a redeployment rather than displacement. This may continue for some time, because there still are possibilities for a further intensification of agriculture. The opportunities for non-agricultural employment are few relative to the staggering size of the agricultural labor force in China.

The spectrum of mechanization of agricultural activities in China is varied. At the one end are the almost completely non-mechanized activities of harvesting, seeding, transplanting, and weeding. In China, weeding is not a chemical but a manual operation of pulling weeds by hand; the weeds then used as animal feed or organic fertilizer are treated as a crop.[76] Clustered at the other extreme of the spectrum are irrigation and drainage and harvest processing. Falling in the middle are the partially mechanized activities of tilling, plowing, and transportation. In addition to machines for harvest processing, water pumps and small or walking tractors are the most ubiquitous items. The latter were first introduced in Japan, then adopted in China in the 1960s after an initial rejection. Large or medium-size tractors, regarded as the centerpiece of mechanization up to the mid 1960s, are rare and concentrated in a few regions. And the ones that exist are used more for transport and haulage than for tilling and plowing.[77] Although this is often cited as an example of waste and inefficiency, that is not necessarily valid; transport is no less important an activity than plowing.

There is a general pattern to the incidence of mechanization in China. The highly mechanized activities are those where machines have an overwhelming technical advantage over beasts or humans, as in irrigation, or where a simple machine can do the job, as in harvest processing. At the non-mechanized end of the spectrum are the activities that require expensive machines and can be performed effectively by humans, such as harvesting.[78] The activities in the middle, such as ground preparation and plowing, are where mechanization in the near future will be concentrated.

Mechanization has been a highly charged political issue in China and, until recently, successive leaderships have been disposed toward a rapid mechanization of agriculture on the East European model.[79] But the actual rate of mechanization invariably has fallen short of politically induced targets. On balance, it is economic rather than political factors that have shaped the course of mechanization, which conforms fairly closely to the early prognosis of Xiang Nan (Hsiang Nan), one of China's top experts on mechanization. In response to the ambitious 10-year plan for the mechanization of agriculture prompted by Mao, Xiang drew up an order of priority. Dividing China into regions by the ratios of agricultural laborers to arable land, he suggested a quick mechanization of irrigation, drainage, and harvest processing in all areas, but confined the mechanization of ground preparation, harvesting, and transport to the sparsely populated regions of the northeast and the northwest.[80] Although denounced during the CR, he was rehabilitated in the 1970s. Rapid mechanization of agriculture figured prominently once again in Hua Guofang's ambitious modernization plan. But, with the shift to family cultivation, the pace and form of mechanization are now determined more by households than the government. The discussion has now shifted to the problem of absorbing the farm labor force into non-farming activities.[81]

China's farming technology still has major gaps and weaknesses. Nevertheless, it compares well with other developing economies, particularly Brazil and India. Moreover, the problems that loom large with respect to industrial technology are relatively minor in the case of farming technology, such as weak links between research and application and unfamiliarity with major strands of technology. The rapid rates of growth since 1978 have displaced the urgent problem of providing enough to eat with the less pressing problem of changing the composition of the diet from grain to meat and dairy products. The crop targets for the year 2000 are well within the reach of the Chinese economy. But, since animal husbandry is a weak component of agricultural technology, the change in dietary composition may not be as rapid as desired by the population and hoped for by the leadership. Opening doors to the outside world has brought immediate and tangible benefits to Chinese agricultural technology; the relative sophistication of its researchers in plant breeding has given China the capacity to readily absorb foreign farming technology; and international cooperation in agricultural research is well developed and extensive with few commercial barriers to its international transmission. The low level of pure research is a problem, but not one that would constrain applied and adaptive research before the turn of the century. That leaves China enough time to remedy its weakness.

CHAPTER ELEVEN

China's Military R&D System: Reform and Reorientation

WENDY FRIEMAN

One of the enigmas of Chinese development since 1949 is the uneven quality of scientific and technological achievements. On the one hand, China has been able to develop, deploy, and refine strategic nuclear weapons and has succeeded in launching satellites, both with relatively little foreign assistance. At the same time, the general level of technology in Chinese industry and the quality of Chinese science at universities and research institutes remained at a consistently low level between the mid-1950s and the mid-1980s. One logical explanation is that science activities to support the military differed in very important ways from purely civilian S&T activities.* This chapter ex-

*The term *military science* in this chapter refers to chemistry, engineering, mathematics, or physics research directed at the development or improvement of weapons or military support equipment, not to the science of war and military strategy.

amines the defense component in Chinese science more carefully than has been possible in earlier works on either Chinese defense or on Chinese science. Have there been significant differences between military and civilian science in China? Do such differences persist? If so, how will they affect China's broader economic agenda and the role China can play in global as well as regional affairs? The chapter is divided into three sections. The first summarizes what is known about how the Chinese defense-S&T system worked and how it differed from civilian science between 1949 and the early 1980s. The second examines how that system, as well as its relationship to the civilian sector, has been changing. The concluding section discusses the implications of those changes.

While attempting to differentiate between military and civilian science in China, it is useful to remember that the imposition of a military mission on a set of institutes and factories in any country is almost certain to alter them in very fundamental ways. A military orientation dictates tight security, restrictions on scientific communication, and limited dissemination of research results. It is likely that military science will be more closely supervised and controlled. A military orientation may also affect the degree and pace of technical innovation. Western studies of innovation in weapons industries suggest, for example, that a military mission can create a climate for rapid advance and technological breakthrough. The Manhattan Project in the United States is an example of one such breakthrough.

Furthermore, the requirements imposed by national security affect the supply-and-demand dynamics associated with innovation. Innovation in the civilian economy results from either the discovery of a potentially new technology (the discovery-push) or the demand for a new capability (the demand-pull). In a military setting, the definition of what is required, or the demand-pull, as well as the definition of what is technically feasible, that is, the discovery-push, can be easily stretched. Thus, proponents of a particular weapons system may underestimate the difficulty of the project and opponents are likely to overplay the difficulties involved. Technical judgments regarding feasibility of a system are inherently subjective to some degree. The fact that military demands can be overstated and technical obstacles can be downplayed results in the growth of a large bureaucracy that has the power to perpetuate itself indefinitely by evoking national-security requirements.[1]

Some features of the Chinese military S&T system, such as the need for tight control over information, can be assumed to exist simply because they are common to weapons-research environments everywhere (although differences in degree are important to recognize). Other characteristics of the military-science system can be explained by conditions unique to China or to developing countries more generally. Finally, to a certain extent, the distinction between military and civilian organizations is artificial and difficult to make. Developments in weapons technology are increasingly dependent on advances in fields such as microelectronics and new materials which are equally important to major non-military sectors. Thus, a growing number of research institutes and personnel work on projects with both military and civilian applications.

CHINESE DEFENSE RESEARCH FROM 1949 TO 1980

Once shrouded in secrecy, China's defense sector has been much more visible since the promulgation of the open-door policy. Articles about new developments in China's weapons factories and interviews with defense-industry leaders now appear regularly in the Chinese press. Discussions of recent reforms are often a source of further confusion, however, since these reforms are difficult to interpret without an understanding of the procedures and organizations the reforms are intended to change. How the defense-science system operated between 1949 and 1980 is of more than historical interest, moreover, because reforms in China are implemented only gradually, and the status quo can be expected to persist for some time after a series of sweeping reforms are first announced. Thus, despite announcements of change, defense science in China today probably resembles the system of the 1950s and 1960s more closely than the ideal toward which its leaders would like to be moving. This section analyzes how China's defense-science system has been operating over the past thirty years by examining four interrelated themes: the diversity of organizations involved in defense-related science; the effect of military doctrine on defense S&T; the role of technology transfer in Chinese defense S&T; and the influence of modern weapons development on national science planning. The section concludes with an assessment of the key differences between Chinese civilian and military science prior to 1980.

Chinese Organizations Involved in Defense-Related Science

The defense component of Chinese science is difficult to assess, in part because it is difficult to locate. Organizations that have played a role in defense science in the past include the Central Military Commission, the General Departments of the PLA, the National Defense Industries Office, the Chinese Academy of Sciences, and the Ministries of Machine Building. Each of these is described briefly below in order to set the stage for analysis of broader issues.

The Central Military Commission (CMC, also referred to as the Military Affairs Commission) reports directly to the Chinese Communist Party (CCP) Central Committee (CC) and has always been the single most important military organization in China. Its members have had enormous influence not only on military but on all areas of national policy. The CMC has been responsible for setting overarching military policy, formulating China's strategic concepts and doctrines, and playing a key role in resource allocation for major defense manpower and hardware programs. The major departments of the PLA report to the Central Committee through the CMC. The CMC's equipment department traditionally played a role in foreign weapons procurement until it was merged into the Commission for Science, Technology, and Industry for National Defense (COSTIND) in 1983. On other matters, the Commission has concerned itself primarily with defining broad policy objectives and has left implementation to the services and the Ministry of National Defense.

The General Staff Department (GSD) and the General Logistics (formerly General Rear Services) Department (GLD) have both played peripheral roles in weapons-related R&D. The GLD has traditionally had factories and research institutes under its jurisdiction. Although GLD facilities have been responsible primarily for manufacturing non-military items, such as uniforms, consumer goods, foodstuffs, and the like to support the military, GLD institutes may have had a role in testing and evaluating new weapons or military support equipment.

Responsibility for coordinating and supervising weapons research and development fell to the National Defense Industries Office (NDIO), the National Defense Science and Technology Commission (NDSTC), the National Military Industry Commission, and the Science and Technology Equipment Office of the CMC. The history and evolution of these earlier bodies is complicated and obscure, but several key points are worth noting.

From 1949 until 1983, the Chinese leadership attempted to separate the administration of defense-related research from that of actual weapons production. The NDSTC, on the one hand, focused on basic research and development for new weapons systems, especially for nuclear weapons. The NDIO, on the other hand, had the responsibility for supervising and coordinating conventional-weapons production in the factories under the Ministries of Machine Building. The NDIO was not be involved in the development of new systems. Despite the organizational separation, it is virtually certain that this distinction was difficult to maintain. Turf battles developed among competing national-level organizations, since, in practice, R&D spill over into testing and production for both conventional and nuclear weapons. The attempt to separate the administration of research and production of weapons encouraged redundancy and inefficient use of scarce resources, and was ultimately detrimental to the military-industrial complex.[2]

The Chinese Academy of Sciences (CAS) is China's premier science organization and is responsible for basic and some applied research across a wide range of disciplines and technical areas. It has a large network of branch academies. Because of its responsibility for funding basic research, it is a major player in determining national science priorities. Although less is known about the participation of CAS in conventional-weapons production, the nuclear program had its inception in the CAS Institute of Nuclear Physics (later known as the Institute of Atomic Energy) in 1957. The Institutes of Physics, Mathematics, and Mechanics were also extremely active in nuclear-weapons research and development. In 1958, CAS established the first of a number of science and technology universities to train students in subjects closely related to nuclear-weapons development. The best known of these, the Science and Technology University at Hefei, was responsible for the design and manufacture of China's first supercomputer, the Galaxy. The Chinese Academy of Space Technology, which now houses the key research institutes supporting the ballistic-missile and satellite programs, originated in CAS but was spun off from it in 1968.

Virtually all China's weapons have been developed and produced in the factories and institutes of five different ministries which are not part of the formal military command structure. These ministries constitute what is usually meant by the term "military-industrial complex." Each has large research institutes within its jurisdiction, and some have training facilities as well. The product lines for the ministries as they were organized in 1979 are listed in Table 1. Although each is now titled according to its specialty, these ministries until 1982 were referred to as numbered Ministries of Machine Building; what is now the Ministry of Electronics Industry used to be the Fourth Ministry of Machine Building. As Table 1 indicates, all the machine-building ministries (MMBs) manufacture a range of civilian industrial and consumer goods in addition to weapons and military support equipment. It appears that under each MMB there are three types of facilities: those that manufacture only weapons, those that manufacture only civilian goods, and those that manufacture both. In addition, several non-machine-building ministries play a role in military production; the State Automotive Industry Corporation, for example, produces military as well as commercial vehicles.

Throughout the 1960s and 1970s, the MMBs were run by PLA officials; the factories were staffed and operated largely by civilians under the direction of the NDIO. The weapons factories are among China's largest facilities; many have more than 10,000 personnel. Besides being large, these units are also extremely self-contained, with their own equipment for processing raw materials and manufacturing components, their own schools, and their own research institutes. US aircraft-industry officials who visited China in 1980 reported that the Shenyang Engine Factory, for example, had facilities to make the basic tools needed in the metalworking and parts-fabricating processes that precede the buildup of sub- and final assemblies. According to one visitor: "They have tool shops that literally make the cutting tools they need to build production tooling and to cut metal. They also make all the screw machine parts they need in the manufacturing process."[3] There is little evidence of a second or third tier of suppliers or subcontractors comparable to those in Chinese civilian factories; all supplies and materials are provided by the

TABLE 1 China's Ministries of Machine Building

Ministry	*Product Areas: Military*	*Product Areas: Civilian*
Ministry of Nuclear Industry	Nuclear warheads; stock of both fission and fusion warheads believed to number 300.	Meters and instruments for measuring radiation; nuclear electronic instruments; uranium survey and mining equipment; optical instruments; mechanical components, e.g., hot cells, air filters, heat exchangers, and valves.
Ministry of Aviation Industry	Jet fighters, F-2 (MiG-15), F-4 (MiG-17), F-6 (MiG-19), F-7 (MiG-21), F-8 (Mach 2 fighter), F-9 (twin-engined MiG-19); ground attack bombers, Sheyang Tu-16, 11-28, Tu-4; transports, An-2, 11-14, Li-2; helicopters, Mi-4, 6 and 8, Aerosaptiale Super Frelon and Dauphin 11; Yak-18 basic trainer; Spey engines, Tumansky R-11 engines; air-to-air missiles.	Agricultural planes, optical recorders, watchmaker lathes, transducers, switches, motorcycles and helmets.
Ministry of Electronics Industry	Avionics, early-warning radar, electronic countermeasures, space electronics.	Earth stations, navigation equipment, computers and peripheral devices, radios, TVs, integrated circuit technology, and consumer electronics, components and materials.
Ministry of Ordnance Industry	Tanks: T-54, T-59, T-34, T-63 Is-2; reconnaissance vehicles: T-60, T-62, PT-76; armored personnel carriers, Types 55 and 56; BTR-40, BTR-50, BTR-152, K-62; antiarmor weapons, 57-mm, 85-mm, and 100-mm guns; SU-76, SU-85, and SU-100; ISU-122 self-propelled artillery, 82-mm, 120-mm, 160-mm mortars; 107-mm, 140-mm rocket-launchers; 75-mm, 82-mm recoilless-launchers, and 37-mm, 57-mm, 10-mm antiaircraft weapons.	High-precision and heavy-duty metal-cutting tools, bicycles, chemical industrial products, electrical appliances, steel and wooden furniture, oil-extraction pipes.
China State Shipbuilding Industry	Submarines, destroyers, frigates, patrol escorts, fast-attack craft, ocean minesweepers, hydrofoils, and infantry landing craft.	Bulk carriers, container vessels, marine engines, factory design and consulting services.
Ministry of Space Industry	Strategic weapon systems: IRBM (CCS-2), limited range ICBM (CSS-3), full range ICBM (CSS-X-4) launch vehicle under development.	Communications, weather and earth resources satellites; telemetry, tracking and control equipment.

state. Finally, it is worth noting that a number of key industrial research laboratories and production were originally under the jurisdiction of CAS and later transferred to the MMBs. Two of these labs, the Lishan and Wanyuan Corporations, both under the Ministry of Space Industry, have some of the most sophisticated electronics equipment in China.

This brief description of the key defense science and technology organizations demonstrates that the responsibility for weapons research and development has traditionally been spread among a number of different ministries, commissions, academies, and institutes. The relationships among these organizations, to the extent that they can be determined, are depicted in Figure 1. CAS, the MMBs, and the NDSTC initially resembled the Soviet organizations on which they were modeled. They were all vertically organized, with tight central control in Beijing over the regional subsidiaries in provinces and key municipalities. Although some horizontal ties exist at the local level, they have always been considerably weaker than the vertical ties on which the entire system depends.

The Influence of Military Doctrine on Military Technology

The day-to-day activities of the organizations described above were dictated, at least in part, by the hardware requirements associated with China's prevailing military doctrine of people's war. This doctrine called for a mixture of denial and retaliation, although over time the balance between these two elements shifted. Denial consisted of absorbing a nuclear attack and "luring" the enemy in deep until he could be encircled and overcome by PLA infantry; retaliation consisted of one or more nuclear counter-strikes.[4] The defense industrial base responsible for supporting this doctrine allocated resources to developing a wide range of systems, but only achieved success in two areas; these were at opposite ends of the technological spectrum. At one end were infantry weapons for use by large numbers of foot soldiers; even at their most sophisticated, these are relatively easy to acquire, reverse engineer, modify, and mass-produce. At the other end were nuclear weapons—large, complicated systems that required a range of capability from theoretical physics to precise manufacturing and integration of electronic components. Other hardware, such as interceptor aircraft, frigates, destroyers, or armored

FIGURE 1 Chinese Defense Science and Technology (pre-1980)

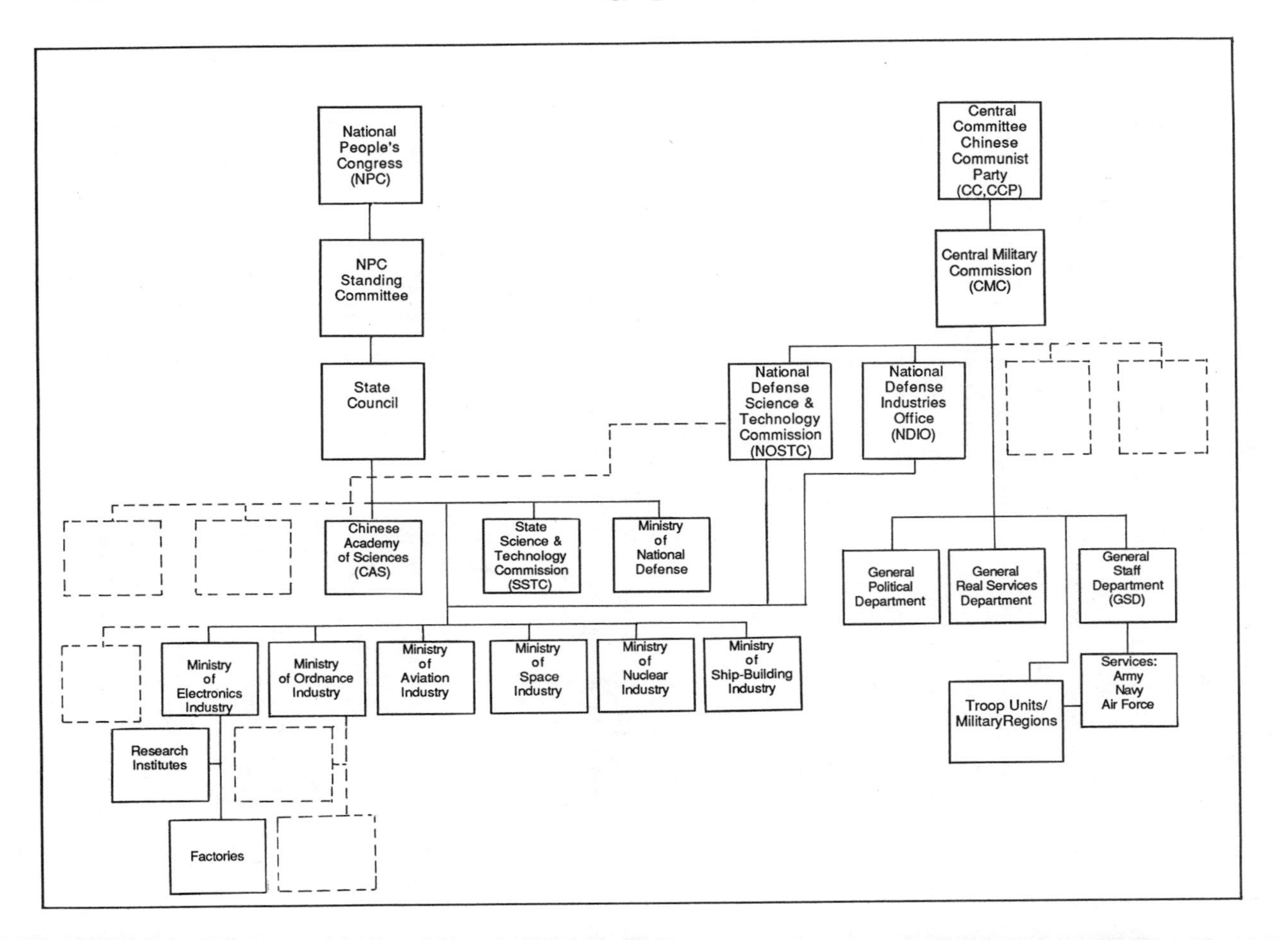

personnel carriers, which did not figure prominently in either denial or retaliation, were necessarily relegated to a lower order of priority. The hardware programs to support the two key military missions are worth examining in more detail.

A conventional-weapons industry to support the infantry existed, albeit in primitive form, when the Communist Party came to power in 1949. Under Soviet guidance, the Chinese reorganized military research and production into two ministries of machine building. After the break with the Soviets in 1960, Chinese military and industrial planners became aware that China weapons manufacturing was disproportionately concentrated in the coastal industrial centers, which accounted for over 60 percent of China's industrial production. This concentration was troublesome for security as well as economic reasons; the weapons factories located near the coast were highly vulnerable to foreign invasion or occupation, particularly from the USSR. In the following years, therefore, the Chinese government established another four Ministries of Machine Building and made a conscious effort to move the weapons industry inland away from the already vulnerable coastal industrial centers. During the early 1960s, in what is now referred to as the "third-line" policy, the government built new factories in cities such as Chongqing, Chengdu, Taiyuan, Shijiazhuang, Loyang, Baotou, Lanzhou, and Xian. Many of these cities today are major centers of defense production.[5]

The decision to move the weapons industry inland was also heavily influenced by the prevailing doctrine, although the concept of people's war suggested conflicting views of the necessity for large-scale weapons-production facilities. On the one hand, Mao maintained that, ultimately, weapons were less important in war fighting than human will and determination. Nevertheless, people's war also required that China's military regions be self-sufficient in production of infantry weapons. The establishment of inland weapons centers eliminated interdependencies and ensured that interior provinces could continue to mass-produce supplies and equipment in the event of foreign occupation of one or more coastal cities.[6]

For the next twenty years, both the coastal and the "third-line" factories continued to mass-produce weapons based on the Soviet prototypes provided in the 1950s. Chinese engineers appear to be extraordinarily adept at tinkering with existing systems to achieve marginal improvements, and, in the area of infantry weapons, particularly rifles and machine guns, those modest modifications were sufficient for Chinese needs. There have been no revolutionary advances in rifle production in the past twenty years, and the Chinese version of the AK-47 rifle is one of the most effective infantry weapons produced anywhere in the world.

The doctrine of people's war initially depended primarily on fairly simple infantry weapons as a way of compensating for China's technological weaknesses. It is possible that, by continuing to embrace this doctrine, the leadership ultimately perpetuated those weaknesses; no priority was attached to more complex conventional weapons research and production because the doctrine did not require it. It is possible that China succeeded in low-technology weapons simply because they were easier to produce. In any case, there was no need to develop systems for more complex scenarios not envisaged in people's war. As a result, the aircraft and tank industries, which needed more than incremental improvement, failed to keep up with their American or Soviet counterparts. Modern technology has dramatically changed the way these systems are designed and manufactured over the past twenty years. Although Chinese standards of craftsmanship are reputed to be high, most Chinese conventional weaponry has changed little in quality or design since the early 1960s. Nevertheless, production continued at very high levels even after the fall of Lin Biao in 1970; Chinese defense production capacity actually tripled between 1965 and 1979.[7] As a result, the PLA now has a massive inventory of obsolete equipment that would be of limited utility in combat against modern American or Soviet aircraft or tanks.

The nuclear-weapons program in the Ministries of Space and Nuclear Industry has a dramatically different history. In contrast to the conventional-weapons industry, the nuclear and space-sciences fields have progressed rapidly through China's own efforts. While not advanced by American or Soviet standards, they enabled China to develop a strategic deterrent adequate for current defense needs. Nuclear-weapons research began in 1957 at the Institutes of Physics and Atomic Energy in Beijing, under the terms of a bilateral agreement with the Soviet Union, by which the Soviets were to "provide China with a sample of an atomic bomb and technical data concerning its manufacture." When the Soviets reportedly tore up this agreement two years later, China was forced to continue the nuclear program without foreign assistance.[8] China exploded a 20-kiloton atomic bomb in 1964 and a megaton-range thermonuclear weapon in the summer of 1967. Thus, the Chinese had made the transition from fission to fusion in less than three years, an impressive achievement for a country whose technological infrastructure had been in a shambles fifteen years earlier. After the initial warhead research and development was completed under the Ministry of Nuclear Industry, the focal point of the program gradually shifted to the Ministry of Space Industry (MSI), which assumed the responsibility for developing delivery and guidance systems.

Since 1967, China has deployed over a hundred intermediate-range ballistic missiles (IRBMs) with a range of approximately 7,000 km, and several intercontinental ballistic missiles (ICBMs) with a range of 13,000 km. The Chinese have also tested a MIRV warhead, upgraded strategic missile guidance and control systems, and tested a sea-launched ballistic missile. The satellite program, which has also made significant gains, is jointly administered by the MSI and COSTIND, the successor to the NDIO and the NDSTC. Despite the fact that the satellites themselves have non-military missions, they are very much a product of the military S&T system.[9]

Thus, China's defense S&T system seems to have performed best in areas related to those weapons required by the prevailing doctrine. This has resulted in impressive accomplishments at both ends of the technological spectrum, and relatively mediocre performance in systems to support a range of intermediate missions, such as troop mobility, close air support, tactical air power, and so forth.

Foreign Technology and National Defense

China's weapons factories as well as the entire defense industrial process were closely modeled after the Soviet system. With the exception of some nuclear weapons facilities, the early PRC weapons plants were established in consultation with Soviet experts, and the ministries of machine building were modeled on the highly centralized Soviet ministries with the same names. Soviet assistance to China in the 1950s was the most massive international transfer of technology in modern history. Between 1950 and 1960, over 11,000 Soviet specialists were on site in Chinese facilities, where they were involved in every phase of military industry from basic research to prototype testing and serial production. The result of this transfer was that, by 1960, Chinese factories had access to the technology for manufacturing weapons that were close—if not equal—to the Soviet state of the art. It was also in 1960, however, that Sino-Soviet political tensions, which had been escalating since the outbreak of the Korean War, provoked the Soviets into suspending all bilateral educational and technical cooperation. In July, the remaining 1,300 Soviet engineers, managers, and scientists left the country, taking with them blueprints, plans, and technical libraries.

The massive transfer of technical assistance from the USSR permitted China to "leapfrog" ahead, accomplishing in ten years what may have taken twice as long to do on their own. As a result of the abrupt pullout of Soviet assistance, however, China could not hold on to the state of the art and fell considerably behind world levels in the years that followed. This taught the Chinese a lesson they would never forget about over-dependence on one source of foreign technology and expertise. Not only was the PRC left high and dry to manage projects conceived and executed under Soviet tutelage, but, once the Soviets pulled out, the Chinese did not have the capability to move ahead. Left with only prototypes, the PRC had no choice but to reverse-engineer them, a process that took years; by the time the Chinese had successfully entered serial production, the equipment had been rendered obsolete by new technologies.

In a sense, the Soviet transfer may have worked *too* well. Soviet assistance was so wide-ranging that China had no need to cultivate a strong scientific talent base capable of designing and engineering the next generation of weapons systems, machine tools, instrumentation, or control equipment. If the Chinese had been on their own from the beginning, they might have been forced to develop an indigenous capability. This is exactly what happened in nuclear science, where Soviet assistance was cut off in the very early stages.

The Role of Modern Weapons in National Science Planning

The preceding discussion suggests that national defense figured prominently in the PRC planning process during the early years of the People's Republic. This was true for several reasons. China's military industries were thoroughly dilapidated after the war with the

Nationalists and badly needed new technology. As a result of the close friendship between China and the USSR, the PRC inherited the Soviet administrative structure, which was itself designed to serve the needs of the military sector. Although this structure was intended to support the production of conventional weapons, Mao developed an interest in nuclear technology as early as 1954. After China broke with the Soviet Union in 1960, moreover, the need to develop a nuclear deterrent was all the more critical, since China was now facing not one but two superpower adversaries, this time without foreign assistance.

The far-reaching influence on national priorities of the desire to develop an independent nuclear deterrent can be seen in the 12-Year Science Plan drawn up in 1956. The top five fields targeted by the plan consisted of atomic energy, radio electronics, jet propulsion, automation and precision instruments, and petroleum and rare mineral extraction and processing. All these are directly applicable to missile production. This science plan was to guide the activities of CAS, the science and technology universities, and the MMBs.

The leadership's strong commitment to nuclear science and technology is also evident in the early establishment of the NDSTC and the NDIO. As early as 1958, China's leadership decided that military research and production were of high enough priority to require coordination and control mechanisms distinct from those that administered civilian science. Evidence suggests that the primary function of the NDSTC, when it was first established, was the development of nuclear weapons, a task beyond the scope and capabilities of the existing Science and Technology Commission. Not surprisingly, the NDSTC assumed considerably more importance after the withdrawal of Soviet scientific and technical assistance in 1959, when the PRC decided to proceed with a nuclear-weapons program independent of foreign assistance. The fact that work on the nuclear program continued uninterrupted during the CR is evident in a series of weapons tests in the early 1980s. Chinese advances in MIRV technology and in sea-launched ballistic missiles were the fruit of the previous five to ten years of research.

Finally, the nuclear industry has been a major employer of scientific manpower: It was reported in 1980 to have between 100,000 and 150,000 personnel in research institutes, schools, factories, and offices; roughly 80 percent of these people are involved with military programs.[10] Given the size of China's total scientific manpower base, these are extraordinarily high figures, which testify both to the importance of the program and to the degree to which resources were reallocated from other initiatives to make the nuclear-weapons effort possible.

The unavailability of detailed defense-spending estimates makes it virtually impossible to speculate on the relative share of the defense industrial budget devoted to the nuclear weapons effort. Nevertheless, the impressive achievements of the nuclear program must have resulted from the allocation of scarce and critical resources to that project. The national commitment to the development of a nuclear deterrent took the form of a "big push" dedicated to achieving that goal. Thus, the most talented scientists, the best equipment, and the most modern facilities were all devoted to the development of nuclear weapons and delivery vehicles. Other defense and civilian science efforts simply could not command the same level of leadership commitment or resources.

Defense vs. Civilian Science Pre-1980: Some Comparisons

The system that existed until the late 1970s therefore consisted of a Soviet-style bureaucracy dedicated to supporting the doctrine of people's war and to developing a strategic nuclear deterrent. The MMBs and the NDIO/NDSTC were successful in mass-producing infantry weapons and in establishing a nuclear program that answered China's needs. In other areas of conventional and nuclear weaponry, however, Chinese industry remained hostage to the Soviet experience and failed to move beyond the technological level of the 1950s.

The conventional wisdom holds that the Chinese defense establishment is "better" than its civilian counterpart. Although neither Chinese nor Western commentators have been specific about this perceived superiority, military science and technology institutes, factories, or projects are presumed to have had access to the best technical talent, the most advanced equipment, priority allocation of raw material supplies, and so forth.[11] These generalizations certainly hold true for Soviet military research and development; given the heavy Soviet influence during the period when China's defense industrial system was first established, it would not be surprising if they were equally applicable to China. It is true that China's weapons-related factories and institutes have some degree of priority in the allocation of raw materials, energy supplies, and trained personnel. The fact that the MMBs are vertically integrated and manufacture all their own components and machine tools has afforded them a degree of protection from the supply uncertainties that plague the rest of the economy. In this respect, the conventional wisdom (and the analogy to the Soviet military) are well founded.

At the same time, the vertical integration that has isolated the MMBs from the rest of the Chinese economy has not helped those weapons factories that badly needed an infusion of new technology. Moreover, the tendency toward isolation and self-sufficiency has drawbacks as well as advantages. On the whole, it seems to have prevented weapons factories from developing a local network of suppliers and subcontractors. Over-reliance on the central allocation system has been a serious problem for most of China's heavy industries, whereas enterprises in the light and export-oriented industries have been forced to establish horizontal ties with local, sometimes collectively owned, businesses. These local suppliers may be less reliable than the central government, but they provide more flexibility (in case products need to be custom-manufactured, or delivered quickly) and diversity, as well as substantially lower prices for components and equipment. Without these local, unofficial sources, the process of technical innovation bogs down in an endless pile of bureaucratic paperwork. In one military electronics facility, for example, a manager wanted to modify the design of one of the factory's products, but could not pursue the idea because he was unable to find a supplier for the customized chips the modification required. Although there was a semiconductor factory in the same compound, this manufacturer claimed that the order was too small to justify a customized chip. The chip manufacturer had no incentive to satisfy the requirements of any customer other than the central ministry in Beijing, which supplied all its raw materials and purchased all its output.

Ultimately, therefore, much of the conventional wisdom about the superiority of military over civilian science in China applies only to the nuclear program, not to all defense industries. The analogy of the Soviet military sector is perhaps valid for China's nuclear-weapons program, but not to the Chinese defense industry as a whole. There is much evidence to suggest that conventional-weapons factories have suffered from many of the shortcomings that have stymied efficient manufacturing and innovation in non-defense sectors. Many weapons factories were for years run by Party committees, with no one person willing to take responsibility for major or even minor decisions. Everyone involved in factory management sought to avoid risk. The Chinese equivalent of the Soviet "chief designer" is not at the plant but at the ministerial level. As a result, the opinions of party cadres at the factories continue to influence decisions on the design and production of weapons, despite the fact that they are often not among the most technically or managerially sophisticated personnel. Even the chief engineer or chief designer in the Chinese system does not have the same degree of authority or autonomy as his Soviet counterpart. As a result, China's defense industrial establishment lacks the product or program "champions," the influential scientists or managers to mobilize personnel, equipment, and funding behind a particular project and energize the innovation process. Only the nuclear-weapons and space programs had strong technical leadership in personalities such as Qian Xuesen, the famous rocket expert educated at Caltech in the 1950s.

By the same token, the conventional argument that weapons production and defense-related research, unlike civilian science, continued uninterrupted during the 1960s and early 1970s, applies more to nuclear than to conventional-weapons research and production. It now appears that the entire military establishment in China was caught up in some fundamental ideological debates and that the military as a whole, including the defense industrial and scientific organizations, suffered serious losses in the 1960s and early 1970s. Even after the reascendance of Deng Xiaoping in the early 1980s, scientists and technicians in the defense sector were slower than professionals in other industries to regain prestige and to make contact with the industrialized world; they have only recently begun to acquire advanced technology and participate in international conferences.

Before discussing how the defense science system established in the 1950s has begun to change, it is worth remembering that comparisons between civilian and military science in China are necessarily oversimplified and based on sweeping generalizations. In reality, there is a wide range of technical sophistication in both civilian and military science. The weapons industry has its "high end" (nuclear weapons) as well as its "low end" (tanks), just as the civilian economy has its strong (textiles) and weak (telecommunications) sectors. Chinese defense S&T is both better and worse than civilian science and technology; it depends on what is being measured. Finally, the areas of overlap, particularly in electronics, are growing to the point where it is becoming more and more difficult to separate military from civilian science activities.

CHINESE DEFENSE RESEARCH SINCE 1980

After the CR and the demise of the Gang of Four, the future role of the military in China was uncertain. By 1980, the PLA had been compromised by its participation in the CR and its leaders have been gradually weeded out of the Party and state government apparatus. The Deng Xiaoping leadership has, furthermore, repeatedly stated that national defense is the last priority of the Four Modernizations. However, the Deng leadership also insists that the modification of military strategy and upgrading of military hardware are critical mid- to long-term objectives. What do these developments mean for defense-related science and technology? This question can be explored by revisiting the themes identified for the pre-1980 period, discussed here in a slightly different order: the role of modern weapons in national science planning; foreign technology and national defense; the influence of doctrine on defense science and technology; and the organizational structure for defense research.

The Role of Modern Weapons in National Science Planning

Although the technology gap between Chinese and American or Soviet military forces has never been greater, the direct influence of the military on national science priorities appears to be considerably less than it has been in the past. Once the driving force behind the allocation of science resources, the military sector is now itself driven by the overriding objective of improving economic efficiency. For the first time, the defense establishment feels the need to justify itself economically. This need is reenforced by the decision of the leadership that, although the Soviet Union is still the key adversary, it does not represent an immediate threat. This has resulted not only in a decline in prestige for the PLA, but also in an increase in the Deng leadership's commitment to a development strategy in which im-

TABLE 2 China's Defense Expenditure (in billion RMB yuan)

Year	1980	1981	1982	1983	1984	1985
Spending	19.38	16.79	17.63	17.71	18.37	18.6
Increase over previous year	-13%	-13.4%	2%	0.4%	1.8%	3.5%

provements in the scientific, educational, and industrial infrastructure must precede acquisition of sophisticated weapons.

One indication of a decreased role for the military in national science planning is the relative decline in defense spending. Clear-cut evidence is difficult to find, since the budget for China's defense-related research and development is not broken out separately and is only indirectly linked to total defense spending. Given the fact that figures on the defense budget probably include some, but not all, defense-related expenditures, even estimating China's total military budget is extremely difficult. Nevertheless, it is instructive to compare the Chinese figures for different years with each other. The defense budget as a whole has been growing only marginally since 1982 to reach a level of 18.6 billion RMB yuan as depicted in Table 2; this figure will probably not change dramatically in the near future.

Even with allowances for reductions in procurement, these figures are extremely low. It is possible that some defense S&T expenditures are subsumed under other categories. A delegation to China was told by the State Science and Technology Commission, for example, that half the national budget for S&T in 1985 was used for military projects. This represents one hidden source of funds for defense R&D. In any case, demands for major new weapons systems, which typically involve large expenditures, are not evident in the announced budget figures.

A second illustration of the defense industry's concern for economic viability can be found in their involvement in weapons exports. Since 1980, the MMBs as well as COSTIND have all established foreign trade corporations and have begun selling weapons to Middle Eastern and other Third World countries. Revenues from foreign weapons sales are reported to have totaled $1.66 billion in 1984 alone and $4.3 billion between 1973 and 1983.[12] Only a handful of Chinese systems meet international standards; these tend to be weapons incorporating fairly simple technology. Furthermore, the Third World weapons market is beginning to get quite competitive, particularly with the entry of countries such as North Korea, Egypt, and Brazil.[13] The Chinese cannot increase sales in this market indefinitely; they will soon encounter a revenue ceiling beyond which they cannot move without dramatically increasing the sophistication of their weapons exports. Nevertheless, the revenues from arms exports, some part of which presumably are available to the defense industries, represent another unannounced source of funds for investment in new Chinese weapon systems.

A third indication of growing sensitivity to profit and loss by the MMBs is their more active role in the non-defense economy. Diversification into non-defense production has received considerable attention in the Chinese press. For the past five years, weapons factories have been using idle capacity to manufacture non-military products—especially consumer goods—needed in the local market. They have also transferred equipment and machinery and provided technology-related services to local industry. According to Chinese sources, by the end of 1982, 315 military production lines had been set aside for civilian products, and nearly 300 of the products had been included in the state plan.[14] Although it is impossible to determine exactly the scale and number of factories involved in this effort, the range of products being produced is quite broad, including consumer electronics, electronic components, chemicals, paints, machine tools, printing presses, construction machinery, and vehicles.[15] Li Peng, newly appointed Premier and head of the Electronics Industry Leading Group, reported in November 1985 that the state had selected 120 military production lines for technical transformation (that is, for conversion to non-defense products) and that production of civilian goods in the Ministries of Machine Building accounted for 40 percent of total output.[16] The leadership has also announced that the Chinese nuclear industry will reverse its program of 80 percent military projects and 20 percent civilian projects to 20 percent military and 80 percent civilian (presumably nuclear-power) projects.[17]

MMB production of civilian goods will, in all likelihood, result not in a deterioration of China's current defense posture or weapons-production capability but in more efficient use of previously idle capacity. For many Chinese manufacturers of conventional weapons, diversification is simply a rational decision in the face of decreasing subsidies from the central government, declining orders for military products, and the opportunity to earn additional revenues from sales of civilian goods.

Examples of technology transfers and technology sharing between military and civilian units may ultimately be more significant than the production by defense factories of civilian goods. Both COSTIND and local governments have established mechanisms and channels that defense factories can use to sell technology and equipment to their civilian counterparts; COSTIND has founded a newspaper to publicize military technologies that may have profitable consumer or industrial applications. An article on military technology transfers reported in 1984: "In 14 provinces and municipalities alone, over 8,000 technological contracts were signed with transactions amounting to 430 million yuan. When all fulfilled, these contracts are expected to create over 4 billion yuan of results for the state each year."[18]

Even if these numbers are less than reliable, the level of activity suggests that weapons factories are implementing some of the reforms intended initially for civilian science institutions. Factory managers in the defense industry are beginning to assume profit-and-loss responsibility, and therefore now have an incentive to manufacture goods above and beyond the levels called for in the state plan. They are engaging in joint ventures, co-production, licensing of technologies, and a long list of other arrangements from which they profit by providing equipment or services to the civilian economy. It also appears that military research institutes will begin implementing the contract system of research funding (as opposed to block grants) now being used by civilian research organizations. Xinhua reported in December 1985: "The present practice of distributing such funds among the institutions according to fixed proportions is now blamed for causing egalitarianism and laxity on the part of researchers. The new (contract) system will be put into effect next year (and) is expected to expand the decision-making power of research institutes and accelerate the technological flow from military to civilian industries."[19]

Finally, it is worth noting that lower defense research and production budgets provide an incentive for the defense industries to consider internal technology sources, which are likely to be more accessible and less costly than foreign sources. The general loosening of restrictions and tight central control, as well as the specific mechanisms established with the objective of arranging military-to-civilian transfers, may ultimately result in more effective civilian-to-military transfers.

Much of what China acquires for civilian purposes can ultimately benefit the military sector. Representatives of the CMC or COSTIND have a voice in deliberations over the purchase of foreign technology, particularly high technology, that may have military applications. Whether or not the military necessarily benefits immediately from all civilian purchases is difficult to know. Still, Chinese business links with foreign aircraft firms are certain to help the military as well as the commercial aircraft industries. (A summary of technology acquisitions in 1985 related to aircraft modernization is provided in Table 3.) By far the largest number of "dual-use" sales can be found in the electronics and telecommunications sectors. Even a partial list of sales and agreements for transfer of electronics technology (also found in Table 3) is a good indication that the Chinese are determined to strengthen their entire electronics infrastructure. The military applications for electronics technology are limitless, and China will probably not be able to pursue them all with equal success. The most significant purchases from a military perspective appear to be of integrated-circuit production lines, as large-scale integrated circuits are critical to military command, control, and communications systems, as well as to cryptology and missile guidance.

These trends all suggest that China's leaders are rethinking the relationship between defense and civilian research and production, looking for ways to achieve a more effective integration between the two, and beginning to eliminate the privileges traditionally enjoyed by the defense sector. This integration appears to be more than simply a short-term policy designed to take up the slack in China's weapons factories. As a commentator in *Jingji yanjiu* points out, effective integration of the two sectors is not only important for economic reasons, but also critical for those times when China needs to rapidly increase defense production: "In times of war, due to the rapid expansion of the military establishment, the urgent need to fit out the vast number of militiamen, and the huge drain on arms and ammunition, vast quantities of arms and equipment, ammunition

TABLE 3 Chinese Purchases of Selected Aircraft and Electronics in 1985

Amount US $ Millions	*Product*	*Supplier*
	Negotiation for purchase of 747, 767 & 737 aircraft	Boeing Co.
	Negotiating to sell Mirage 2000 Aircraft	Dassault-Brequet of France
	3 wide-body A310-200 airliners	Airbus Industries of France
4.00	Signal-processing integrated circuits	Matsushita Electronics Corp. of Japan
84.31	Joint production of micro-wave marine radio equipment	K.K. Oyo Gijutsu Kenkyujo of Japan
25.00	Design Engine for Silicon plan	Stearns Catalytic Corp.
12.00	Prog. Controller Computers	Gould Inc.
6.00	Semiconductor equipment	Kayex Corp/General Signal
21.00	56 Mainframe comp., 120 micros 200 printers	Honeywell Inc.
	Production line, equipment & technology for PC's	Olivetti of Italy
	Set up factory to produce 100,000 micros/year	Olivetti (US) & NA (Hong Kong)
	Set up fiber-optic cable production	Yurugawa El Indust Plant of Japan
	Technology, IC boards & soft-ware to code Chinese characters	Eastern Computers, Inc.
	6 Jetranger III helicopters	Bell Helicopter Textron Inc.
20.00	Modern Y-7 & Y-12 aircraft prototypes for production	Hong Kong Aircraft Eng Co
2.00	Six image-processing systems	Dipix of Canada
	Technology to produce mag-netic measuring machines	Keihin Densokko Co. of Japan
12.00	Integrated circuit production equipment	Matsushita Electronics Corp.
	Installation CAD Center	Louis Berger Int.
	Technology & equipment to produce spectral analyzers	Fluke Co.
	Technology & equipment to produce signal generators	Aritsu Electric Co. of Japan
	Porcelain fine-tuning capacitors, technology & equipment	Hitec Corp. of Japan
25.00	Design engine for silicon materials plant	Stearns Catalytic Corp.

and necessary accessories, as well as military supplies, will be desperately needed. Only by mobilizing the entire industrial force to step up production will we be able to ensure wartime needs."[20] These observations are drawn from a long article which examines in depth the type of relationship between military and civilian economies found in developed countries, including the United States and Western Europe. How these analyses of foreign industries will translate into Chinese policies, of course, remains to be seen. What is clear is that the questions are being raised by the Chinese in the context of the limited reforms that have already been implemented.

The Influence of Military Doctrine on Modern Weapons Development

The Chinese leadership appears to have reached a consensus that military strategy and military hardware both need to be updated to reflect the conditions of modern warfare. This point has been driven home repeatedly and reenforced by combat experience in Vietnam and by exposure to sophisticated Western weapons on shopping expeditions in Europe and the United States. Since large-scale acquisition of modern weapons will have to wait, the near-term focus will be on doctrine and strategy. An overhaul of Maoist military thinking, once unimaginable, is politically possible now that Mao's legacy has been thoroughly reviewed and reevaluated. Not surprisingly, there is widespread agreement that the concept of people's war needs substantial modification if it is to serve Chinese needs between now and the end of the century. What is less clear is how to change that doctrine or what to replace it with. The way in which this issue is debated and resolved will dictate the activities of the defense S&T establishment for many years to come.

The near-term solution is a concept referred to as "people's war under modern conditions." This concept has not been fully developed or articulated, but its outlines are beginning to emerge in articles and speeches by defense leaders. It is also possible to make some educated guesses about developments in doctrine by looking at recent organizational reforms and military exercises. The limited evidence will be summarized briefly here, with an emphasis on doctrinal reforms most relevant to weapons development. Based on what information is now available, it appears that these reforms will have lasting significance for the entire defense S&T establishment.

Many of the doctrinal changes now evolving are in part a result of the PLA's withdrawal from Chinese politics. The concept of a professional rather than a political army, consistent with the broader economic goal of improving efficiency and instituting rational bureaucratic procedures, has resulted in a reassessment of the relative role of men and weapons in warfare. Combat experience as well as common sense have persuaded China's military leaders that human will and determination, however critical to military victories, is a necessary but not sufficient condition on a modern battlefield. Weapons are equally critical. People's war, as explained above, did result in a defense industrial establishment that churned out a considerable amount of military hardware, but the industry was skewed in the direction of either extremely complex nuclear systems or large quantities of relatively simple weapons to support guerilla units. Now, by contrast, the nature of warfare is assumed to be extremely complex, requiring more than a nuclear deterrent counterbalanced by large numbers of infantry weapons in case deterrence doesn't work.

The need for more diverse and complex weapons results from the leadership's willingness to abandon key tenets in Maoist military doctrine. Discussions of ground-based warfare, for example, now focus on positional warfare, improved logistics, and high degrees of mobility, rather than guerilla tactics and protracted war. This means that the infantry will no longer be the backbone of the PLA in many combat scenarios. In addition, recent combined arms exercises indicate that the PLA is planning for contingencies in which sea, air, and land forces will have to operate in concert. This trend is reflected in heightened interest in modern command, control, and communications systems. The PLA is also preparing for "strategic air raids" that never played a role in people's war.[21]

Equally important, the Chinese are reexamining the role of nuclear weapons in modern warfare. People's war stipulated that one of the likely forms of attack on China was a nuclear strike followed by a ground invasion. Now, the PLA is considering the possibility that it will have to respond instead to a tactical nuclear strike, and has been staging military exercises to explore the use of tactical nuclear forces in battlefield situations.

Finally, the balance between denial and retaliation referred to earlier is shifting in favor of denial, a change that requires enhanced survivability and accuracy in China's intercontinental weapons.[22] The strength of China's commitment to the nuclear-weapons program, already discussed at some length, has not changed, and is now directed at improving the performance of the sophisticated delivery vehicles.

Thus, although the defense industry is de-emphasizing weapons production and operating at reduced funding levels, new Chinese thinking about military doctrine has been reiterating the weaknesses of current PLA inventories and calling for massive and comprehensive improvements in military hardware. Such improvements are simply not affordable in the near term. Tactical aircraft, tanks, frigates, helicopters, anti-tank weapons, anti-aircraft missiles, and communication hardware are only a few of the kinds of systems that need a major infusion of new technology. There is relatively little question as to what the defense S&T system needs to do to support the emerging doctrine; the answer is "Everything." The difficult question is how priorities will be determined in a period of finite human and financial resources.

Foreign Technology and National Defense

The Chinese are keenly aware that the task of modernizing military hardware is too ambitious to attempt without some degree of foreign assistance, and, since 1980, the PLA has been debating which technologies and which systems to buy from abroad. The military establishment has spawned a collection of defense procurement organizations to study foreign weapons and foreign defense industries and make key import decisions. These organizations are staffed by technically competent military personnel and provide the PLA service arms access to international sources of modern military technology.

Despite widespread speculation that the PLA was only "window shopping" and had little intention of making any purchases abroad, the PLA has imported both finished weapons systems and military technology. The Chinese have bought tanks and infantry weapons from foreign suppliers, and are negotiating the purchases of foreign helicopters, engines, and aircraft. In addition to direct sales, which represent a near-term technological fix more than a long-term solution, China is also acquiring the technology and expertise necessary to manufacture state-of-the-art weapons indigenously. A co-production arrangement with Rolls Royce to manufacture Spey engines and the Dauphin helicopter project ran into difficulties due to poor management and inadequate metal-processing equipment. Nevertheless, other transfers from Europe and Japan of casting technology for jet-engine housings, aircraft wings, brakes, bearings, and turbine blades have also been negotiated. The United States is prepared to sell technology which will be used to upgrade Chinese F-8 fighters. Pilkington brothers (UK) has entered into an agreement to build a factory in Shanghai and license float-glass technology, reportedly to be used in optical systems for military aircraft. (Major transfers of military equipment and technology since 1972 are listed in Table 4.) The Chinese engineers and technicians involved in these programs are likely to benefit from the exposure to foreign equipment, processes, and management methods, even if the road has not been entirely smooth. Future military technology-transfer agreements are likely to enjoy more success than those in the past.

Although Chinese technology purchases to date will not have a revolutionary effect, incremental improvement, as evident in a number of conventional Chinese systems, has already had noticeable results. The value of incremental change in electronic warfare can be seen in the appearance of the T-69 main battle tank, an upgrade of the T-59, which includes an infrared searchlight and a laser range finder, presumably purchased from the Israelis. The resulting system may not be as effective as its Soviet counterpart, but the project enabled the PLA, with relatively small capital investment, to alter substantially the performance characteristics and utility of this outdated tank. As the PLA begins to acquire electronic "eyes and ears," electronic improvements are being made to a number of Chinese systems, including aircraft, destroyers, and armored personnel carriers. Other incremental modifications to conventional systems include improvement in gun stabilization and ammunition quality. Because weapons-technology purchases are costly, foreign-exchange constraints place a limit on how much can be imported. This may be alleviated somewhat by the capacity of China's weapons industry to earn hard currency by exporting Chinese weapons to Third World countries. Representatives from foreign defense contractors have speculated, in fact, that the Chinese intend to use foreign technology

Table 4 Chinese Acquisition of Foreign Military Hardware 1972–1985

Date	*Supplier*	*Item*	*Weapon System*
1972	Thompson-CSF	Sonar	Super Frelon SA-321J helicopters
1972	Thompson-CSF	Sonar	Super Frelon SA-321J helicopters
1972	Lamparo	Data-processing unit	Super Frelon SA-321J helicopters
1975	Rolls Royce Spey	Skyranger ranging radar	F-7 fighter
1980	Aerospatiale	Dauphin helicopter including 670HP Turbomeca Ariel I-C turboshaft engine	
1983	GEC Avionics computer system	Display and weapon aiming	F-7 fighter
1983–1986	Crouzet	Nadar navigation system	Super Frelon SA-321J helicopters
1984	Sikorsky	Transport helicopters	
1984	FEC	Skyranger ranging radar	F-7 fighter
1984	GEC Avionics	3400 communications radio	F-7 fighter
1985	General Electric	LM 2500 gas-turbine engines	Frigates or destroyer ships
–	US Consortium	Avionics package	F-8 fighter
1986	Creusot-Loire	100MM dual-purpose gun mounting	Destroyers

to upgrade weapons designated for export before applying it to Chinese systems.

These developments illustrate a very different pattern of technology acquisition from that of the 1950s. Keenly aware of the dangers inherent in reliance on a single technology source, China is talking to, and buying from, a wide range of foreign vendors. If diversification is carried too far, however, the Chinese will encounter a different set of problems. A piecemeal approach, attractive because in theory it will enable the MMBs to buy one of everything and reverse-engineer each prototype, means that the Chinese will in many cases not receive the training and technical consulting generally included in large-scale packages. Unless standards are developed and enforced, the military establishment will end up with a large number of diverse systems, all incompatible with each other, which it cannot operate or service.

The experience of other developing countries in acquiring modern weapons may offer some insight into the trade-offs China is facing. In the rush to make up for several decades of neglecting military technology, the PLA may over-compensate in the other direction. The Chinese are attracted to high-prestige items, such as advanced fighters and sophisticated tanks. Weapons assume a symbolic value far beyond their utility in battle; threat perceptions can easily be inflated or reconfigured so as to require sophisticated systems. Thus, it is not surprising that the Chinese have expressed an interest in military systems and technology well beyond what is required or even advisable, given the current state of China's industry and infrastructure. Other countries, such as Egypt and Iran, have successfully deployed sophisticated foreign equipment even after foreign technical assistance was withdrawn, but they have acquired those systems in packages rather than in bits and pieces from different suppliers.[23]

Moreover, the fascination with modern "bells and whistles" may cause the PLA to ignore other facets of warfare, as dangerous a mistake as downplaying the importance of modern technology. Studies have shown that the level of technology per se is not a critical determinant in the outcome of wars fought by Third World countries.[24] Much more important is tireless attention to the mechanics of logistics and supply and the availability of basic, "low-tech" weapons and ammunition. For the moment, in the absence of abundant spending power, the Chinese military leadership appears to be addressing many of these non-hardware issues.

Organizational Change in the Defense Science and Technology System

The doctrinal and economic reforms discussed above are also reflected in organization and personnel changes in China's military establishment. These changes are reviewed briefly in the text, and the new organization is depicted in Figure 2. First, senior leaders of the CMC who have been retired have not been replaced; thus, the number of people in important leadership roles has declined since 1983 from 9 to 6. Among the retired officials is Nie Rongzhen, a military man who was intimately involved in the scientific work underpinning the nuclear-weapons program. Nie had been chairman of the National Defense Science and Technology Commission (NDSTC) from 1958 to 1975, as well as chairman of the Science and Technology Commission until 1967. The only other comparable individual now in the CMC is Zhang Aiping, Minister of Defense, and also former chairman of the NDSTC, but Zhang is fairly senior. He was relieved of his Central Committee membership in September 1985. The remaining CMC members are military men who have risen through the command structure. Only four of the CMC leaders are both Politburo and Central Committee members, an indication that military representation in the Party's non-military organizations is at an all-time low.

A second change concerns the management of defense production. In February 1983, the NDSTC and the NDIO were replaced by the Commission for National Defense Science, Technology and Industry (COSTIND). COSTIND is a larger, more powerful unit than any of its predecessors, and is heavily involved in selecting foreign technology and equipment for purchase or co-production; it must approve all transactions involving foreign technical assistance to the PLA. Its structure includes the Foreign Affairs Bureau, the Research and Development Bureau, and the Science and Technology Committee and has a vertical system of its own with bureaus and offices at the provincial and municipal levels.

Very little is known about COSTIND personnel. The Chairman, Ding Henggao, was virtually unknown before he was appointed as

FIGURE 2 Chinese Defense Science and Technology (1987)

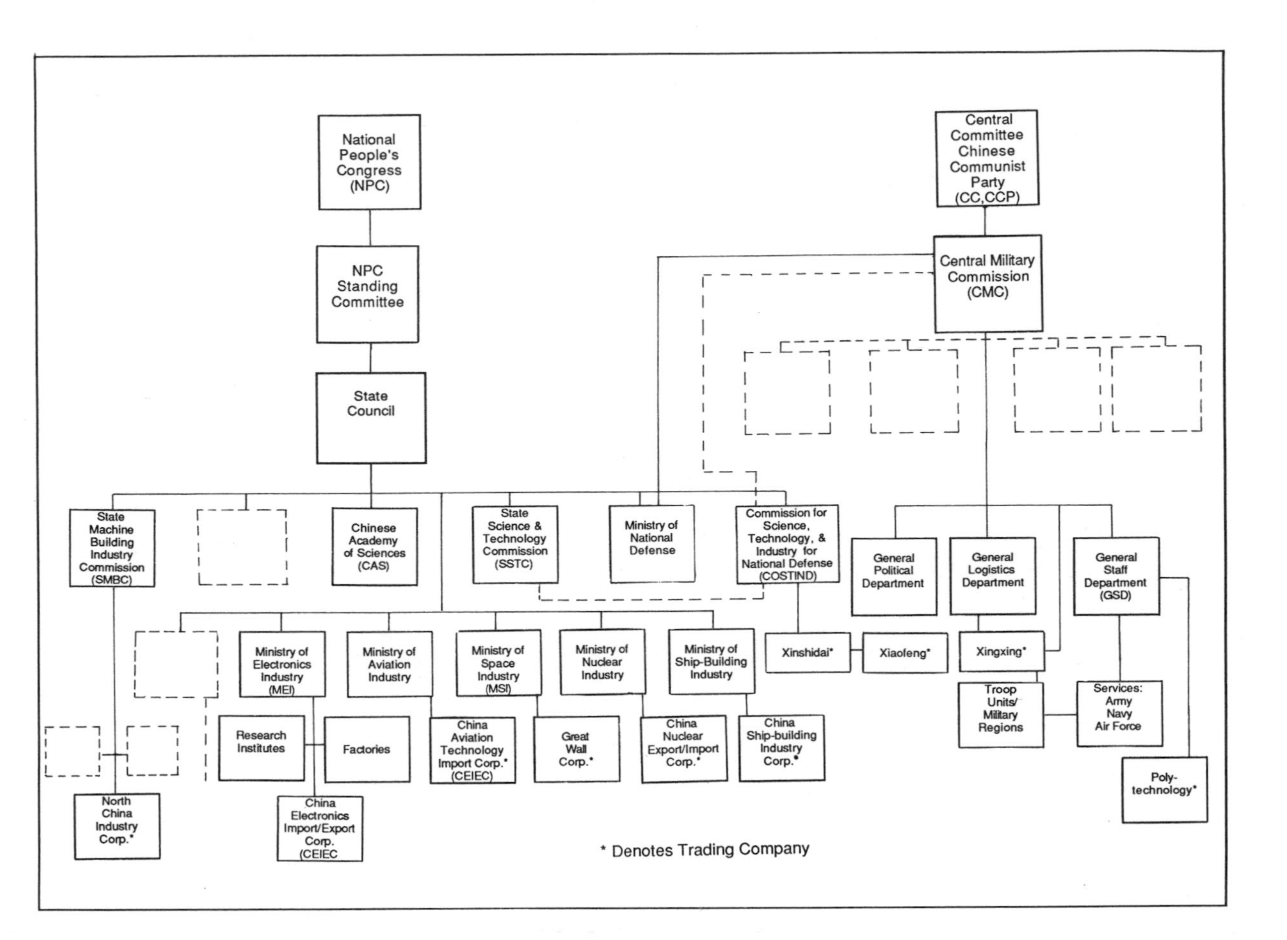

an alternate member of the Central Committee in 1985. He is the husband of Nie Li, an accomplished scientist who is on the Science and Technology Committee of COSTIND as well as the State Leading Group for Electronics. Nie's professional reputation is enhanced by the fact that her father is the recently retired Nie Rongzhen, the founder of the original NDSTC. It seems that the new COSTIND staff consists primarily of technical experts who are part of the uniformed military but do not appear to have strong ties to the service arms.[25] COSTIND reports directly to the State Council, whereas the NDSTC and the NDIO were part of the military bureaucracy.[26]

Another sign of impending change in Chinese defense technology after the CR appeared with the renaming and replacement of the ministers of the machine-building industries between 1978 and 1983. Traditionally, the ministries responsible for weapons production were under the supervision of senior PLA officers, who often lacked the technical background required to move a new generation of weapons and military-support equipment. Beginning in 1978, ministries were gradually put under the command of career bureaucrats with no apparent ties to the military, many of whom are part of a new technocratic elite.

Biographical information on the ministers is scattered and incomplete. Jiang Xinxiong, who became the Minister of Nuclear Industry in 1983 at age 52, graduated with a degree in Mechanical Engineering from Nankai University in the 1950s and then was assigned to work at a nuclear power plant in northwest China. He made important contributions to the implementation of nuclear safety standards and is thought to have participated in the development of China's atomic bomb in the early 1960s. In 1979, he was appointed director of the nuclear power plant and, in June 1982, was appointed Vice-Minister. Jiang is a Central Committee member with no apparent ties to the military command structure.

Li Tieying is another example of the new breed of ministers in the machine-building industries. He was born in 1936 and studied solid-state physics at the Charles University in Prague. Returning to China in 1961, he worked in three different research institutes in Shenyang before being appointed Secretary of the Shenyang Municipality in 1981. He was elected an alternate member of the Central Committee in 1982, and a full member in 1985, when he replaced the former Minister of Electronics Industry, Jiang Zemin.

In 1987, the former Ministry of Machine Building (responsible for heavy machine tools and control systems for both military and civilian applications) was merged with the Ministry of Ordnance Industry to form the State Machine-Building Industries Commission. This commission operates much more like a ministry than a commission. Its newly appointed Minister, Zhou Jiahua, was born in 1928 and is a former Minister of the First Machine-Building Industry, where he was in charge of all non-defense machine tool production. He is also a former deputy director of both the NDSTC and the NDIO. Zhou has a university education and served in the 1960s as the director and chief engineer of the Shenyang No. 2 Machine Tool Plant.

Although it is difficult to identify any changes in the operation of these two ministries that can be attributed directly to the appointment of civilian ministers, nevertheless, the men in charge of weapons production in the 1980s have formal training in physics, mathematics, or engineering, and appear to have been promoted steadily through the ranks on the basis of solid performance records. The changes at the top of the ministries suggest that China's leaders are genuinely committed to reorienting the defense military-industrial complex, but the direction of the new orientation is unclear. It is possible that civilian ministers were appointed to decrease further the role of the military in the state government, or to encourage the conversion of space in older weapons factories to consumer or civilian production. Another explanation is that these ministers were appointed to strengthen the scientific and technical underpinnings of China's defense research and production.

Chinese Defense vs. Civilian Science and Technology Post-1980

The Four Modernizations program has brought to the defense S&T system a new orientation, reflecting a different set of national priorities than those of the previous three decades. The uncertainties about the relationships between civilian and military sectors before 1980 are compounded for the current period by inadequate information about the extent to which many of the reforms have actually been implemented. Nevertheless, it is possible to draw some tentative conclusions.

First, the defense industry does not appear to command the same position of leadership in defining national, scientific, or economic priorities that it did in the 1950s and 1960s. Instead, it is following the lead of the non-defense economy. Reforms in the defense industrial system are intended to increase overall economic efficiency and professionalism; these are objectives that are not exclusively military or scientific. The payoff for implementing these difficult organizational changes, which have certainly alienated large parts of the bureaucracy, will, however, be better organization and a more effective process for weapons research and production. Since many of the problems in the pre-1980 period were a function of managerial rather than technical shortfalls, the reforms represent a large step forward.

Second, there is now a clear effort under way to integrate the military and civilian sectors of the economy. In the pre-1980 period, the transfer of resources was a one-way street from civilian to military, with the lion's share of resources going to the nuclear-weapons effort. The goal of the reform program is to establish a two-way flow, focusing initially on transfers from the military back to the civilian.

Third, and clearly related, the old concept of a "big push" in one specific area has been temporarily shelved. The nuclear-weapons effort still enjoys certain privileges, but the reallocation of resources to more commercial activities, such as communication satellites and nuclear power, suggests that it does not command the same level and quality of resources that it used to. There is a distinct possibility that the "big-push" concept will reemerge; some of the scientists involved in the nuclear-weapons program are advocates of this approach and are in powerful government positions. One example is Song Jian, Chairman of the State Science and Technology Commission.

Fourth, doctrinal changes will have a profound effect on weapons research and development, but the leadership has yet to codify a new doctrine and identify clear-cut hardware priorities. Although the defense-related research process may be improving, there is no consensus as to what tasks that process should be used to accomplish. This may be one explanation for the temporary absence of a "big-push" program.

Finally, in many respects the defense sector, specifically the conventional-weapons industry, is still behind the civilian sector in technology, management, and innovation. The organizational changes proposed will take time to implement, and reforms came to the defense sector several years after they were adopted by many non-defense industries.

IMPLICATIONS FOR THE FUTURE

China's defense science-and-technology accomplishments to date have been achieved either through entirely indigenous research efforts (the nuclear and space programs) or through massive transfers of foreign equipment and technology (the conventional-weapons program). Neither approach seems well suited to current conditions. The conventional-weapons approach has numerous drawbacks. First, not one country is prepared to offer assistance comparable to that of the USSR in the 1950s. China is not prepared to pay for comparable levels of imports and expertise. Moreover, the PRC is wary of over-reliance on a single source of technology. Wholly indigenous research programs, on the other hand, have been successful only in cases where the leadership made them a top priority. While it is possible to do that for one or two programs, it is not possible to make everything a priority. Until questions of strategy and doctrine are fully resolved, there is no one compelling motive to force another major technological breakthrough analogous to the development of a nuclear deterrent in the early 1960s.

Because the immediacy of the threat has been downgraded, and because there is no single overriding military problem that can be remedied with a hardware "fix," the defense S&T establishment has been asked to hold out for the long term and pin its hopes on the success of organizational and civilian economic reforms which will "trickle across" to help the defense sector. Defense S&T organizations are concentrating now on making process improvements, with the expectation that, in the future, they will be responding to requests for much more diverse and complicated hardware. If these requests and the capital to support them materialize, the defense industry will be in a formidable position to make another "big push," which will result in quick and dramatic improvements in force posture. Such improvements will be troublesome to the United States, not to mention the Soviet Union and other Asian countries. However, given the growing overlap in military and civilian research, production, and procurement, compounded by organizational uncertainties, indications of meaningful change in the interim may be difficult to detect. The most likely areas for rapid progress are those in which transfers from the civilian sector can play an important role. Military aircraft, electronics, and telecommunications will be among the first to benefit from an infusion of modern technology.

In the meantime, the strategic-weapons program will retain some degree of priority, and the MMBs will pursue incremental improvements in conventional weapons within the budget constraints. Modest changes in the existing hardware can tip the balance with adversaries such as Vietnam and India, even if they do not fundamentally alter China's force-projection capabilities.

In the future, it will be even more difficult than in the past to distinguish defense-related S&T programs from purely civilian efforts. However, there is still a large defense S&T establishment whose activities will make important contributions to the broader modernization effort: in the near term, by refraining from large expenditures and assisting the civilian sector; and, in the long term, by applying its streamlined organization to the development of more complex weapons, which are certain to improve China's military capabilities and force posture.

PART FOUR

Technology Transfer

CHAPTER TWELVE

Technology Transfer and China's Emerging Role in the World Economy

DENIS FRED SIMON

China's efforts to become a more active participant in both regional and global technology and economic markets promise to have important implications for the evolving international system in the coming decades. The more visible role in international relations being sought by Beijing is particularly noteworthy in view of its almost continuous isolation from the mainstream of the world economy and S&T system during the last three decades. Given its large population, its vast reserves of untapped natural resources, and its rapidly expanding industrial and technological base, China's growing presence will be felt in a number of ways. Most important, not only the quantitative dimensions of China's impact will be significant; the potential qualitative implications of the rise of the PRC could be even more far-reaching—affecting a broad spectrum of present and future regional and global relationships.

Beijing's "open door" to trade, foreign investment, managerial expertise, and imported technology constitutes the primary vehicle through which China will make its presence felt. From the perspective of both buying and selling, the present open-door policy is designed to act as a catalyst for the country's overall modernization program. Through increased exposure to advanced-country management methods, production techniques, and technical know-how as well as the dynamics of foreign markets, and so forth, the PRC's leadership hopes that Chinese enterprises will become more productive and efficient, thereby enabling them to offer a more competitive assortment of products in key overseas markets, such as the United States, Western Europe, and Japan—along with being better able to serve growing domestic demand.

The acquisition of foreign technology, in particular, is viewed as the key element in China's strategy for entering world markets. In fact, technology transfer has become something of a compulsion among Chinese leaders, who see tremendous opportunities for China to leap ahead through the application of foreign equipment and technical knowledge. This is not to suggest that China has abandoned its goal of greater technological self-reliance. To the contrary, foreign technology will be used to enhance indigenous efforts, thereby reducing the time, cost, and risk involved in modernizing existing enterprises, developing new products, and improving the quality of existing items. In spite of the somewhat uneven approach to the import of foreign equipment and know-how since 1978, major efforts remain under way to secure access to a variety of key product and process technologies. And, while the leadership in Beijing may be concerned about excessive expenditures of foreign exchange as well as problems of "spiritual pollution," the fact remains that it has not retreated from its overall commitment to acquiring foreign technology.

Most of the advanced industrialized nations have been supportive of China's expanded participation in the world economy and technology markets. From the perspective of foreign firms, the lure of the enigmatic "China market" has been a major driving force behind this support. While some companies may want to purchase Chinese raw materials or utilize China as an export platform, the majority of companies are ultimately drawn by the PRC's potential as a market for capital equipment and technology or as a market for consumer goods, assorted commodities, and services. In many cases, given the PRC's almost insatiable demand for technology, some foreign firms have been willing to "leverage" selected aspects of their product and process know-how for greater access to the Chinese domestic market.

Underlying much of the foreign participation in China's modernization program, however, has been a much more provocative set of premises. One of the primary assumptions behind the prevailing support provided by the home-country governments of many of these firms has been that a secure, economically modernizing China would be much more conducive to global stability than a China beset by a host of intractable development problems. By providing support to Deng Xiaoping's modernization drive through technology transfer, many Western leaders have hoped to consolidate the continued dominance of a pragmatic leadership in the post-Deng era. Accordingly, leaders from the United States, Japan, and Western Europe have taken a number of substantial steps to expand Chinese access to technology, such as relaxing export controls and readjusting the Coordinating Committee for Multilateral Export Controls (COCOM) restrictions, as well as offering attractive financing to promote Chinese development.[1]

The central argument of this chapter is that a substantial amount of learning has taken place in China with respect to the importation of foreign technology. The Chinese pattern of behavior since 1978 supports the contention of Haas and others, who have asserted that, through the process of learning and the sharing of common norms, countries will come to see greater benefits in collaboration than in confrontation.[2] Yet, despite Chinese progress in this area, many uncertainties remain regarding China's increased participation in the global economy, arising, in most cases, from its uneven performance in using foreign technology. These uncertainties exist inside as well as outside China. Moreover, within a dynamic world economy, major changes taking place in the nature of manufacturing technology and production organization have raised many questions about where the key sources of competitive advantage for China as well as other emerging nations may be found. In some respects, for example, China's greatest resource, its heretofore relatively untapped reservoir of inexpensive labor, may become increasingly irrelevant as a major asset in certain key industrial sectors in terms of establishing the country's competitive position.

China's steady demand for unincumbered access to advanced foreign technology has already raised complex foreign and domestic policy issues (military and economic) for the West, Japan, and a number of other nations in the East and Southeast Asian region. In a world that continues to be plagued by problems of protectionism and increasing competition, China's precise choice of technological and economic priorities could affect, in a potentially significant fashion, global-industrial, employment, and trade patterns as well as investment and lending practices. Under such circumstances, China's emergence is likely to create both conflicts and opportunities as well as pressures toward shifting international economic and political alignments. In this regard, China's trading partners and competitors, particularly in East Asia, will become increasingly sensitive to (both positively and negatively) and take a more keen interest in the success or failure of various dimensions of the Chinese modernization program, especially in the area of technology assimilation and innovation.

TECHNOLOGY AND THE CHANGING FACE OF THE INTERNATIONAL ECONOMY

The 1960s and 1970s were a period of rapid growth and change in terms of all facets of international economic and technology interaction. The face of the global economy was transformed with the emergence of Western Europe and Japan as a counterbalance to America's previous technologically dominant position. In addition, Grunwald and Flamm in *The Global Factory* have noted that, during this period, the operations of American multinational corporations (MNCs) seem to have switched, on a fairly large scale for the first time, to overseas production of manufactured exports for the US domestic market, with Asia and Latin America being the two largest areas of concentration.[3] In most cases, American firms sought out cheap-labor production sites as a means to compete with the rising Japanese presence in such industries as consumer electronics and textiles. This approach stood in sharp contrast to the actions taken by Japan, which viewed overseas production primarily (though not exclusively) as a means to address local market opportunities as well as third-country markets.[4] Moreover, as O'Neill suggests in *The Technology Edge*, Japanese firms were also busy investing in expensive, large-scale automated production equipment at home—which provided Japan with higher levels of productivity, reliability and quality, and lower costs as requisite economies of scale were attained.[5]

The 1980s has also been a time of significant change, especially in the technology area. Whereas it was once possible for places like Taiwan and South Korea to enter into the international market through participation in low-level assembly-type operations, and offering relatively low wages, this is no longer the case.[6] Technological advances in transportation and communications, which were in large part responsible for the movement of American industry overseas, are now enabling some of those same industries (in whole or in part) to move back onshore. As the recent report by MIT entitled *The Future of the Automobile* points out, "The advantages of low wages in the less developed countries do not offset the quality and coordination handicaps, the country risk, and the use of many more hours of labor for most production steps."[7] A similar argument can be made for segments of the microelectronics industry, where automation and other related changes in manufacturing—all spurred on by the demands of greater technological complexity and precision—are making it possible once again to fully integrate design, production, and testing in one location.[8]

Now, one could argue that what we are really talking about is the emergence of so-called high-technology industries in the industrialized nations and that the growing emphasis on these "advanced" industries will allow some of the so-called "traditional" industries with their mature technologies to move steadily to the developing world. While this may be an accurate perception in some respects, actually current technological changes are making selected segments of so-called traditional industries once again viable in the advanced nations. Moreover, as Robert Reich shows in *The Next American Frontier*, these traditional industries serve as the gateways to new ones.[9] Driven by technological breakthroughs, American industry has begun to take traditional sectors such as steel, chemicals, textiles, and automobiles and restructure them toward higher-valued, more sophisticated businesses like specialty steel and chemicals, precision automobiles, and so one. The introduction of flexible-system production into American industry—a system that merges the once separate business functions of R&D, engineering, purchasing, manufacturing, and distribution—will underlie the ability of American companies to compete on a global basis at present and in the future.

In effect, the globalization of critical industries such as automobiles and electronics has changed the nature of the attraction as far as overseas production and sourcing locations are concerned. The prerequisites for participation in these global industries have undergone a major shift in terms of management, technology, production, and personnel. As far as technology transfer is concerned, a country needs to have technology, that is, a basic technological infrastructure, in order to get more technology. At the same time, as noted in a recent article in the *Harvard Business Review*, there continues to be active consideration of the ways to bring "lost" industries back to the United States. And, while it would not be possible or desirable to bring all American industry back home, these pressures are continuing to build, and we have already begun to see a response in terms of domestic technology development.[10]

Today, overseas production and sourcing are no longer seen as primarily a short-term response to a simple wage-rate phenomenon.[11] Corporations with worldwide sales are no longer able to operate simply as multi-domestic firms. Rather, they must be global in scope and orientation, choosing technologies, designing products, selecting manufacturing locations, and establishing marketing arrangements on a worldwide basis. Within this context, the decision to locate a production facility abroad or to transfer technology to a foreign recipient must be based on a strategic set of considerations. While factors such as good will and politics are important, most firms are looking for locations where high value-added can be realized, where global economies of scale can be efficiently and effectively attained, and where the local infrastructure will help facilitate global integration and linkages, that is adequate port facilities, air links, communication capacity. Most important, the transaction costs of engaging in a manufacturing or technology-transfer relationship must not be excessive—in terms of time or of money.[12]

The managerial requisites for participating in such a complex network of economic and technological relationships have also changed. As Abernathy et al. noted in their *Industrial Renaissance*, "The skillful implementation of technology matters as much as does the quality of technical design."[13] It is possible, for example, that the requirements of economies of scale necessitate the establishment of a fairly sophisticated production segment in an overseas location. It is at this stage, however, where progress along the learning curve in the past will begin to yield future benefits for places such as Brazil, Singapore, Taiwan, and South Korea—all of whom have achieved substantial levels of technological progress and have become *key* players in the global networks of many large multinational firms.[14] Thus, while Reich and others may be correct in suggesting that sophisticated machinery may be movable to overseas locations, there exists a clear preference hierarchy in terms of the characteristics of possible and viable partners. The use of more complex and precision equipment, for example, means that the service and maintenance requirements will increase as will various aspects related to the "software" side of the production process—all of which exist in sufficient quantities in relatively few locations outside Japan and the Western industrialized countries. In other words, while standardization may facilitate continued overseas activities by foreign firms, the imperatives of closer coordination and precision may call for a set of managerial skills and capabilities generally absent or in short supply in most developing countries.

These shifts in the nature of production and the sources of competitive advantage have important implications for China, because they could inhibit Chinese objectives in a number of significant areas, particularly with respect to the acquisition of foreign technology—whether through equity-based investments or more arms-length-type licensing agreements. In order to assess China's ability to play a more strategic role in international product and technology markets, it is important to focus on a number of key questions:

(1) What are China's technological priorities?

(2) What role does foreign technology play in China's strategy for technological development? What is the relative balance in terms of emphasis between indigenous R&D efforts and foreign technology imports? What are the key channels and vehicles for obtaining foreign technology?

(3) How effective has China been at acquiring and utilizing foreign technology? What factors have been important in determining the effectiveness of technology import for overall development?

(4) In what areas is China likely to make its presence felt most directly in terms of its participation in foreign markets? Is China strategically situated to become an important actor in the evolving inter-

national division of labor of the 1990s as a result of its technological progress?

TECHNOLOGY TRANSFER AND CHINA'S TECHNOLOGY MODERNIZATION STRATEGY

China's strategy for modernizing its domestic technological capabilities is multifaceted, reflecting a combination of top-down, market-oriented, and horizontal policy initiatives. In essence, the current drive is characterized by the strong emphasis on organizational reform and structural change—though it must be acknowledged that the Chinese have not entirely backed away from the "big-push" approach to technological advance. Nonetheless, the degree to which the leadership is prepared to initiate fundamental change in the S&T area is significant. This is best reflected in the March 1985 Central Committee decision on "Reform of the Science and Technology Management System." The document is significant because it spells out a broad array of modifications regarding the funding of science and technology activities as well as the treatment of technical know-how and the procedures for managing scientific and technical personnel. It also serves to complement a number of other initiatives introduced over the past few years in the area of economic reform, for example, the production-responsibility system, and overall industrial policy.

From the perspective of technology transfer, the reforms hold particular significance.[15] Three of these initiatives stand out. Most important is the growing emphasis on the technical transformation of enterprises.[16] Total investment in technological transformation and equipment-renewal for state enterprises during the 7th Five-Year Plan is projected to be 276 billion yuan (RMB), which constitutes an 87-percent increase over such investment during the 6th Five-Year Plan.[17] This decision to give attention to the renovation and reorganization of existing plants and facilities reflects the recognition that the "software" side of industry (plant layout, scheduling procedures, inventory control, and so on) may be just as vital to increasing productivity as the acquisition of large quantities of foreign equipment. As such, China has de-emphasized its previous focus on whole-plant purchases as the major form of so-called "technology import"—in the belief that this acquisition vehicle did not provide adequate levels of technology transfer—and is now stressing the acquisition of "know-how" and basic design data in its relations with foreign companies.

It goes without saying, however, that technology imports in the past did contribute to the development of such critical industries as chemical fertilizers and steel. Yet, the fact remains that, while the Chinese were able to obtain greater quantities of the products they desired, very little direct transfer of design capacity took place. In many cases, even the simple task of retrofitting a basic energy-saving device on a synthetic ammonia plant must be done by the original supplier. A recent article in *Intertrade* (April 1987), commenting on China's import experiences with chemical-fertilizer projects, noted that some foreign firms "are skeptical about Chinese design capability and progress of the project(s), which would inevitably affect the quality of the final product." The same article went on to suggest that "some foreigners still do not trust the ability of the Chinese and insist that key parts must be imported and the patented technology for manufacturing the equipment must be purchased."[18]

The second key element has been the linkage of the economic reforms with efforts to stimulate technological progress.[19] As Fischer argues in his chapter of this volume, the ongoing movement toward greater decentralization of decision making, although fraught with a number of problems and uncertainties, promises to improve, to a greater degree than is presently true, the process of technology selection and may even increase demands for foreign technology.[20] In the past, many factory managers decided to eschew the potential benefits of new technology (in many cases, both product and process technology) because of the additional burdens that integrating a new piece of equipment or procedure might create.[21] With increased autonomy and responsibility for profits and losses, as well as reduced bureaucratic red tape, Chinese leaders hope that production units will become more attentive to their technological opportunities as well as their technological choices and options—especially since the costs of waste and inefficiency will increasingly have to be directly borne by the unit itself.[22]

The third area that is significant is the effort to build closer links between the civilian and military sectors in order to facilitate the transfer of knowledge and capability from the latter to the former.[23] Heretofore, China's defense sector has generally been the recipient of greater financial resources, larger numbers of more qualified technical personnel, and more sophisticated equipment, instrumentation, and so forth. Industries such as electronics and nuclear energy, for example, were primarily "military-oriented," and contributed very little to the civilian economy—were, in fact, probably a drain on limited personnel and technical resources.

Since the early 1980s, this has begun to change, primarily in the belief that the technological and managerial leadership of the defense S&T and industrial base can help spearhead more rapid advances in the civilian sector, and perhaps even stimulate productivity and efficiency gains in the defense sector as well. Consumer electronics has now become a critical part of the electronics industry; nuclear scientists are now devoting their attention, in part, to development of a civilian nuclear-energy program. Various mechanisms have been established, such as the working relationship formed between Tianjin municipality and the National Defense Science, Technology, and Industry Commission, to facilitate greater civilian-military interaction. In essence, China's leadership appears committed to moving away from the highly compartmentalized system it borrowed from the Soviet Union; its new model appears to be more akin to the so-called "military-industrial complex" most frequently associated with the United States.[24]

These changes in both organization and emphasis—the emphasis on technical transformation, the linkage between S&T reform and economic reform, and the effort to build closer links between the civilian and military sectors—cannot be viewed in isolation from China's own set of specific technological priorities. The recent *White Paper on Science and Technology* issued by the SSTC identities a number of key areas where China hopes to make rapid and sustained progress.[25] In some cases, market-driven forces will be largely responsible for promoting technological advance, for example, in light industry; in other cases, strong central guidance will be maintained, as in microelectronics. Current priority fields, discussed in an October 1986 issue of *Liaowang*, include (1) transportation and communication, especially telecommunications; (2) agriculture and food processing; (3) energy, including coal, hydropower, and various power-transmission technologies; (4) electronics, especially integrated circuits, computers, and software, for example, CAD/CAM/CAE; and (5) manufacturing technologies—across the board.[26] These choices reflect both the desire to enter into new technology-based industries (the "High-Technology Development Program" announced by the SSTC in March 1986) as well as modernize traditional industries and the country's economic infrastructure to reduce foreign imports and expand exports.

CHINA'S TECHNOLOGY-TRANSFER STRATEGY

Propelled by a desire to gain expanded access to foreign technology, the Chinese have set in motion a series of acquisition programs to support their overall modernization. During the 6th Five-Year Plan (1981–1985), according to statistics provided by the SPC, China signed over 1,300 technology-import contacts with foreign firms worth a total value of US$9.7 billion. Of these, 127 projects were considered "backbone" projects, meaning that they were directed at developed or expanding high-priority sectors, for example, Baoshan Iron and Steel Complex, the Beijing Dongfang Chemical Plant, and the Jiangxi Copper Plant.

As in the past, the basic framework of the program for the 7th Five-Year Plan has been laid out in a list of 3,000 key items designated by the former State Economic Commission (now part of the State Planning Commission) for purchase during the 1986–1988 period. As noted, these items place the import of technologies to support the technical transformation of enterprises in the forefront—with the State Planning Commission being the lead organization in this effort. Most of the focus is on the import of *know-how*, with emphasis on technologies in the electronics and machinery area. While technology imports are not confined to this 3,000-item list, the list does appear to contain the priority items.

The salience of the so-called "3,000-projects" list can be seen in a recent Chinese assessment of the impact of those technologies imported under this program during the last three years of the 6th Five-Year Plan. According to the evaluation, which was made in late 1986, "the 3,000 projects have enabled some Chinese industries and products to leapfrog technically, significantly narrowing the gap with the advanced nations. Ten percent of the products of the machinery industry today reach international standards of the late 1970s and early 1980s. In electronics, the manufacturing of color TV sets, videocassette recorders, tape recorders, and copiers has grown and matured. More than 30 percent of [our] electronic products are now on a par with the best of the world in the late 1970s and 1980s, up from 15 percent in 1982."[27] While perhaps somewhat over-optimistic in its tone, the assessment reveals the critical value that such a program could have if properly supported and managed, especially since, as the source suggests, "these projects require limited outlays, have a short construction period, pay off quickly, and yield good economic results."[28]

Since 1979, we have seen the emergence of a large number of Chinese organizations doing business overseas as well as the proliferation of local foreign-trade entities within the country at large. Each of the production ministries as well as many of the provincial and municipal governments have their own import-export arm, parts of which are responsible for foreign-technology acquisition. A good example is the Shenzhen Scientific Equipment Export Service Corporation, whose main function is to provide relevant Chinese organizations with information and data on the latest developments in international science and technology. Chinese organizations have also begun to proliferate in Hong Kong, each one seeking to take advantage of Hong Kong's special access to the West. No less than 15 provinces and several of the key municipal governments have formerly registered firms in the soon-to-be-former British colony.[29]

Foreign investment abroad by Chinese organizations has also become an attractive mechanism for gaining access to foreign technology. China has not only made investments in timber in the United States, for example, but it has established an equity position in a computer firm in New Hampshire (Santec) and a machine-tool firm in New York (Autonumerics). The Chinese see these as technology listening posts as well as opportunities to train technical personnel.

Of course, these types of formal organizations only begin to touch the surface. One must also consider China's efforts to attract foreign firms to establish equity-based investments in the country. Between 1979–1986, over 6,150 contracts were approved for foreign investment; 2,300 have involved equity joint ventures, 3,700 are contractual joint ventures, 120 include joint exploration, and 120 are 100 percent wholly foreign-owned projects. Total foreign investment committed surpassed US$16 billion by the end of 1985, though only US$4.6 billion has been utilized. So far, Hong Kong firms account for about 80 percent of all investments, with the United States second and Japan third. Chinese leaders see foreign investment as a means more easily to bring about the transfer of key production and managerial know-how to Chinese industry—though both Chinese and foreign observers acknowledge that many questions remain about the effectiveness of several existing joint ventures for acquiring technology. China has put in place an entire legal infrastructure to encourage investment by companies from abroad. In addition, four "special economic zones" were created in South China, not only to construct export platforms or generate employment but, more important, to provide a channel for technology transfer.

A series of regulations on technology transfer have also been introduced.[30] For example, a patent law was announced in April 1985, in large part to alleviate foreign concerns about violations against the integrity of their technology and know-how. Regulations also have been introduced that maximize Chinese access to core technologies, but whose principal aim is to limit the amount of control foreign firms can retain over the application of technology once transferred to a Chinese counterpart. Announced in May 1985 by the State Council (24 May 1985), the rules restrict the use of so-called restrictive business practices by foreign firms in China, such as limiting the geographic export area of products manufactured with the imported technology. In August 1985 and in December 1987, the State Council also promulgated another set of measures designed to institutionalize the procedures by which technology-import projects are reviewed in order to better ensure that proposals are adequately evaluated before final approval is granted. This latter pronouncement was part of the overall recentralization of technology-import process under which a "certificate of approval" will have to be obtained from the Ministry of Foreign Economic Relations and Trade before a contract can be implemented.

Acquisition mechanisms have also been created at the enterprise level with the establishment of so-called "technology-introduction or import departments." In many cases, these departments were created on an ad hoc basis to handle the process of technology import. Subsequently, they have been formally institutionalized and are now a permanent part of the Chinese enterprise. Their primary value is that they constitute a critical mass of individuals responsible for all facets of an import-related project. In some instances, in addition to having a small core staff, there is a rotating staff of technical and financial specialists that move in and out of the office as needed during the inception, negotiation, agreement, and implementation phases of particular projects. Enterprises are now responsible for preparing in-depth feasibility studies to not only justify the need for the purchase, but also to indicate their capabilities and resources for effectively implementing the new technology.

CRITICAL ISSUES IN CHINA'S TECHNOLOGY-ACQUISITION SYSTEM

It is clear that China has begun to benefit substantially from the growing presence of foreign technology in the local economy. In the electrical-appliance area, for example, product quality and designs have been greatly improved.[31] A good example is the Yingkou Washing Machine Factory in Liaoning province, which, after extensive cooperation with a Japanese firm—including the completion of two large-scale technical upgradings and the setup of 22 computerized production lines—now produces 7 models of washing machines.[32] In addition, according to one study of imported technology and equipment which have been put into operation in Beijing, Shanghai, and Tianjin, each yuan in investment in imported technology has increased output value by 2.5–2.8 yuan.[33]

Nonetheless, in spite of the large-scale mobilization of organizational and personnel resources aimed at the acquisition of foreign technology, the Chinese have encountered a number of serious problems in this important effort. Many of these were spelled out in a revealing series of three articles published in the overseas edition of *Liaowang* in May 1986.[34] Written by Cao Jiarui, a former Ministry of Foreign Economic Relations and Trade (MOFERT) official responsible for overseeing China's technology-import program, the articles point to a plethora of difficulties, many of which have limited the impact of foreign technology transfers to the PRC or at least led to a significant waste of scarce financial and technical resources.

Among the problems cited by Cao, four stand out. First, there is *insufficient coordination* among the various organizations responsible for various facets of China's technology-import program. China's bureaucracy continues to be plagued by competition and parochialism, which make themselves felt not only at the national level but also in the relations between the central government and the provinces and key municipalities. At times, for example, relations between Shanghai and Beijing, have tended to be more competitive than cooperative, leading to inefficiency, delays, and congestion in administration. Relatedly, a second problem involves the *consequences of decentralization* vis-à-vis the decision-making process for technology imports. While the central government has granted various localities certain levels of financial decision-making autonomy regarding technology imports (a US$5-million threshold has been set similar in thrust to the foreign-investment levels of autonomy), the fact remains that this autonomy can be and has been circumscribed in a number of ways by Beijing. Thus, on some occasions, local authorities have gone too far afield from Beijing's dictates in rendering decisions on projects, while, at other times, they have lacked the authority to move ahead on important and highly desirable projects.

A third problem cited by Cao is the *excessive duplication* of technology and equipment imports. One area where the problem has been extensive has been the import of color-television production lines. Estimates suggest that China has imported over 100 color-TV lines, creating a production capacity far exceeding projected domestic demand. And, given current difficulties with quality control and product design, most of the excess production cannot be exported. The severity of the duplication problem is exacerbated by the jurisdictional sensitivities of units who belong to different administrative hierarchies. Few organizations are willing to share items, such as imported computers—many of which tend to be seriously underutilized in terms of their overall capacity.

According to one source, "It is hard to tell the quantity, types, and nature of technology China has imported since 1979."[35] The severity of the situation also appears to be a result of poor organizational coordination. Cao suggests that "the management system for technology import lacks overall control and planned guidance. . . . There are nearly one hundred departments that have the right of examination and approval. . . . And after local autonomy was expanded, the various supervisory departments lost control of various trades." Even though technology imports have drastically increased since 1978, a technology-import information-management system has not been created, thus precluding the exchange of information and data among potential buyers.

The fourth problem highlighted by Cao deals with *poor assimilation capabilities*. Much attention has been focused on the acquisition of technology, including the problems of export controls and COCOM, that seemingly prevented China from purchasing the technologies it desired and needed. With the significant relaxation of restrictions on technology exports to China that took place in December 1985, an appreciable number of China's problems in this context have disappeared. What is clear, however, is that, with much of the attention focused on acquisition problems, in many cases Chinese officials and enterprise managers have not paid adequate attention to technology absorption.

In some instances, Chinese purchasers, fearful of buying obsolete technologies, have sought state-of-the-art items, only to find that they lack the expertise, maintenance and service capabilities, and supporting infrastructure to utilize fully the imported technology or equipment. In other cases, purchasers of technology have been apprehensive that their foreign-exchange holdings might be gobbled up, and thus they have sought to spend their foreign currency as soon as possible. Under such pressures, they have frequently failed to conduct adequate feasibility studies or carefully evaluate technological alternatives. Moreover, as Cao notes, "China's current technological imports reflect the will of the high-level government departments or officials rather than the market reality or latent needs."[36] Thus, heretofore, there has been little incentive or direct pressure to be successful or effective at assimilating foreign technology.

It must be noted that the *Liaowang* articles are not unique in their criticism of Chinese performance in technology acquisition and assimilation. At all levels, serious reservations have been expressed about the effectiveness of current practices and methods for utilizing foreign technology. One significant critique has also been provided by Zhang Aiping, retired Minister of National Defense, in a December 1985 article in *Hongqi*, directed at China's poor assimilation performance. He talks about the three "unifications:" in planning, coordination, and organization—which have been missing in China's experience but are required to do a better job in the assimilation process. More significantly, especially given Zhang's former position and his previous role as head of the NDSTC, he expresses strong concerns at the general tendency that has emerged over the past few years to overemphasize the role of technology imports to the neglect of indigenous R&D and S&T capabilities. The article reflects the fact that China has not abandoned its quest for greater self-reliance and indigenization of imports, and that the debate about the precise role of technology imports has not been resolved.[37]

The new pieces of legislation regarding technology transfer mentioned above were apparently formulated for the purpose of standardizing existing technology-import procedures, tightening up control over the import process, and ensuring more careful evaluation of specific acquisition proposals.[38] They are part of an overall Chinese effort to add more discipline to the acquisition process and thereby improve the prospects for effective and efficient application of imported technology. The major provisions of the technology-import contract regulations include: (1) prohibitions on the use of so-called "restrictive business practices" by foreign firms wishing to sell technology to China;[39] (2) requirements concerning the levels and appropriateness of the imported technology; it must, for example, meet 1 of 8 criteria as far as its contribution to economic modernization is concerned; (3) the need for guarantees by the foreign supplier that the technology is not defective and will allow the recipient to meet its stated objectives; (4) provisions limiting the term of the contract to ten years; and (5) stipulations that MOFERT or its stated representative must review and approve the contract. While the regulations still leave many questions unanswered (for example, what obligations does the foreign firm incur by virtue of its acceptance of the "guarantee" clause?), they still go a long way, especially from the Chinese perspective, toward resolving some of the inconsistencies in PRC technology-import practices across the board.

Similarly, the procedures for "examination and approval of technology-import contracts," revised and reissued in December 1987, represent an attempt to clarify the decision-making criteria and review process for potential Chinese recipients as well as foreign technology suppliers. Under these regulations, technology-import contracts must meet supplementary criteria (in addition to those spelled out in the May 1985 provisions) in order to be approved by PRC government officials. For example, there is a quality guarantee not merely for the technology itself, but also for the products manufactured with the technology. This raises a number of critical questions for foreign technology suppliers, since there are strict limits placed on their control over the various inputs and the source of those inputs to be used in the production process. The procedures also stress the use of the concept of "reasonableness" in assessing various provisions of the technology-import contract; for example, price and payment methods should be reasonable. While this may provide room for negotiation and flexible interpretation, it also leaves much to the discretionary judgment of the Chinese bureaucracy—a situation that could leave many firms quite uncomfortable.

The reasons for this discomfort are plentiful. What has come to the surface over the last several years is that, far from being a command economy operating in a coherent, well-coordinated fashion, China is a country where bargaining, negotiation, and internal competition tend to have a greater influence on decision making than dictates from the central government in Beijing. Chinese managers interested in acquiring foreign technology must often proceed through a maze of bureaucratic obstacles and negotiations in order to obtain the needed foreign exchange, raw materials, and infrastructure support to implement a project.

As a result, the decision-making process regarding technology import must not be viewed in terms of the purposeful-actor model that suggests the image of a system fully mobilized to get the best deal from the foreign supplier and to take the most efficient route to effective assimilation. Rather, the technology-import process in China tends to be characterized by great uncertainties and lack of clarity regarding the criteria for project selection and choice of recipient. Moreover, the process can be radically uneven in terms of cross-provincial treatment and frequently leaves the proposed Chinese counterpart to the foreign supplier just as frustrated and confounded as the foreign firm itself as the process of project approval proceeds through the multiple steps of review in the bureaucracy. Unfortunately, even though decentralization has proceeded to a great extent in terms of general economic decision making, the reality is that most of the autonomy of local managers regarding technology-import decisions remains circumscribed by higher officials who, through their control over foreign exchange, maintain continued authority over the factory's "external" economic and technology relations. This is not to suggest there are not "successes" or that the process is always so cumbersome; things do work and are improving. Rather, the key point is that problems of mixed signals and lack of predictability can be just as pervasive to those within the system as to those outside of it looking in.

CHINA'S IMPACT ON THE WORLD ECONOMY

On the positive side, the gradual integration of the Chinese economy with the world's major market economies—stimulated and supported in large part by the transfer and more effective use of foreign technology—could serve to make China increasingly interdependent with important segments of the global economy. Given China's factor endowments, the Chinese economy could eventually become an important site for the production of some key labor-intensive components or products, a supplier of key natural resources and minerals, or even a source of R&D support for foreign corporations.[40] The evolution of Sino-Japanese, Sino-West European, and Sino-American economic and technology relations could take on long-term significance in this context. For example, China could conceivably become the next battleground for the playing out of US-Japan competition—with the PRC being the key element as partner or market in this competition.[41]

From the Chinese perspective, the availability of technology has become the quid pro quo for entrance into the PRC domestic market. More specifically, Chinese officials are on record as indicating that those foreign firms who are more forthcoming on technology transfer will be granted greater access to the local market.[42] And, in the future—to an extent that is even more the case than at present—technology transfer will become the "cement" holding China's relations with the West together. From the Western perspective, the steadily growing links between China and the rest of the world seem designed to make present and future Chinese leaders more aware of the real and potential benefits to be derived from behaving in a responsible and pragmatic fashion in the years ahead.

In the commercial area, in particular, while the promise of the China market has remained largely unfulfilled and major sales have been slow to develop and difficult to finalize, the Chinese penchant for wanting to link up with the West continues to be strong.[43] Given this fact, and assuming that further progress is made in overcoming present absorption limitations, China, in all likelihood, will steadily increase its demands for manufacturing know-how and managerial expertise from the West. Opportunities for the licensing of technology, co-production agreements, joint ventures, and assorted training programs may grow beyond present levels as China attempts to improve in a drastic fashion the quality and marketability of its products. In the process of linking up with the West, as China expands its reliance on Western standards, measurements, designs, spare parts, and managerial procedures, it will become extremely difficult and costly for the Chinese easily to extricate themselves from the network of relationships they have developed with the industrialized world—though it may be acknowledged that no real guarantees exist to prevent China from bearing such economic costs in order to promote what they perceive to be more imperative political considerations.

At the same time, and perhaps more important, Chinese progress in certain facets of its modernization program could prove both politically and economically unsettling to selected industrial constituencies in the West and the regional economy of East Asia. According to development strategy articulated by Beijing in the recently announced 7th Five-Year Plan, the Chinese hope to finance a large proportion of their technology and equipment acquisitions through rapid and sustained expansion of exports. In spite of recent controls on overall imports of goods and equipment, China's leaders recognize that export growth and not import cuts are the key to more rapid and sustained modernization. While petroleum may become a major source of these export earnings in the long term, the recent decline in the world price of oil (as well as a number of other factors) has led the Chinese to focus more on consumer goods, light industrial products, and selected capital goods as the core of their export drive. As China's industrial base improves, and its products become more desirable in foreign markets, the Chinese economy could gradually, albeit steadily, prove to be an important source of economic competition to the various sectors in the ASEAN nations and the "newly industrialized countries" as well as selected parts of the United States, Western European, and Japanese economies.

In areas such as textiles, machine tools, and light industrial products, for example, the Chinese have already begun to make substantial progress. During the 6th Five-Year Plan (1980–1985), the textiles industry was a principal source of foreign exchange, earning US$17.2 billion in foreign exchange (exports rank first in the world in tonnage); the volume of textiles exports increased 64 percent—averaging an increase of over 10 percent each year.[44] In Shanghai, China's leading area of textile exports, the textiles industry has been a main target for that municipality's technology-import *and* technical-transformation programs. Through the contribution of foreign technology (since the 1970s the Ministry of Textiles Industry has imported advanced technology valued at more than US$2 billion), China has been able to upgrade *significantly* the quality of its textile products and move into higher value-added segments of the industry.[45] Between 1981 and 1986, the textile industry spent RMB 14.5 billion and more than US$700 million on renovating factories.[46] Moreover, consistent with the past experiences of textile companies in South Korea, Taiwan, and Hong Kong, they have also reaped substantial benefits from interacting with foreign buyers and designers who have imparted know-how related to style and color to the Chinese in their efforts to increase their purchases from PRC textile producers.

New organizational amalgams have already appeared in order to better coordinate the technological modernization and promotion of exports in the textiles industry. A series of industrial associations has been formed to establish a union among spinning and weaving, printing and dyeing, tailoring, and exporting. Similarly, new efforts are being made to link research with production, such as in the case of the Changzhou Chemical Research Institute, which linked up with a number of local enterprises to create new and improved products and processes.[47] According to one commentator in the January 1985 issue of *Guoji maoyi*, raising the quality and style of Chinese textiles is imperative because (1) about 75 percent of the markets served by PRC clothing exports are in the Western developed nations; (2) the prevailing consumption patterns in the United States require greater attention to quality, new designs, and more fashionable products; (3) the technological capacity to meet these requirements is now in place in China and needs to be utilized; and (4) higher-quality products tend to yield greater returns and more foreign-exchange earnings.[48]

The Chinese textile industry is becoming more conscious of the notion of fashion.[49] Replacing the existing drab materials with new color materials, transplanting foreign designs to China to replace existing styles, and getting items to the market while they are still in vogue are all objectives that are present within the strategic thinking of textile producers. In fact, the absence of key clothing accessories in appropriate colors in the domestic market has forced many firms to import needed items. For example, about 58 million meters of zippers will be produced for the Beijing municipal market in 1986, even though 10 million meters are already stockpiled. Many of the zippers sitting in Chinese warehouses are not available in multiple colors or are made of metal and not suitable for some styles of garments. Similar problems have existed for linings, packing materials, and so forth.[50]

China's recent technological progress in industries such as textiles should not be taken to portend an inevitable clash between China's modernization program and the economic prosperity of the West or East Asia. Structurally, particularly with respect to the situation in East Asia, while a combination of growing protectionism and competition may impose limits on easily achievable market shares in Western countries, the current efforts by the Asian NICs to move away from their labor-intensive, light-industry orientation could provide some relief for China.[51] Furthermore, interesting complementarities could develop among some of these countries, such as the evolving relationship between South Korea and the PRC.[52] Nonetheless, should these countries encounter problems in their programs to move into more skill-intensive, technologically sophisticated product lines, the presence of Chinese products could bring about an intense economic rivalry with strong political overtones.

From a political perspective, China has reiterated its intention to maintain its independence in the realm of foreign policy, avoiding any movement that would bring it too close to either superpower or make it excessively dependent on any one nation. One hallmark of China's open door since 1978 has been its willingness and desire to engage in political, economic, and technology relations with many nations simultaneously, including those of the Eastern bloc, the Third World, and the West. In fact, Beijing continues to see itself as a leader and protector of the Third World—a role that could conceivably bring it into growing conflict with the United States and other Western nations on a host of global economic and technology issues.

Beijing's efforts to ease tensions with the Soviet Union, reflected in the recent signing of an economic and technical cooperation agreement with the USSR, also raise some potentially sensitive issues.[53] Along with Western concerns about the possible diversion of American, Japanese, or West European technology to the USSR (and possibly North Korea), questions may emerge about how viable the strategy of "interdependence" really is under circumstances of renewed Sino-Soviet cooperation.[54] It is clear that there are at least three major stumbling blocks (Kampuchea/Vietnam, the Sino-Soviet border, and Afghanistan) to overcome as well as a number of other problems to confront before technology relations can even begin to approach the level of the 1950s. Yet, the fact remains that Chinese concerns about foreign-exchange spending, a slowdown in the economic reforms, or their dissatisfaction with the self-perceived paucity of technology transfers from the West could lead them increasingly to turn to the Communist block if they cannot get what they want from Western countries on the terms they like.[55]

In a sense, China's growing participation in the world economy, engendered through its technology-transfer relationships, will leave the leadership confronted with a fundamental dilemma. Chinese leaders will be faced with the increasingly difficult task of trying to balance the trade-offs between maintaining their "independence" of action and reaping the benefits that come from closer economic and technological integration with the advanced industrialized nations. This is especially true in industries such as telecommunications. What promises to compound the difficulty in achieving an acceptable balance is the fact that the decentralization of decision making in the country has fostered the more visible articulation of local interests. As such, no longer (if it ever did) can Beijing speak as the monolithic authority in China; local interests and capabilities must be taken into account. This effort to balance local and central perspectives most recently has been manifested in the *attempt* to tighten up central control over foreign trade and technology imports.

The penchant for wanting to import technology has also run up against concerns from within a number of economic circles in China regarding protection of domestic industries.[56] Even though reducing foreign-exchange spending has been a main motivating factor in the clamp-down on imports, the fact remains that the excessive influx of foreign products, ranging from automobiles to integrated circuits and microcomputers, has caused apprehension about the ability of various domestic sectors to grow in the face of competition from foreign products. Out of this apprehension have come the roots of a typical import-substitution policy.[57]

In the early 1980s, for example, the domestic microcomputer industry in Shanghai, as well as in other parts of the country, was literally destroyed by the too-rapid import of foreign microcomputers. While many of these were imported in kit form for assembly and recorded in the annual statistics as "production," they contributed very little to the overall technological advance of the country's computer-development programs. The same argument could be made for China's integrated-circuit industry, which heretofore has suffered from poor yields and serious reliability problems. Designers of China's counterpart to the IBM/PC/XT, the Great Wall 0520, have decided to use mainly imported components because of the poor reputation of domestic ones.[58] Fearful of never getting out from under the exigencies of foreign availability and quality pressures, Chinese leaders have imposed controls on imports of certain foreign-made ICs, while at the same time pushing to expand acquisition of foreign IC-production equipment to improve the quality and technological level of locally made chips. These efforts will be focused on four key computer and electronics bases: Beijing, Guangdong, Shanghai, and Jiangsu (Wuxi).

The Chinese have also gone to extremes with respect to their policy on television production. Anxious to meet a growing domestic demand for television sets, but also seriously committed to establishing a presence in the world television market, the Chinese imported over 100 production lines for manufacture of black-and-white and color televisions as they have curtailed the import of TVs from abroad.[59] Under the long-term trade agreement with Japan, for example, 14 of the 24 complete plants imported were television-production plants. Estimates suggest that, once all of the plants come on line, annually China will have over a 90-million-set capacity. Significant improvements have already taken place in the basic quality of Chinese-manufactured TVs; the mean time between failures (MTBF) has steadily increased to 15,000 hours. In many cases, however, even those involving joint ventures, such as the Fujian-Hitachi partnership in Fujian, the cost structure and the overall quality of the sets still may preclude export, let alone sales at a competitive price.[60]

In many respects, what the Chinese are facing is a twofold problem. On the one hand, they are coming up against the constraints within their own system. Previously mentioned assimilation problems combined with a host of factors associated with low labor productivity, poor quality control, and ineffective management make it difficult to produce a package of competitive exports at this time—except in product areas that are the object of intense competition and protectionism. On the other hand, the various efforts by the Chinese to attract foreign know-how and expertise that could help overcome many of these problems have been slow to materialize because of continuing questions about China's investment climate, its commitment to protect proprietary information, and a series of related issues.[61] In reality, there are many other attractive investment alternatives to China where the local skill levels, managerial capabilities, and investment environment are all less problematic.

China's concerns about gaining unincumbered access to foreign technology are focused mainly on the United States and Japan—though in recent months there does appear to be a growing emphasis on Western Europe. Since 1982, the United States has been criticized by Chinese leaders for its maintenance of export controls on the sale of advanced technology to the PRC. China's leadership has viewed the problems as "government-derived," with the implicit assumption that, if and when a substantial relaxation in export controls were to occur, American firms would be ready, willing, and able to arrive in China with technology and know-how in their arms for the taking. Yet, as evidenced in an editorial in the 20 April 1987 edition of *Beijing Review*, even though there has been a steady relaxation in controls on the sale of advanced technology to the PRC, the Chinese continue to criticize American policy.[62] The real problem is that many American firms remain unconvinced about the widely heralded China market—and thus do not want to offer their technology. Moreover, among other firms, a sense of "techno-nationalism" has emerged; concerns about "creating another Japan" in the future inhibit the pace and extent to which technology transfers will be forthcoming.

China's problems with Japan appear to be even more serious, especially in view of the fact that Sino-Japanese trade, which totaled US$13.86 billion in 1986, was almost three times as large as Sino-US trade (US$5.99 billion). [63] The PRC's trade deficit with Japan, which reached about US$5.13 billion, was more than twice the size of its deficit with the United States (US$1.06 billion). While some improvement occurred in 1986 and 1987, the problems are still severe and remain a source of political friction. These problems were exacerbated by the Toshiba case, in which Tokyo froze a number of Sino-Japanese deals as part of the penalties imposed on Toshiba for selling unauthorized technology and equipment to the USSR. China has criticized the Japanese for being unwilling to transfer any advanced technology; Japan, it is claimed, is only interested in selling *products* to the Chinese market. Similarly, until this year, Japan's level of direct investment in China was one of the lowest—a surprising state of affairs, given the vast network of trade and aid relations established between the two nations. Only after the imposition of various trade barriers by the Chinese have Japanese firms begun seriously to consider making appreciable equity investments in China.

PROSPECTS AND CONCLUSIONS

In many respects, China's desire to establish its presence in the world economy through the use of imported technologies is occurring at a very unfavorable time. First, as both the recent example of the ill-fated Fujitsu-Fairchild agreement and the Toshiba technology-diversion case indicate, national-security concerns along with concerns about unauthorized sales of technology have created a mentality of control in the United States and some other countries that has spilled over into China's own quest to gain greater access to advanced know-how and equipment. While gradually changing, this is still a problem for China. The recent Sino-American confrontation over China's alleged sale of Silkworm missiles is also relevant in this context, the main result being that the US will not be forthcoming on technology transfers when it believes that its national-security interests are not being served by Chinese behavior.[64]

Second, increasing global competition has made some firms shy about releasing their technology to potential competitors. At the present time, these firms do not see China as playing an important role in this dynamic competitive situation. If a firm is looking for a high value-added production site and a steady, secure value stream, it will not find it easily in China. As suggested earlier, the impact of design and cost-reduction efforts by US manufacturing firms, in particular, has been that, while labor costs are important, they are no longer the major cost factor; material and component procurement has become more critical. Therefore, the decision to locate offshore is being viewed from a radically different perspective than in the past.

Third, rising protectionism in the West has raised questions about the viability of a strategy designed to rely on exports as a major source of growth, especially when the targeted markets are already highly crowded by exporters such as Japan, South Korea, Taiwan, Hong Kong, and so on. Neither the US nor the EEC is prepared to absorb large quantities of Chinese goods. And, fourth, as noted, due to the introduction of technological changes such as greater automation into such industries as automobiles and electronics, the need to seek out low-cost labor manufacturing sites by MNCs has begun to decline. Moreover, the financial and technological barriers for new entrants into such industries as microelectronics are steadily increasing as the pace of technical change further accelerates and the cost of building new, modern facilities climbs.

These factors have combined with the PRC's internal constraints to help slow down Beijing's access to technology and equipment and to diminish some of the significance of China's own progress. Yet, they have not stopped progress. Along with advances made in the textiles industry, the Chinese have been busy responding to current changes in global technologies by investing and expanding research into such fields as biotechnology, new materials, and robotics. Much of the information Chinese organizations gather about developments in these fields comes from an extensive network of information-collection organizations that are part of certain large enterprises, corporations, and ministries. China's newly structured Ministry of Machine-Building Electronics Industry, for example, has a very active "information-collection" effort under way that swallows up data about electronics from Taiwan, South Korea, Hong Kong, and Singapore as well as from the industrialized world.

China's robotics program, although still behind the West's, has already begun to make ample progress.[65] The first robotics R&D center under the Institute of Automation of the CAS in Shenyang is being built. China's apparent aim is to build low-priced simple robots for raising worker efficiency and reducing labor intensity. In addition, it will also develop special-purpose robots, including movable robots to work in dangerous environments. Here again, foreign technology has already played a role through the import by the institute of underwater-robot manufacturing technology from Perry Offshore, Inc., of the United States. Related research is being supported by the Ministry of Space Industry and carried out by Harbin Engineering University. The Fenghua Machinery Plant in Harbin helped produce China's first arc-welding robot.

The Chinese have also indicated their intention to establish a presence in the international satellite market—a market heretofore dominated by NASA of the United States and Arianespace, a West European space consortium.[66] China has launched 18 satellites of various types over the last three decades. Its first telecommunications satellite was launched in 1984. The second one was successfully launched in February 1986. An agreement has apparently been reached between Sweden and the PRC to launch a communications satellite (Mailstar) for the state-owned Svenska Rymdatiebolaget. One of China's main competitive advantages in the commercial end of the satellite business appears to come from its lower insurance rates through the People's Insurance Company.

These two examples of the use of technology for commercial purposes reinforce the enigmatic quality of China and the difficulty of predicting when and how the PRC will make its presence felt in the international economy. While the country continues to have trouble producing the right color zippers for traditional garments, it is also undertaking complex technological challenges such as robotics, biotechnology, space launches, and so forth. Obviously, in the short term, China is most likely to make its presence felt in the lower end of the technological spectrum as far as the international division of labor is concerned. Through a combination of technological push and market-led innovation and renovation, however, a number of Chinese enterprises will increasingly possess the tools to offer a competitive group of products for export. The greater attention being paid to standards by the central government and its efforts to impose strict requirements for meeting these standards are part of the "push" side of the equation. In addition, the recent decision by the central government to provide direct and indirect subsidies as well as assistance to those factories that are successful in meeting export requirements will further enhance the prospects for China to establish a presence in select market niches.

Of course, numerous problems will persist. The prospects for foreign investment will remain largely unimpressive, and, as a result, the value of this mechanism as a vehicle for technology transfer will be somewhat limited. The biggest gains will be in terms of the experience Chinese workers obtain from being employed in a work environment where quality control, discipline, productivity concerns, and so forth are strongly emphasized. This will leave the Chinese frustrated, especially since they feel they have gone a long way toward adapting to the needs of the foreign business community. Similarly, the SEZs will continue to experience difficulties, mainly because expectations have been set too high and they are not structured to serve the functions of facilitating technology transfer. As this author has indicated in his forthcoming study (*Taiwan, Technology Transfer and Transnationalism*), the experience of export-processing zones in Taiwan, South Korea, and elsewhere reveals that diffusion of technology within and from the zones takes time. Without much greater opportunities for free labor mobility, especially for engineers and technical personnel, the most central mechanism for the movement of technology and know-how out of these projects and into the larger economy may be missing. And foreign-exchange controls are likely to place some real limits on China's ability to buy large quantities of technology, though, if predictions about Chinese greater willingness to borrow in international finance markets do come true, this situation might change somewhat.

In the final analysis, it will be the interface with the international market along with the interaction between foreign buyers and Chinese firms that will provide the greatest source of technology transfer. Yung Whee Rhee et al., in a study entitled *Korea's Competitive Edge*, suggest that this sustained contact with foreign firms in the international marketplace served to improve the understanding of foreign market structures and characteristics among Korean firms.[67] This went a long way toward strengthening the ability of these firms to survive and prosper in such markets. And, while nothing can substitute for a good overall economic and technology strategy (like Japan's) to support an export drive, in reality, the more the Chinese learn about the intricacies of the markets outside their country (as well as within their own country) and attach themselves to a dynamic "learning curve," the better they will be able to compete.

Foreign technology will continue to provide the linkage through which China interacts with the world economy. As a key buyer for technology in the international market, the Chinese promise to become more selective as their level of sophistication and their ability to do comparative shopping improve. As their experiences with Volkswagen and American Motors indicate, they have learned much from the history of the Brazilians and Mexicans in dealing with multinational auto firms. Moreover, working with such organizations as the World Bank and the United Nations Development Program, they have learned about the critical value of feasibility studies, market surveys, and so on. This all suggests that substantial learning has taken place, and that gaining continued and perhaps expanded access to foreign technology on more beneficial terms is likely to remain a high-priority item on the agenda of Chinese leaders for the foreseeable future. The question about the likely impact of all of this technology flowing into China is also to remain a prominent one.

Of course, a failed modernization program, where foreign technology is not effectively employed, could have strong, negative consequences for China as well as for East Asia. Ironically, the attempt to alleviate this possibility may lead to the undoing of some of China's present reforms; as many other countries are finding, strong statist policies may be needed to succeed in a world where expanded government intervention, especially with respect to technological development, is leading to a more neo-mercantilist environment. Moreover, rather than proceeding in a broad-based fashion and seeking to achieve a comprehensive "great leap forward," better industrial and technological targeting in order to achieve a more optimal use of resources will become increasingly necessary, as will the acceptance of a more modest pace of modernization in all areas.

CHAPTER THIRTEEN

Acquiring Foreign Technology: What Makes The Transfer Process Work?

ROY F. GROW

When Song Weimin visited the 3M Corporation in Minneapolis, he noticed things that the other members of his delegation from Shenyang had not. All of the delegation was fascinated by the company's equipment and machinery–the vast array of scientific paraphernalia that abounds in most of 3M's work sites. One of Song's colleagues whispered, "If only we had 1 percent of what is here we could change our entire province."

Song was impressed by the equipment, but he was even more excited about some of the methods used by the corporation to acquire new product and process technologies. He had talked with several of the firm's "corporate scientists"–individuals given special authority to design projects on their own, purchase the equipment they needed, and use the results as stepping stones to even more complex projects. He noticed that 3M had developed a corporate culture that gave free rein to the innovative attitude of these scientists, making such an attitude an integral part of the process by which the corporation acquired new technologies. As one of the corporate executives explained to him, 3M's philosophy was to let their bright and committed corporate scientists seek out and bring into the corporation the new skills, formulae, and equipment—the technologies—that they saw as relevant to a particular project they had in mind. The underlying logic of the philosophy, said one scientist, set 3M apart from those companies that acquired new technologies in a more traditional fashion and helped to account for the firm's technological "cutting edge."[1]

What Song saw at 3M brought to mind a situation that concerned him: How was it that new foreign technologies—machines, scientific formulae, management strategies—were arriving in China in such an uneven manner and having such an uneven impact? He knew of a number of success stories: factories that were beginning to produce world-class products with Japanese machinery, farms increasing their output by factors of three or four with Dutch strategies, hospitals using American electronic gear to bring dramatic improvements to their patients. But he also knew of a number of failures: computers sitting idle under plastic covers, equipment for a planned communications project gathering dust in a Shenyang warehouse, steel-factory renovations in Anshan that couldn't quite get off the ground. Song wondered whether the principles that explained technology acquisition at 3M might be relevant also for China.

The literature on foreign technology transfer offers several explanations for the varying degree of success that foreign technologies have met with in China. Most of what has been written emphasizes the importance of the "mesh" between the incoming technology with the requirements of the receiving side.[2] Kojima, for example, has argued that technology transferred from Japan to a developing country such as China is often more "appropriate" than the more sophisticated technology from the United States.[3] Other analysts, such as Wells, outline the differences between capital-intensive and labor-intensive technology transfers and maintain that the latter are more relevant to Asian markets with their large labor pools.[4] Still others speak of the importance of central-government decisions.[5] These and other examinations of technology transfers share a concern for what happens once technologies have been ordered and put in place.[6]

What Song learned at the 3M Corporation offers another possibility—that one of the major factors determining the success or failure of foreign technology-transfer projects involves the *process of acquisition.* The *procedures* by which foreign technologies find a home in the new Chinese economy may be more important than such factors as appropriate mesh and sophistication of the technology.

In the course of more than ten years of work investigating more than 65 Japanese and 40 American firms involved in technology-transfer projects in China, I have interviewed several hundred corporate and government officials who have been closely involved in these projects. Most of these interviews took place over a sustained period of time, and in a number of cases I worked as a consultant for the companies or became part of the corporate planning process. Altogether, the firms in my study were involved in almost 250 different projects with Chinese end users.[7]

I have found that China's acquisition of foreign technologies is shaped by (1) a multi-tiered *process* that includes a number of distinct and separate stages, and (2) a group of *key players* whose activities shape the workings of the different stages. An important part of whether the foreign technology transfer process works or not in China depends on how well the key players interact with each other and how successfully they negotiate each stage of the acquisition process.[8] The decision of the Chen County Canning Plant near Dalian in Liaoning province offers an interesting case study that illustrates the interaction between players and process in a technology transfer in China. The *process* and *players* of this case are in many ways typical of other technology-transfer projects I have studied. The same kinds of hurdles and obstacles encountered by this project were present in virtually every other episode of foreign technology transfer I have examined.[9]

THE CHEN COUNTY CANNING PROJECT

Chen county lies 20 miles from Dalian. The county is in the middle of one of Liaoning's most productive agricultural regions, and the area's farms are known for their excellent harvests of vegetables and grain. The Chen County Canning Plant, however, was in rather bad shape. The plant dated from the 1950s when Soviet engineers had attempted to design a food-processing project for the area. A good part of the plant's output was apparently destined for Soviet Siberia as part of the repayment schedule for earlier Soviet loans used in China's "rehabilitation" projects.

The plant had become one of the important employers in the county by the 1980s and supplied canned vegetables to four separate distributors in different areas of China. The plant itself employed several hundred workers, and a good many more individuals were involved in the transportation of raw materials to the plant or in the shipment of the final product to other parts of the country. Most important, over the course of three decades a good part of the county's agricultural output found its way in one form or another into the plant's canning operation. The canning plant was one of the focal points of the region's economic activity.

The plant's processing techniques, however, had not changed much from the original technologies introduced in the 1950s. The system involved the acquisition of locally grown vegetables, preparation of the vegetables by hand, and a washing and heating process during which the vegetables were pushed through by hand and then sealed in large metal containers. The machinery that moved the system was in frail condition—breaking down every three or four days. More disturbing, the wasted resources—both raw and canned—were enormous. The manager, Mr. Wang, thought that perhaps 30 percent of all the cans were unusable because of improper seals, temperature controls that often were too low, lack of adequate sterilization, or other problems. A foreman in charge of vegetable procurement estimated that as much as 50 percent of the raw vegetables acquired by the plant were unusable in the canning process owing to poor harvesting and transportation techniques. The factory, if it was going to continue operation, needed an infusion of new equipment and procedures.

Deng Xiaoping's new agricultural policies had a major impact on the county. As Chen county implemented a form of the family-responsibility system, there were significant production increases. In most categories of crops—but especially in sweet corn and other vegetables used in the canning process—the area's output began to jump dramatically in 1983. Farm incomes increased correspondingly, purchasing power was up, and shops were moving a great many items from their shelves. Still, a sense of uncertainty and vulnerability hung over the area. The uncertainty was a residue of some of the unpleasant "aggressive" political activity, primarily the result of the actions of some young Party workers in the previous decade. There was a feeling in the county that this sort of "unreasonable" power might return and perhaps the economic good times might be fleeting, especially if the area's major employer—the canning plant—could not be improved and made more efficient and competitive.

Although Mr. Wang was committed to finding some way to refurbish the canning operation, the new planning and allocation rules provided no source of funds for such an undertaking. He had discussed his hopes with a group that included several of the more successful farm families in the area, county officials, several members of the old commune administrative organization, and Mr. Li, the county magistrate.

From their discussions, they concluded that to make any significant changes in their operation, they would have to find outside help. Magistrate Li reported that one of his officials had attended an agricultural conference in Dalian, where he had been impressed with the attitude and expertise of several of the Dalian officials he had met. They seemed an energetic group, committed to a process of change and development and to aiding groups in the greater Dalian area that were looking for solutions to specific problems. Perhaps in Dalian, Li suggested, there would be some help or advice for Chen county's problem.

The Dalian Development Program

Dalian has been a city on the move since the early 1980s. An energetic group of Dalian officials had become the nucleus of a push to make Dalian's greater municipal area the hub of Liaoning's commercial activities. In contrast to other areas, this group did not emerge from any single bureaucracy or agency of the municipal government. It consisted of a rather loose coalition of individuals scattered throughout some of the municipal organizations, research centers, trading corporations, banking institutions, production organizations in and around Dalian, and some provincial ministries.[10]

Those who came together were attracted more by each other's attitudes and training than they were by anyone's formal position. The group included the following:

Scientists and technicians, especially those who had studied abroad. Since the early 1980s, Liaoning and Dalian sent many students to study in foreign countries, and the returning students were becoming a force for change in a number of different agencies.

Economists and statisticians who had adopted new tools and used increasing analytic power in their work in the economic bureaucracies.

Managers trained in new techniques of decision making, organization, cost accounting, and production technology. Large numbers of Liaoning's factory and agricultural managers have studied either in the United States, Japan, or the Soviet Union. Even larger numbers are currently being trained in China by foreign university and corporate teams.

Purchasing organizations and representatives who had become an extremely important part of Liaoning's general economic scene in the first half of the 1980s. Most of the province's firms, bureaucracies, and service organizations now employ their own purchasing representatives to act on their organizations' behalf or in conjunction with some larger provincial and national purchasing organization.

The individuals from the above groups were scattered throughout the city's ministries, municipal organizations, research centers, trading corporations, banking institutions, and production organizations, and often they assumed the coloring of the institutions they represented. But they increasingly stood on one side of a generational "fault line" that separated them from some of the older bureaucrats identified with the "problems" in the province.

Many of these men shared the belief that Dalian could become the Northeast's premier commercial center. In their discussions, they had listed a series of priorities that could push Dalian toward economic prominence: a new technological base, increased domestic and international trade, new management techniques for some of the area's major employers, and new bureaucratic structures such as those of Japan's Public Policy Corporations. Such a program, they believed, could make Dalian a focal point for the region's economic development.

How to turn these visions into reality? The Dalian group had worked hard to stimulate projects in a number of different areas; they pushed the new Shenyang-Dalian highway system and lobbied for the harbor renovations that could be undertaken with Japanese support. They also had run a series of programs designed to inform groups about the new economic opportunities that existed in the city and surrounding areas.

The actions of the Dalian group were directed less toward the implementation of new projects than toward conferences and sponsored technical seminars that might *lead* to the new projects. Some members of the group also joined a larger Chinese entourage that traveled aborad–most notably to Japan in 1983 and the United States in 1984.[11]

Chen county's Magistrate Li traveled twice to Dalian where he visited the Dalian branch of China's national import/export corporation for cereals, oils, and foodstuffs (CEROF) and had several meetings with two agronomists who were members of this informal Dalian group. The agronomists then visited Chen county and walked through the canning plant. They agreed that the canning plant could become more important to the region, but only if the entire harvesting and canning process were overhauled.

Two Foreign Proposals

Two foreign firms–the American Food Coporation from the United States and Marubeni from Japan– were important to Dalian's initiatives and, eventually, to the Chen County Canning Plant project. Each firm submitted a proposal that would have a potentially major impact on the economy around Chen county. Each firm's proposal called for the transfer of equipment and technologies that might have long-term consequences for Dalian's general development. Each proposal outlined a different course, and the decision that was ultimately made provides some clues about the variable nature of the foreign-technology-transfer process.

The American firm's initiative was largely the work of one man–Richard Spencer–a senior vice-president of the American Food Corporation.[12] The corporation is and has been one of the United States' leaders in virtually every form of food processing. Its perennial problem was finding a steady and reliable source of raw materials for its canning operations. Food-processing companies often use a "follow-the-sun" strategy–moving their sourcing operations from one area to another as crops ripen that can be used for canning.

American Food was involved in a minor way in several other areas of Asia, most notably Taiwan and the Philippines, and had some food-processing operations in these areas. But China seemed a much different sort of region. Spencer knew a bit about the country and believed that its wide range of geographic and climatic conditions raised some intriguing possibilities for a food-processing company. The thought of supplying canned goods to China to meet the changing food demands in such a market made China even more attractive to him.

When Spencer first went to China in 1983, he talked with officials in a number of central ministries about buying vegetables in China. The officials were all cordial, if somewhat distant. He reported that most of these officials did not know quite what to think about either his company or his notion of purchasing vegetables in China for sale on the international market. At each office he visited, he came away with a different set of recommendations for his project.

Even though Spencer was a senior vice-president at American Food, he still regarded himself as a "field man," and he prided himself on the fact that he could spot an ideal site for growing a particular vegetable from a mile away. He liked nothing better than walking through fields, using his practiced eye to spot potential sites for new crops. He needed no tests or measurements to spot such sites—he simply recognized them intuitively and had a track record to confirm the soundness of his intuitions. He believed that, if he could actually walk around a number of farm areas, he could find some good sites for the vegetables he needed.

At Beijing CEROF, Spencer talked to an official who knew of the Dalian group and its interests in developing the region. The CEROF official suggested that a trip to Liaoning might prove interesting to Spencer, since the province possessed the sort of soil and climate similar to those in the great American corn belt. The official put Spencer in touch with the Dalian group. He liked what he saw of the Dalian area, particularly the part of Chen county that he was shown. It was a good site for several types of vegetables needed by American Food, and he believed that a strong pilot project could be begun there. During his first day in Chen county, Spencer talked for over eight hours with Chen county officials and Mr. Wang, the canning plant manager, about his ideas. He proposed improved seed stock, different harvesting and storage methods, and some new transportation facilities.

Spencer's negotiations within American Food are a story in itself.

American Food had not committed itself to any sort of a China project—that was Spencer's idea. Many in the company questioned the logic of having a company without much foreign expertise begin an operation in a land so alien. The most common question posed was, "We have enough sources as it is—why take on headaches in a new and unknown market?" Spencer responded that he saw the Dalian-area project as part of a larger strategy for entering the Chinese market with other American Food products. Some of the vegetables would be sold in China as well as transported to and sold in the United States. He finally received a sort of tacit approval from his company's officials and was allowed to pursue the Chinese negotiations further. But he faced some tough bargaining sessions during the next months—both at home in his company and in China.

At Marubeni, the Japanese trading company that was engaged in a worldwide exchange of commodities of almost every type, there were also plans for the area. The trading of different products at Marubeni is handled on a divisional basis—a separate division for each category of product. Marubeni operates by giving its divisional managers a considerable amount of latitude in buying and selling products around the world. Periodically, the heads of all of the divisions and financial units meet in Tokyo to exchange information, discuss each division's financial position, and coordinate future activities.

At these meetings in 1983, several of the divisional managers were concerned that increased American protectionism toward certain goods and commodities, such as automobiles, farm equipment, and motorcycles, might spread to other manufactured goods where Marubeni had greater stakes. Marubeni also was an agent for American agricultural goods imported into Japan. While the American situation had some ominous implications for Marubeni, it also held out the promise of opportunity. Why not approach the China market with the idea of replacing some of Japan's agricultural imports from the United States with Chinese agricultural products in exchange for opening Chinese markets to those Japanese manufactured goods that were handled by the trading company?

Marubeni explored this possibility by focusing on two agricultural products. The first was corn—especially feed corn for Japan's cattle industry. China's Northeast, where Marubeni had long historical ties, appeared to offer an ideal site for purchasing such corn. Not only was the area one of China's most productive; it had the added advantage of the Dalian port, a logical locus for a point of export to Japan. Even though the port was notorious for its month-long delays in getting ships in and out of its harbor, Dalian was much closer to Japan than Baton Rouge, Duluth, or Seattle. The long wait in Dalian's ports might be more than compensated for by the shorter distance and lower fuel costs. And, if the Dalian port could be improved—perhaps with Japanese development loans—Chinese corn might indeed become a substitute for some part of the American supply.

A second project promised to be more difficult but more lucrative: raising fresh fruit in China that could be harvested, processed, and rushed to the Japanese market. The Japanese demand for fresh fruit had grown considerably since the mid-1970s, and had broadened to include almost every type.

A Marubeni analysis pointed to strawberries as the most likely product for a first test. Demand for the fruit was increasing, the plants themselves took little time to mature (compared to, say, grape vines or cherry trees), and there was as yet little competition in their production. But who would grow the strawberries? Marubeni had a number of variables to consider in the calculation. The Chinese sites had to be near good transportation, and the work force had to be reliable and capable of using hothouse technologies. Local enthusiasm for the project also had to be strong.

After four months of study, the Marubeni team came up with a series of general guidelines for their project. The strawberries would be transported to Japan by air, so proximity to an airport was important. Hydro-technology would be used, so a dependable source of water was required. First-stage processing (washing, and perhaps quick freezing) would be done on site in China to preserve the quality and taste of the product, so the necessary technologies would have to be transferred to China, probably from a Marubeni operation already underway in Central America or the Philippines. Using these general guidelines as a basis, Marubeni analysts identified three Chinese regions that seemed appropriate for the project: Dalian, Tianjin, and Canton, each of which offered different combinations of location, climate, and proximity to transportation. Marubeni's strategy was to contact officials in the *provincial* bureaucracies and have the provincial bureaucracies set up meetings with local officials and managers.

In Liaoning, Marubeni met first with Shenyang CEROF. The prospect of Japanese corn purchases was especially attractive to Shenyang since the province was a heavy supplier of both sweet and feed corn. Shenyang also agreed to help Marubeni tour the Dalian area for their strawberry project, even though the relationship between many Shenyang and Dalian agencies had become strained as a result of Dalian's aggressive development policies. With Shenyang's help, Marubeni officials traveled to Dalian and toured the surrounding farm areas where they talked with local Chinese farmers. When the Marubeni team forwarded their final recommendations to division headquarters in late 1983, Chen county was listed as one of two possible locations in Liaoning.

MAKING THE TECHNICAL DECISIONS

The actual decision to adopt or use a specific new foreign technology, commodity, or production process is most often made by Chinese end users. In the Chen County project, the most important decisions about foreign technologies were made by Manager Wang, Magistrate Li, and their staffs. They weighed the possibilities of different projects, balanced them against existing arrangements, and made their decision in terms of the expected consequences for their important constituencies. They did not make general decisions to "change" or "modernize" but, rather, the *specific* choices between one technology rather than another. Should it be a Sanyo or a GE cat-scanner, a Honeywell or Mitsui industrial-grade air-conditioning system, a Marubeni or American Food food-processing technology?

Spencer's visit to Chen county in 1983 created a great deal of interest and excitement there. Following his visit, a series of meetings that brought together Magistrate Li, several county officials, some farming families, and the manager of the canning plant, Mr. Wang, were held in the community assembly hall to discuss the general outline of the American Food project. Manager Wang believed that the county's economy had a number of different needs. The most obvious was machinery for the canning plant itself and new processing technologies. He knew that Chen County's technology wasn't even close to that of several other exporters of canned goods. He had also read of several American companies interested in China projects, Del Monte and Beatrice being the two most prominently mentioned in

CEROF literature. Indeed, he had seen Del Monte canned fruit and been impressed with the taste and appearance of the product.

Manager Wang and Magistrate Li also believed that canning technology was not enough. A whole new range of inputs was needed—seeds, harvesting machinery, transport vehicles. Chen County's high rate of wastage and the comparatively low quality of its products could certainly stand some change. Wang had several discussions with Dalian officials, and together they decided to propose a combined growing and canning operation that could produce an export-class product. If such a facility could be created, it would generate employment and a more balanced, long-term economic outlook for the community. Dalian CEROF had asked Beijing CEROF for information on American Food, and Beijing had forwarded brochures, reports, and advertisements from American magazines. They learned that the company was an American leader in a number of food-processing areas, had a powerful track record in high-quality production, and was highly advanced in canning and other food-processing technologies. American Food technology thus could become the basis for renovating the entire range of the Dalian area's agricultural production and food processing.

Wang decided to approach American Food with two proposals. First, the company might consider using the county as a source of vegetables for worldwide consumption. Second, the company might consider the county as a possible site for a canning operation of their own— run as a joint venture with the Chen County Canning Plant. The quality might even be high enough to make the Chen County plant competitive with the planned Del Monte operation in Canton.

Spencer made an extended visit to Chen County in early 1984 and held several meetings with Wang, who found Spencer easy to get along with (for a foreigner) and genuinely interested in Chen County's modernization. Spencer liked to wander across Chinese fields, wasn't afraid of getting his shoes dirty, had farming experience himself, and was asking the right questions about Chinese rural life. But, as Wang talked with Spencer, he was aware that their agendas were different. Spencer proposed a relationship whereby American Food would help Chen County set up trial projects for growing several vegetables needed by the company. American Food would supply know-how and direction, management expertise, and new seed stock; the company would also commit itself to purchasing a certain amount of the output. Chen County would supply land, transportation, labor, on-site management, and would handle all the bureaucratic regulations with Chinese agencies. If the project worked, Spencer suggested, American Food would then consider a second stage—a joint venture for a full-fledged canning operation. Spencer also raised several questions about entry into the Chinese market, asking about distribution channels and trademark protection.

The American Food offer was not quite the integrated planting and canning operation envisioned by Magistrate Li, Manager Wang, and Dalian CEROF, but it was a start. If what Dalian CEROF said about American Food was correct, the American Food proposal could become the basis for a larger project in the future.

The Marubeni project was a different story. Shortly after the first discussions with American Food ended, Shenyang CEROF informed Magistrate Li that a Marubeni team would again visit the area. The Japanese team was still interested in the strawberry project and wanted to begin some preliminary discussions. Wang and two county officials met with the Marubeni team and discussed the details of the project over a four-day period. What Wang learned both excited and worried him. The Marubeni project seemed even more substantial than what American Food had in mind and held out the promise for important gains for the area: permanent installations, new technologies, and an almost guaranteed market for the product in Japan.

The Marubeni project, however, had some traps. On the technical side, the strawberry project seemed more complex than the American Food plan—perhaps because the strawberry-growing technology itself was alien to the area. Marubeni's demands—water, transportation, expertise, tight time schedules—also seemed more burdensome. Then there was the organizational question: Who would manage the strawberry project? Spencer had talked directly with Wang and seemed to view him as the focal point of American Food's effort. But Marubeni was talking about a *new* installation, perhaps making it necessary to bring in an outsider to manage an operation that might even be independent of the Chen County Canning Plant. Finally, on the personal side, was the unfortunate fact that many of Chen county's citizens remembered the very difficult occupation of the area before 1945 and still regarded the Japanese with some degree of suspicion. Nevertheless, while the drawbacks were great, the project held out the promise for real benefits to the area, especially if the American Food project did not materialize.

NEGOTIATING THE CHINESE BUREAUCRATIC MAZE

Since both the American Food and Marubeni projects were potentially attractive but also somewhat worrisome, weighing their relative merits was an extremely difficult task.

The initial discussion of all of the issues involved Wang, his staff, and those most closely involved in the canning plant's operation. Magistrate Li sat in on several of the discussions, but, since the canning plant was the county's largest single employer, and its staff the most familiar with the technologies under debate, Magistrate Li believed Wang to be the most qualified to evaluate the proposals. Occasionally, outside experts and officials were consulted for advice, but most of the mulling over of these options was done by Wang and the Chen county officials. Since they were to be the primary users of this particular set of technologies, they would decide its relevance for their particular needs.

Negotiating within the Plant

Wang's first talks were with his own people—primarily the foreman and engineers in the plant itself. Most were already familiar with the general outlines of both the American Food and the Marubeni projects and had developed some strong positions about each project.

One group, made up mostly of the younger technical staff-people, saw the long-term possibilities of the relationship with American Food. They were intrigued by the access to Western expertise that the project offered. They envisioned a large canning operation, as well as the chance to learn about a whole range of new food-processing technologies. The American Food project also had the support of Chen County Canning Plant's procurement people—those responsible for actually acquiring vegetables for the existing canning operation. They, too, were attracted by the benefits of access to American Food technologies, although in this group they tended to focus more on what they could learn by being part of American Food's "sourcing" operations—the growing and harvesting of export-class vegetables.

There was, however, an undercurrent of worry in all of the discussions. The primary concern revolved around the employees' questions about their own limitations: Would American Food introduce methods far beyond the capabilities of the plant's personnel? And, if so, would their jobs be threatened?

Since it could provide a near certainty of an ongoing strong, stable Japanese market, Marubeni's primary appeal was economic. But the Marubeni project had little support within the plant, because virtually everyone saw enormous problems for the plant if the project were adopted. The technology was new and alien to the area, control over the operation itself might fall primarily to Japanese technicians, and the drain on the region's resources would be great. Magistrate Li had a different concern. He worried that Shenyang's support for the Marubeni project could clash with Dalian's interest in the American Food proposal, putting Chen county in the midst of the growing Shenyang-Dalian rivalry.

The Marubeni project was based almost completely on a set of initiatives developed *outside* the county. As a result, the Marubeni project did not meet most of the hopes that many in the county had for their area, and was seen as a threat by some. By contrast, the American Food project took shape as a result of prolonged negotiations with the Chen county officials and Manager Wang and seemed to mesh more with the hopes and plans of the area's leaders. Almost everyone—from Wang to his foremen—believed that if only one project were to be attempted by the plant, it should *not* be the Marubeni strawberry project. Within the plant, then, there was a decided tilt toward the American Food proposal.

Negotiating with the Outside World

The discussions with local constituencies *outside* the canning plant—the farmers and laborers who would be most affected by the new project—were the most difficult. The American Food project called for a substantial amount of acreage to be planted in new vegetables. Since Wang estimated that as much as 10 percent of the county's farmland would be affected in some way, the project raised basic questions: What land was to be used? How would it be paid for? And what guarantees were there if the new (and untested) seed failed? Wang proposed a loose amalgamation of farms: some that would sign contracts and would be paid in cash at harvest time, and others already under contract to the canning plant and for which additional labor would be hired.

Wang faced some difficult demands. Those with land wanted a definite commitment before they moved to a new crop. Those without land feared that a successful American Food project would make their chance for a piece of land in the future more remote. But there were also families who saw a part of their future, as well as that of their children (who yearned for the greater opportunities afforded by city life), tied to a successful canning operation that would hire more workers and provide a higher standard of living.

Wang had to negotiate a host of additional and equally difficult issues with "outsiders," such as:

-Water supplies for emergency irrigation of the vegetable crop

-Fertilizer for both the corn and the asparagus

-Tractors that would break the ground in a new way, to be paid for by American Food

-Gas and oil for the tractors

-Electricity for the new washing and packing procedures

-New buildings, such as sheds, for the tractors and temporary storage for the harvested vegetables

-Labor rules for newly hired workers: wages, job assignment authority, and the right to hire and fire

-Land consolidation permits, to allow the joint tilling of separate pieces of ground

-Repair of the road leading to the canning plant

-Vehicles and transit permits for transporting the vegetables to dockside in Dalian.

Another potentially charged situation came when Wang briefed his existing customers about the American Food project. The most delicate point centered on the canning plant's continued reliability as a source of canned vegetables if the new project indeed materialized. Wang had to be very careful, because to lose his existing customers would be dangerous, especially if the American Food negotiation fell through. Wang could offer these distributors enhanced quality and variety that would make their products competitive with those of some of the southern units, but he could not answer some difficult questions, such as those about cost. He simply did not know what his products might cost in the future. Above all, Wang believed that the new project would benefit all concerned and would cause little disruption to his meeting supply deadlines for his longstanding customers.

The process of talking out all of these issues with so many different agencies and officials tested Wang's skills to the utmost. His hope was to keep the situation fluid by not pinning himself down to a specific commitment or need. His basic strategy was to persuade some of the officials he was dealing with of his own vision of the area's future by holding out the promise of general benefit—greater income and employment and access to Western capital.

To gain minimal consensus, Wang had to engage in five bargaining processes, almost simultaneously. First, there were the talks within his plant, which had to be managed carefully if the plant itself was to continue to run smoothly. Second were the discussions with the Chen county residents over their interests and wishes. Third, was the bargaining with American Food over their proposal. Fourth were the delicate and complex negotiations with both upstream and downstream users. Finally, there was management of the potential Shenyang-Dalian tension so that his project would not be caught in the crossfire. Bringing the five negotiations together was an intricate juggling act.

There were moments of real despair for Wang, when he was certain that all of the negotiations would collapse. The differences in the Chinese and American perspectives showed most starkly during the contract negotiations. While Wang was trying to keep his own bargaining situation as fluid as possible, American Food was talking about a contract that actually spelled out all of the details on a piece of paper. Wang was interested in building a personal working relationship that could be adjusted as the details of the new situation took shape; Spencer needed guarantees, numbers, and commitments he could take back to American Food for the corporation to weigh both in absolute terms and against possible sites in other parts of the world. There were also smaller problems. American Food wanted crops, but for how long? One season or five? Would they guarantee a purchase price? What if American Food folded their operation? Who would then be stuck with the commitment Wang had made to the Chen county residents? And could American Food give him the guarantees he needed without pinning him too tightly into a contractual straightjacket that would inhibit maneuvering with his other suppliers and users?

GAINING APPROVAL FROM "HIGHER-LEVEL AUTHORITIES"

Most Chinese enterprises looking for foreign technologies have some sort of relationship to local, provincial, and national agencies and bureaucracies. Some of these agencies are influential and can have a life or death impact on the proposals under consideration. Others are little more than one-room offices with a new sign on the door (perhaps wishful thinking on the part of an official somewhere in the bureaucracy).

In none of the enterprises I studied did *one* agency control *all* aspects of an enterprise's life. Most often the situation is one in which a number of agencies and organs have a limited influence over some aspect of an enterprise's needs. In a new project for a steel complex, for example, CAS might have some say in the assignment of scientific personnel; the municipal power authority might affect the outlook for new electricity needs; a provincial ministry might issue transit permits for equipment; a Party organization might negotiate work rules. The days of central resource and service allocation are fast fading, and a good manager learns how to bargain with many different groups and agencies at the same time.

Such was the case for Manager Wang and the American Food Corporation proposal. Six months after the proposal had been submitted, Wang and his team had worked out a rough agreement with American Food and had discussed the agreement with the canning plant's upstream suppliers and downstream users. The agreement called for American Food to deliver 900 pounds of new seed, show local users how to plant and harvest the crop, pay for fertilizer and other inputs, and take delivery of the vegetables upon harvest at Dalian.

Dalian CEROF seemed content with the terms of this agreement and enthusiastic about the potential for future growth in the Chen County-American Food relationship. One official believed that, if American Food could be enticed into a more general joint venture, the Chen County operation had a good chance of competing with some of the other canning operations elsewhere in China—especially in the Northeast market, and possibly even for export to Japan and Soviet Siberia. Everything that Chen County had agreed to fit nicely into Dalian's plans for the area. Shenyang was less happy. Marubeni had been interested in the Chen county site. Wang's agreement with American Food had complicated Shenyang's discussions with Marubeni on the corn issue, and Marubeni was now talking with groups in Heilongjiang.

Wang, on behalf of the canning plant, could have initialed the agreement himself, and American Food would have been satisfied. Because there was very little in the way of Chinese funding up front, he did not have to bring any of the area's financial institutions into the process. And, since he had worked out a rough agreement with most of his upstream and downstream constituents, no further negotiations seemed necessary. But American Food made one final demand that was most difficult for Wang. The company wanted "exclusive rights" to all the vegetables of that particular variety that were grown in the province. No other Chinese unit in the province would be allowed to sell similar products to any other foreign firm.

This American Food demand brought Wang into a complex negotiation with "higher-level" bureaucracies in Dalian, Shenyang, and Beijing. The discussions with Dalian went easily. While Dalian CEROF could make no such guarantee to American Food about other areas of the province, they certainly seemed to have little difficulty in approving the request in principle.

Shenyang was a different story. The tension between Shenyang and Dalian had deepened at virtually every level of interaction. This tension was especially pronounced after Dalian received its special-economic-zone status in 1984, a move that seems to have caught many Shenyang officials by surprise. Shenyang CEROF was also upset over the handling of the Marubeni strawberry project which, they believed, provided a more substantial basis for economic development in the province than the relationship with American Food, especially in light of Marubeni's desire to purchase large amounts of Liaoning corn. Since Wang needed Shenyang's agreement for the exclusivity provision requested by American Food, he found himself in a very difficult situation. Shenyang CEROF simply refused to sign off on the American Food request for exclusivity in the province. Regardless of whom he approached, or which channel he used, the answer was always, "We have no authority to issue such an agreement."

Thus Wang found himself in Beijing. After almost two months of trying, with Dalian's help he received an appointment with one of Beijing CEROF's deputy directors, who saw the logic of Wang's and American Food's request and was sympathetic. However, he was apparently unwilling to come down on one side or the other in the Shenyang-Dalian struggle. The best he could do was to issue a somewhat nebulous statement which, when translated into English, gave the appearance—but not guarantee—of exclusivity. Wang could only hope that the statement would be acceptable to American Food.

The procedures for obtaining an official imprimatur for a project are some of the most difficult for outsiders to comprehend. Understanding this process means distinguishing between substantive decisions that produce results and administrative approval that confers legitimacy. In the Chen County project, Manager Wang's decisions about *specific* technologies were separate from Dalian CEROF's approval of the *general* project. Wang negotiated specifics; Dalian ratified his general choices and stayed away from the technical discussions; and Beijing "took note" of the general process. It is easy to confuse or blur these steps. Many foreign businessmen (especially Americans) believe that individuals and organizations that oversee some part of the general administrative process actually make the decisions about the new technologies themselves.

There is in many Chinese technology-transfer projects a rough distinction between decisions about specific technology on the one hand and ratification and acquisition on the other. For example, in Anshan, another city in the northeast, technical decisions about new machinery are made at the level of plant superintendent and ratified by the larger Anshan directorate. The machinery is then acquired through a series of bureaucracies, and MOFERT's CUTC. Similarly, the Canton Hospital Association makes decisions about new technologies, using a committee structure that includes doctors who will actually use new medical equipment. The purchases are funded through a consortium arrangement that funnels funds from several hospitals into a joint account, and then purchases the new equipment through a provincial-level health ministry.

The key for the successful Chinese enterprise manager is knowing the point of access. *Guanxi* (connections) continue to be an important bureaucratic fact of life in China, but there is more at work here than a simple "Whom do you know?" or "Who is your patron?" Successful negotiations require knowledge of the *levers* of influence—knowing why a bureaucracy or individual bureaucrat might want to act in a particular manner and getting a grip on the questions of "What's in it for me and mine?"

The most difficult waters for a Chinese user of foreign technology to navigate are those where competing agencies have some say in the disposition of a project. Manager Wang's negotiations were complicated because he was caught between the Shenyang bureaucracy and Dalian CEROF. China's economic reforms increase competing demands on local project managers. But they also offer multiple points of access to what otherwise might be a monolithic and impenetrable bureaucracy. When Wang was frozen out of Shenyang, he used different channels to reach Beijing.

IMPLEMENTING THE TRANSFERS

The agreement with the American Food Corporation was signed in mid-1985. Dalian CEROF, acting on behalf of the Chen County Canning Plant and Beijing CEROF, initialed the agreement for the Chinese side. But the agreement was only the beginning. The make-break point for any technology-transfer process occurs during the actual implementation, when the differences between expectation and reality are faced directly.

The track record for technology-transfer projects in China is spotty. Projects range from successful new laser and food-processing technologies to some real disasters in heavy-assembly operations. The failures generate the most agonizing appraisals. Chinese managers, for example, sometimes become so entranced with new machines and processes, and so wrapped up in making them work, that they are late in discovering the possible negative consequences of the new processes, such as higher costs, interrupted schedules, or unhappy personnel. At the other extreme are projects that get bogged down in a buck-passing misperception of where a project's difficulties lie. Chinese end users, faced with unexpected difficulties, tend to reject the notion that a problem might lie on site or result from something they neglected in the negotiations. Instead, such managers may blame delivery shortcomings or suppliers who fail to live up to contractual obligations.

Even successful technology-transfer projects often generate agonizing reappraisals. Does one judge success by the rapidity with which results are obtained or by long-term potential? Are economic considerations, such as return on investment and return on sales, the sole criteria by which a project's success is judged, or are social considerations (number of people employed) equally important? Almost always there are unintended consequences and unforeseen ripple effects in a technology transfer that can shake even the most confident manager. Lives of real people in real places—often the economy of an entire county or province—can be dramatically altered by the introduction of new ventures and new technologies, as the story of Mr. Wang and the Chen County Canning Plant illustrates so vividly.

There were some immediate and tangible benefits from the Chen County project. American Food proved to be a good teacher, especially in the technologies needed for growing the new vegetables. The company provided new varieties of seeds, advice on fertilizers, information about harvesting techniques, and aid in barreling the harvest for shipment abroad. The nuts-and-bolts part of the new agreement seemed to be working so effectively that Wang unexpectedly took a risky gamble. Since American Food seemed to want the vegetables so badly, he increased the amount of acreage to be seeded in new vegetables by 30 percent. If the crop could be sold, it would provide a great deal of new income, which Wang pledged to Dalian CEROF. He in return received a loan to construct the first building for his cherished new canning operation and began construction before the first harvest was complete.

When additional income did not materialize, Wang quickly ran into the realities of the American corporate world. Since there was a "glut" of that particular vegetable in 1985, American Food could not use as much as the Chen County operation had grown. While American Food would stand behind its commitment to buy the quantity contracted for, the "world price" of that vegetable had fallen dramatically in the previous several weeks. Moreover, American Food was not certain that it could use the crops from Wang's additional acreage.

Not being able to dispose of all they produced was a novel experience for Chen County. Their problem in the past had always been just the opposite—more was wanted. To be paid lower than expected was an even greater shock. How would they take care of the expenses?

The questions came quickly from Chen county farmers and residents. Didn't American Food know that they had an "obligation" for the county's welfare? Who would pay for the expenses associated with the excess production of vegetables such as the water and the gasoline that came from other sources? Who would pay for the new building? Some of Wang's downstream users asked whether Chen County's overproduction meant that the products they expected for their distribution networks would still be on time?

Wang's set of relationships seemed constantly on the verge of shifting from underneath him. He wondered if the arrangement with American Food would hold and if it was really worth all of his time and energy. He had no real clout with the corporation if it chose to back out of the relationship. He had only a contract that might take years to enforce, *if* he had the time or expertise. He had discovered that dealing with Western corporations and using their new technologies created a number of ripples in his pond that threatened to tip over his old ways of doing things, and might even drown him. Was he really in control of events or were events in fact controlling him?

SUCCESSFUL AND UNSUCCESSFUL CHINESE TECHNOLOGY-TRANSFER PROJECTS

What does the Chen County project say about the processes by which foreign technologies are acquired in China? What factors are most responsible for success or failure? Why do some projects succeed and others fail?

Details of technology-transfer projects vary enormously. Some projects, such as auto-assembly operations, steel-modernization projects, and turnkey projects, are large enough to come under the direction of Chinese "higher authorities." Others, such as seismographic-equipment-testing projects, bicycle-assembly operations, food-service and hotel developments, are small and do not involve high-level agencies.

Those foreign-technology acquisition projects that succeed almost always pass through a decision-making process that takes place in identifiable stages:

First, decisions are made about the *specific* technologies needed on the "demand" side and those that can realistically be offered on the "supply" side. Most often these decisions are made by the organizations most directly involved, the Chinese enterprises seeking foreign technologies and the foreign firms offering them.

Second, consultations and negotiations typically take place on two levels: discussions within the Chinese enterprise, as various groups came to grips with the implications of the technological innovation; then discussions between the Chinese enterprise and its constituencies about changing requirements necessitated by the new technologies (such as more electric power or fuel) and the productivity to expect from this technology.

Third, the above decisions have to go through a process of authorization by "higher authorities." It is here that decisions are reviewed and new arrangements ratified. Finally, the users put their new technology to work and evaluate the results.

If any stage is missed or completed in a partial or slipshod fashion, the acquisition process almost invariably falls through, regardless of the appropriateness or "fit" of the new technology.

Yet, as the Chen County project demonstrates, the stages in the process are only part of the story. Another important key to success lies with the *actors* in the process and how they play their parts. At the core of the entire technology-acquisition process are three distinct groups. First, there are the managers and chief engineers of the 500,000-odd Chinese enterprises in the Chinese system. It is they who oversee any decision to use foreign goods and technologies and who usually decide about the *specific* nature of the new technologies. Second, there are the staff personnel in the local and provincial regulatory agencies, planning organizations, and administrative bureaucracies, who retain a limited jurisdiction over the flow of new commodities, technologies, and personnel. Chinese enterprise managers and chief engineers must consult with members of this group in their search for foreign technologies. Third, there are the bureaucrats and state planners who work at the national level of complex political controls and fuzzy lines of authority and who view the impact of new technologies from a more general perspective. Successful acquisition and implementation of new foreign technologies depend on the kind and style of interaction that occur among these three groups.

Yet, whether one examines process or players, it is the end users of the new foreign technologies—the production enterprises, service organizations, and state agencies—that play the most significant role in the process of foreign-technology acquisition. And within these units, it is the decisions and actions of *managers* that determine the workability of entire acquisition projects. How they work through the problems associated with the new technologies, analyze the benefits and costs, and bring the various players together generally spells the difference between success and failure.

As Chen County's Manager Wang illustrates, the role of the manager is complex. The best managers try to deal with active and often competing needs and desires both within their units and with those outside groups whose cooperation is necessary to the unit's work. Ripple effects from the new technologies extend far beyond the basic production unit. The need for raw materials and parts from other units combined with changes in the delivery schedules caused by new production processes make negotiations complex. If production schedules are disrupted or if prices change or quality is altered, users must be consulted and their approval obtained. Chinese managers also must deal with political authorities in the bureaucracies, both national and local, that shape the Chinese economic scene. Competing demands among these different bureaucracies and agencies place the manager in a precarious position. He is both the most vulnerable and most important cog in the entire technological-acquisition process.

How typical is Chen County's Manager Wang of successful managers of foreign-technology acquisition projects? Chinese managers who successfully move through the steps of foreign-technology acquisition usually demonstrate some combination of the following abilities:

They recognize the advantages of change. The impetus for change in Chinese units most frequently originates with some sort of unexpected crisis in the enterprise—the failure of a machine, a new demand for a different sort of production, a serious financial situation. Sometimes it occurs when an employee or local official returns from a year abroad full of new ideas. Managers successfully acquiring new technology know how to turn necessity to advantage and seize the opportunity when changes are forced on them.

They understand the potential disruption that a new technology can cause. Since a new technology may alter working relationships within a unit and therefore cause dissension, it is the Chinese managers who have mechanisms for dealing with changed circumstances or new inputs who move most easily through the different stages of the acquisition process. The mechanisms for coping with change are not remarkable. They include procedures recognized by organizational theorists worldwide: staff meetings, worker-manager forums, and formal channels for the submission of suggestions. Occasionally they also include more informal strategies such as after-hours get-togethers and group retreats. In China, as elsewhere, the real importance of these mechanisms is that they convey a sense that the concerns surrounding crisis and new situations are taken seriously.

They build good relationships with technical subordinates. While the manager is usually responsible for overall coordination of a project, many Chinese enterprises put a chief engineer in charge of technical decisions. A key variable in the success of foreign-technology acquisition projects is the nature of the relationship between the general manager and his chief engineer(s). When managerial and technical perspectives do not mesh, the possibilities of conflict are magnified, and discussion within the enterprise can become fragmented and lacking in focus. When the manager works easily with technical personnel, the acquisition process proceeds more smoothly, regardless of the "fit" of the new technology.

They are sensitive to the needs of major constituencies. Most Chinese organizations seeking to acquire foreign technologies do not exist in a vacuum: Upstream suppliers and downstream users are also affected by technological changes. Units that move most easily through the acquisition and implementation process work closely with all their constituencies, involving them early on in many of the discussions and negotiations.

They are relatively autonomous. Managers who are not too closely wedded to any one government agency or ministry seem to have greater leeway to make the changes required by new technologies. Managers with a strong patron at "higher levels" have initially greater access to foreign technologies by virtue of their patron's position. But, in most cases, the patron also keeps a tighter rein on such an enterprise, thereby limiting the manager's discretionary power to bargain and maneuver.

They maintain multiple lines of communication. Lines of authority and control for Chinese managers have changed dramatically since the late 1970s. As the process of central planning and allocation gives way to a more diffuse system of operation, Chinese enterprise managers are facing a more fluid decision-making process. The most successful managers have a multiplicity of contacts on the local, provincial, and even national levels. Many have built lines of communication with individuals and agencies that in the past would have been out of bounds to them because such communications would have crossed bureaucratic jurisdictional lines. These multiple channels of communication help successful managers in obvious ways: They provide information, expertise, and access. They also give managers options: If one channel is closed, others can be used. Having multiple points of access, however, is like having a double-edged sword. Just as it increases managers' points of access to others, it also increases the ways in which others can try to influence the managers' actions. It is a rare Chinese manager or chief engineer who, when seeking technology from Western or Japanese sources, has not been contacted and pressured by more than one Chinese bureaucracy.

They are entrepreneurial. Along with the increase in the points of access (both for and to the manager), comes a correspondingly greater amount of time spent looking for a path through the Chinese bureaucratic "maze." Success for managers nowadays necessarily involves the ability to maneuver between and among the different government bureaucracies. Chinese managers who are most effective in acquiring foreign technologies are entrepreneurial types who play multiple points of access and control off against each other and take a certain joy in the process.

Manager Wang, who demonstrated all the above characteristics, was very successful at acquiring foreign technology. But he gambled when he decided that producing more crops than he had originally contracted for with American Food would be beneficial. His decision reflects the potential pitfalls that even successful managers may face once they have acquired foreign technology. Since he was accustomed to a Chinese system that guaranteed a market for whatever was produced, American Food's surprise at what he considered a successful application of their new methods caught him off guard.

Rosabeth Moss Kantor argues that at the heart of technological revolutions is a *personal process* in which change is valued and allowed to proceed in an orderly fashion.[13] The findings of Peter Drucker lend support to Kantor's argument and suggest that the most important element in the process of change and development turns out to be the human component rather than structure or finance.[14] They both argue that the decision and acquisition stages are crucial to the success of technological innovation. If a project is not handled successfully during these first stages, it will almost certainly run into real difficulty later.

We thus come full circle back to Mr. Song and his observations about the role of the corporate scientists at the 3M Corporation. In Minneapolis, Song saw a system that emphasized process and players as much as "appropriateness" of technology or its relative level of sophistication. Similarly, in the new China, the decisive elements in successful technology-transfer projects involve both process and players. Responsibility falls increasingly on the actual end users of the new technologies–the Chinese production enterprises, service organizations, and state agencies. These units exist in a rapidly changing world of shifting priorities and volatile bureaucratic alignments. How their managers handle the pressures and requirements of technology acquisition increasingly shapes the larger direction of the Chinese economy itself.

"We aim to acquire by 1985 a comparatively advanced force in research in computer science and build a fair-sized modern computer industry. Microcomputers will be popularized, and giant ultra-high-speed computers put into operation. We will also establish a number of computer networks and data bases. A number of key enterprises will use computers to control major processes of production and management."

–Vice Premier Fang Yi,
Speech at the National Science Conference (March 1978)

"Twenty-seven units in Shanghai are directly engaged in the scientific research, production, application, and service of computers. They come under the vertical jurisdiction of [several different] ministries . . . and relate horizontally to [several different] bureaus. . . . Each unit has a certain strength, but they cannot coordinate their actions because they are administered by different 'grannies' and have different sources of income."

–Shanghai Vice-Mayor Liu Zhenyuan (1984)

"A lack of coordination in research, production and management has led to stockpiling of computers and low usage. According to official statistics . . . 40,000 microcomputers are stocked in warehouses owing to lack of trained personnel to operate them. . . . The [overall] utilization ratio of microcomputers is only . . . around 15 to 20 percent. . . ."

–China Daily, *11 January 1986*

CHAPTER FOURTEEN

DOS ex Machina: The Microelectronic Ghost in China's Modernization Machine

RICHARD BAUM

A Chinese microelectronic revolution of some note has taken place in the 1980s. Since the turn of the decade, domestic Chinese production of integrated circuits (ICs) and microcomputers has on average doubled annually; in the same period, computer imports have risen almost a hundredfold. By early 1986, China had manufactured almost as many personal computers as the Soviet Union; computer boutiques in Peking and Shanghai were selling the latest in 16-bit foreign and domestic micros; a computer dating service had opened in the Chinese capital with the blessings of the local Communist Party branch; and Chinese college students were relaxing between classes, playing "Flight Simulator" and "Space Invaders" on fully equipped IBM PC-XTs in modern computer labs. For all this surface glitter, however, China's computer revolution remains in virtual infancy, its progress severely constrained by a number of technical and developmental problems. Consequently, though the computer's potential impact on Chinese society is enormous, its actual consequences to date have been quite limited.

The first Chinese-made computers were bulky vacuum-tube models, copied from Soviet prototypes supplied in the 1950s. By the mid-1960s, Chinese engineers had designed and developed a second generation of transistorized machines; and, in the early 1970s, the first Chinese IC-based third-generation computer was successfully trial-produced. Despite rapid advances in basic technology, by the close of the 1970s Chinese computer R&D remained an estimated fifteen years behind that of the West, with the gap being most apparent in the areas of IC production, external storage, input/output peripherals, and applications software.[1]

In 1979, there were an estimated 1,500 computers installed throughout China.[2] The vast majority of these machines– including the country's main production-line models—were based on foreign designs and prototypes, reverse-engineered in Chinese laboratories.[3] Most computers in the Chinese inventory were large, expensive mainframe or minicomputers dedicated to numerical data processing and scientific computation; a smaller number were utilized for industrial process control, primarily in the defense sector.

CHINA'S MICROCOMPUTER BOOM

The onset of the Chinese computer boom in the early 1980s coincided with the spread of a worldwide revolution in microchip technology.[4] As small, low-cost 8-bit desk-top computers and microprocessors became readily available on the world market, the PRC's computer inventory jumped 5-fold in just two years—from 2,300 units in 1981 to more than 10,000 in 1983. At the same time, Chinese manufacturers began to incorporate imported semiconductors and subassemblies into their finished products. As a result, China's domestic computer output registered an 8-fold increase between 1981 and 1983, from less than 700 units annually to almost 6,000.[5] From mid-1984 to the spring of 1985, the microelectronics boom gained fresh momentum as Deng Xiaoping's newly expanded "open-door" policy (*kaifang*) resulted in a fresh wave of hard-currency imports and high-tech joint ventures. By the time the wave subsided in the last half of

1985, there were an estimated 100,000 computers in the country, the vast majority of which were desk-top PCs—approximately one-third of which had been domestically manufactured or assembled.

So rapid was the buildup of China's microcomputer inventories that, by the beginning of 1986, some 40,000 PCs were stacked in warehouses throughout the country "owing to lack of trained personnel to operate them." Beijing alone reportedly had 20,000 computers on hand early in 1986—almost 10 times more than the entire 1981 Chinese national inventory.[6] The dramatic rise in China's computer stocks from 1980 to 1985 is detailed in the Tables 1 and 2 below.

By 1985, the 16-bit architecture of the IBM PC had emerged as the micro industry standard in China, displacing the older 8-bit Apple II. In that same year, the bulk of the PRC's microcomputer production consisted of final assembly of partially knocked-down, imported components and subassemblies. Average output of finished machines at each of the country's 111 computer-assembly plants in 1984 was less than 1 computer per day, as compared with almost 1,000 a day at Apple's automated Macintosh plant in Silicone Valley.[7]

Along with the outbreak of microcomputer fever there has occurred in China a less well-publicized—but no less significant—advance in the field of large-scale integrated (LSI) technology, used in the manufacture of so-called "supercomputers." LSI research began in China in 1975, culminating in the successful trial manufacture of the PRC's first two fourth-generation supercomputers in 1983—the "Galaxy" and the "757."[8] Although China's LSI program is still in the experimental stage, and Chinese computer technology continues to lag several years behind the West and Japan in such important areas as manufacturing techniques, software, and peripheral devices, the gap has clearly been narrowing.

The rapidity with which PC fever spread to China in the early 1980s, and the considerable enthusiasm with which Chinese state planners, managers, and technical intellectuals welcomed the dawning of the new microcomputer age,[9] stood in marked contrast to the more ambivalent, conservative response evinced by Soviet authorities. Despite sharing similar Marxist doctrines and Leninist institutions, China and Russia have diverged markedly in their basic approaches to the acquisition, allocation, and end-use of computer technology.

TABLE 1 PRC Computer Inventory, 1979–1985 (selected years)

Year	*Micro*	*Mini*	*Mainframe*	*Total**
1979	(negl.)	900	(600)**	1,500
1981	700	n/a	n/a	2,300
1982	n/a	2,300	1,300	n/a
1983	(7,500 ±)	3,000	(1,600)	12,000 ±
1985	(90,000 ±)	n/a	n/a	100,000 ±

*Excludes single-board computers
**Numbers in parentheses interpolated from correlative data

Sources: Various sources cited in note 4

TABLE 2 PRC Domestic Computer Output and Total Inventory (selected years)

Year	*Annual Domestic Output**	*Total Inventory**	*% Domestic*
1981	600	2,300	26
1983	5,500	12,000 ±	46
1985	12,000	100,000 ±	12

*Includes all computers wholly or partially manufactured or assembled in China, excluding single-board computers

Sources: Various sources cited in note 4

THE SOVIET RESPONSE

To state the contrast most sharply, the Russians have generally stressed large-scale, centralized, institutionally based computer operations and applications, while the Chinese have shown a greater interest in smaller-scale, inexpensive, dispersed systems.[10] In large measure, this difference stems from the PRC's comparatively recent entry into (and hence relatively low sunk costs in) computer research–giving China the latecomer's advantage of being able selectively to bypass intermediate stages of R&D; equally important, the greater Chinese willingness to experiment with smaller, decentralized computer systems and applications may also reflect the two countries' increasingly divergent economic and administrative environments.

A Soviet penchant for technological "gigantomania" was clearly evident in the first phase of the USSR's "Unified Computer System" in the 1970s. Under this plan, Soviet microelectronics R&D was geared to the production of large- and medium-sized mainframe computers for primary use in national defense, economic planning and data-processing, scientific research, and industrial process control. By the end of the decade, at least three respectable mainframe models were reportedly being batch-produced at Soviet manufacturing facilities in Byelorussia and elsewhere. Also, by the end of the 1970s, the Soviets had designed and successfully trial-produced two so-called "supercomputers," reportedly capable of speeds of 12 and 125 MIPS, respectively, and possessing hard-disk storage capacities of 100 and 200 megabytes (MB). At the same time, a certain portion of the Soviet technological R&D budget was for the first time allocated for the design and development of mini- and microcomputers; and, by the early 1980s, at least four desk-top models were being produced in moderate quantities, with speeds ranging from 200,000 to 500,000 instructions per second and on-board memories ranging from 4 to 256 KB.

Despite impressive advances in Soviet microprocessing technology, only about 50,000 desk-top microcomputers were produced in the USSR from 1980 to 1985–compared with more than 3 million in the United States and roughly 30,000 in China. Most of the Soviet PCs were used for educational purposes, while the vast majority of "hard" computing tasks continued to be performed on large, institutionally operated mainframes located in some 3,000 Soviet "computing centers," 1,300 production control facilities, 2,000 industrial enterprises and conglomerates, and 290 central and republic ministries and ministerial branch agencies throughout the country.[11] The single largest consumer of small computers in the USSR thus far has been the school system, where the machines are used to teach computer literacy. Under the current Soviet Five-Year Plan, published in 1985, half a million PCs are scheduled to be put into Russia's 60,000 secondary schools by 1990.[12]

Computers in the USSR have always been regarded essentially as producer goods rather than consumer goods. Therefore, very few microcomputers have ever been sold on the open market. A related reason for the relative weakness of private demand for microcomputers lies in the structure of the Soviet economy, with its highly centralized production planning and its unified system of command-driven technological R&D. Within this institutional framework, the vast majority of computer-assistable functions—including scientific calculation, economic planning, information storage and retrieval, and industrial process control—are performed in relatively large, public or collective organizational units rather than in small, individual, or private settings.

Too, the USSR's Leninist political system imposes substantial limitations on the ordinary citizen's access to—and transmission of—information. Soviet leaders are extraordinarily sensitive to the problem posed by the unauthorized dissemination of dissident ideas and viewpoints. Since every stand-alone word processor has the potential to become a *samizdat* press, Soviet policymakers are constrained to restrict private citizens' access to PCs, printers, communications software, and memory-storage devices.[13]

THE CHINESE RESPONSE

While Soviet leaders have pursued a relatively conservative policy, China's post-Mao leaders have been less cautious and risk-aversive. Where the Russians have kept computers under tight central bureaucratic control, the Chinese have—at least since the early 1980s—been more willing to relax administrative restraints in the interest of promoting the nation's modernization goals. And, while China's leaders have by no means eschewed the acquisition of large and medium-sized computers for scientific research and industrial process control, they have clearly been more receptive to date than their Soviet counterparts to the more diverse applications, cost economies, and greater flexibility of the desk-top PC.

Since the early 1980s, the PRC has pursued a bold, sometimes freewheeling "open policy" of foreign-technology acquisition designed *inter alia*, to procure, as quickly and cheaply as possible, up-to-date microelectronics production techniques, equipment, and know-how from the industrialized West and Japan. Unlike the Russians, who import—often covertly—large amounts of computer hardware while avoiding more potentially "corrosive" forms of technology transfer, China has openly embraced virtually all forms and carriers of microelectronic technology—from private computer boutiques, computer fairs, and local PC users' groups to foreign franchise distributorships and software-writing joint ventures.

Despite xenophobic protest from domestic critics who periodically warn against the dangers of "slavish devotion to things foreign," Deng Xiaoping has opened China's borders to a steady stream of eager foreign investors and technology suppliers and has granted prospective Chinese end-users ready access to available Western and Japanese microelectronics technology. Coupled with a significant (if somewhat unsteady) decentralization of investment decision making in China, Deng's "open-door" policy has made possible a degree of local technological flexibility that would be all but inconceivable within a more orthodox Soviet-style centralized command economy.[14]

As a result of increased local decision-making autonomy, streamlined procurement practices, and liberalized foreign-exchange regulations adopted in the early 1980s, a large—and at times uncontrolled—influx of foreign-made computers and computer components took place between 1983 and 1985. During this period, a number of venturesome Chinese officials and aspiring entrepreneurs, often employing Hong Kong merchants as middlemen, began importing large quantities of Taiwanese and Hong Kong-made Apple and IBM PC look-alikes into China—with as many as 20,000 of the unlicensed copies reportedly reaching the Chinese market in a single month.[15]

For many years, China's efforts to gain access to foreign computer technology were hampered by a series of elaborate U.S.-sponsored COCOM export controls designed to prevent Communist countries from acquiring state-of-the-art microelectronic equipment and know-how. Prior to 1985, for example, 8-bit computers equipped with more than 256K of main memory, flat-screen video monitors, or bubble-memory systems could not be sold to China by COCOM members without prior authorization, while 16- and 32-bit machines could not legally be sold at all.[16] Then, in 1985, at the urging of the Reagan Administration, U.S. export controls were eased somewhat to facilitate licensing of selected non-strategic technologies (including certain lap-top computers and 16-bit micros) for wholesale export to China. However, the new "supermicros"–including the 32-bit Apple Macintosh and the high-performance 16-bit IBM PC-AT–remained on the banned list.[17]

Spurred on by newly liberalized U.S. and COCOM export policies and by a $200-million educational development loan from the World Bank, the wave of foreign computer imports crested in China in 1984–1985. As individual Chinese provinces, ministries, municipalities, factories, schools, universities, research institutes, and corporations began to negotiate their own individual contracts with foreign computer manufacturers and suppliers, one immediate result was to place more and more people in close physical proximity to the enormous information-processing power of computers. Some Chinese schools and universities were virtually inundated with free (or heavily discounted) foreign machines "dumped" by firms eager to gain a foothold in the burgeoning China market. Industry giants such as IBM, Wang, and Hitachi set up permanent field offices in China, some reportedly staffed by as many as 100 employees.[18] By mid-1985, it was not uncommon to see computer labs in Chinese universities equipped with one or more late-model foreign minicomputers linked to a local network of a dozen or more PCs. Nor was it uncommon to see Chinese students relaxing in such labs, playing casual video games such as "Flight Simulator."

On average, from 1979 to 1985, the number of computers in China roughly doubled each year. Projecting such exponential growth a few years into the future, we would arrive at an order-of-magnitude estimate of roughly 1.6 million computers (over 90 percent of them micros) by the end of the present decade–an increase of massive (and indeed highly unlikely) proportions. Despite the improbability of such long-term projections,[19] some industry analysts–encouraged by PRC computer imports worth a reported US $300-400 million in 1984 alone–have visualized a wide-open market for microcomputers in China. Several American and Japanese manufacturers have responded by flooding key target groups inside the PRC—schools, universities, and government agencies—with sizeable donations of free hardware as well as with product exhibits, seminars, service centers, and even franchise outlets.[20] The number of Sino-foreign joint ventures in the area of computer electronics (and, most recently, software development) has also risen dramatically since the early 1980s.[21]

GHOSTS IN THE MACHINE

Despite the steady upward trend of recent years, it is highly unlikely that the phenomenal growth rates of the past half decade can be sustained—or even approximated—over the next several years. For one thing, the Chinese government in the spring of 1985 began severely to curtail hard-currency expenditures on imported consumer durables. As a result, Western computer sales to China slacked off considerably in the second half of the year.[22] Another constraint on growth is the lack of competent technical personnel. One authoritative Chinese source, noting that a great many of the PRC's current stock of 100,000 computers remain idle for lack of qualified programmers, operators, and maintenance/repair personnel, wrote that "to solve the personnel shortage problem, even with a very optimistic estimate it would take China at least five years to train 100,000 qualified people to work on those computers."[23] Currently, China is graduating qualified computer technicians and programmers at the rate of around 10,000 per year.

The extremely low personal income of the vast majority of Chinese people (average income of China's 100 million urban blue-color and office workers in 1985 was around US $350/year), coupled with the extremely high price of PCs sold in China, militates strongly against any major surge in the Chinese home-computer market. The average selling price of a fully equipped, dedicated IBM PC Model 5550 in China in early 1985 was around US $15,000 (equivalent to more than 30 years' income for the average Chinese college teacher or mid-level manager), while a locally manufactured 16-bit IBM PC-XT clone, the "Great Wall" model 520A, sold for about $7,000. A Chinese copy of the Apple II+ (the "Venus") was available for around $2,500, while Taiwanese-made Apple knock-offs were reportedly priced below $1,000. A locally-cloned Z-80 CP/M machine constituted the bottom of the microcomputer line, and could be purchased in Peking for as little as $400, still more than a full year's income for the average worker.[24]

To a certain extent, the prohibitive cost of purchasing a foreign-made computer in China reflects the regime's deliberate attempt to pursue a policy of high-tech import substitution. In order to promote Chinese co-manufacture (or assembly) of finished computer products—and thereby to stimulate the process of transferring needed manufacturing technologies from abroad—the PRC in 1985 imposed a tariff of 100 percent on all fully assembled computer products brought into China (up from 50 percent in 1984), while at the same time retaining substantially lower import duties of 25 and 7.5 percent, respectively, on partially pre-assembled and completely knocked-down units.

Despite repeated efforts to raise Chinese industry standards for domestically produced PCs, foreign computers continue to be in much greater demand than their local look-alikes—notwithstanding the substantially lower prices of the latter. The workmanship, reliability, repair, and maintenance record of Chinese-made machines remain generally poor—far below current Taiwanese, South Korean, or Hong Kong standards. For this reason, few Chinese end-users, offered a choice, opt for domestic products.[25] Indeed, the overwhelming majority of mini- and microcomputers seen by visitors in Chinese government offices, enterprises, universities, and research labs in recent years have been of foreign manufacture; much the same also holds true for mainframes.[26]

A rare glimpse into the frustrating world of Chinese computer end-users was provided in a poll conducted by the fledgling China Computer Users' Association in October 1984. Members of the Association were asked to rate 38 computer products—28 domestic and 10 foreign-made—according to 6 criteria: product quality, technical competence of company personnel, written documentation, spare parts availability, repair and maintenance service, and warranty reliability.[27] Not one of the 38 products evaluated in the survey was rated "good" by consumers in as many as 4 categories; only 8 were judged good in 3 categories.[28] By far the most widespread source of consumer dissatisfaction the survey revealed was in the area of written documentation, in which only one native product—a Suzhou-made clone of the Nova 1200—rated a high mark.[29] Second worst overall ratings were in the category of spare parts, where only 8 of the 38 rated products were deemed satisfactory. Best overall performance was in the category of manufacturer's warranty reliability, wherein 23 of the 38 products received high marks.[30]

The results of this survey reveal a Chinese computer market plagued by lack of adequate technical support, written documentation, spare parts, and after-sales service. Compounding these difficulties are a number of endemic problems such as poor software support (reflecting a severe shortage of off-the-shelf programs capable of processing Chinese characters); insufficient numbers of trained computer operators, programmers, and technicians; inadequate and irregular electric-power supplies (making expensive constant-voltage transformers and battery back-ups virtually indispensable for reliable computer operations); poor telephone transmission quality (rendering high-speed data transfers via modem all but impossible); and high concentrations of particulate air pollution in most Chinese cities (making an air-conditioned environment a virtual necessity for trouble-free computing). All these factors combine further to reduce the salutary impact of the micro revolution on the prospective Chinese end-user.

In addition to these various developmental obstacles, still other difficulties confront the Chinese in their effort to leap forward into the age of mass microelectronics. Amusing—and often poignant—anecdotes abound concerning the use, misuse, non-use, and abuse of computers in China.[31] A recent review article summed up the highly-checkered Chinese "state of the art":

> China's computer industry is far behind. The largest domestically produced RAM chip in wide use holds just 1 kilobit. . . . Central processing chips are early 8-bit designs. . . . Virtually all microcomputers now made in China . . . use imported parts for most of the critical components. . . . Prices are high. . . . The profusion of incompatible models afflicts China just like everywhere else. . . . Hardly anyone in China is familiar with applications software. . . . Nearly all users write their own programs in BASIC or FORTRAN; even statistical programs running on the mainframes servicing ministries in Peking are written locally in BASIC. Duplication and triplication of effort are the norm. Users rarely communicate with one another, much less share their work, even if they are in the same city. . . .

> In offices and factories, an extreme shortage of competent managers often makes the few available micros ineffectual. . . . State-run businesses are little concerned with efficiency or profit and loss, so decision-making is much less analytical than in capitalist countries. . . .
>
> Some users, lacking manuals and access to knowledgeable help, don't even know how to start up their computers. . . . All too often, expensive imported equipment is purchased without allowance for maintenance and repair; mainframes designed for twenty users can be reduced to a handful of working terminals as disk drives fail, and frequently no hard currency can be found to buy replacement parts. Even worse is the presence of off-beat equipment that has been foisted off on innocent buyers."[32]

All this is compounded by a highly bureaucratized, rigidly compartmentalized, organizationally redundant Chinese R&D system that makes interagency communication difficult at best and renders the nationwide coordination of computer planning, research, and production virtually impossible.[33]

Given such constraints, it would be unwise to render predictions of a trouble-free future for Chinese computerization. Yet the microelectronic revolution is almost certainly irreversible; the computer—no matter how mixed the blessing—is here to stay; indeed, it has already left a visible, if embryonic, imprint on Chinese society.

CHINA'S EMERGING MICROCULTURE

If foreign reports and travelers' tales provided accurate guides to how computers are used in the PRC today, one might easily infer that video games, dial-a-match computerized dating services,[34] automated herbal tonic dispensaries,[35] and other such electronic ephemera currently dominate the applications landscape. And it is incontravertibly true that large numbers of Chinese microcomputers—probably most of them—have been wastefully employed in what are essentially trivial pursuits or, even worse, have served as expensive, inert paperweights. Yet, beneath the surface of China's apparent micro-boondoggle lies a more important, if somewhat less sensational, reality.

Sectoral Allocation and End Use

According to official data, China's end-use environment is dominated not by computer cupids or Space Invaders, but by dedicated industrial, scientific, and statistical applications. A 1982 survey of 3,500 installed mainframe and minicomputers revealed that 36 percent of the machines were used primarily in scientific computation, 35 percent in data processing, and 25 percent in industrial process control (civilian and military). At the other extreme, less than 3 percent were dedicated to applications such as information storage and retrieval.[36]

In terms of sectoral allocation, almost 85 percent of China's mainframe and minicomputers were installed in civilian work units, with 14.8 percent used in national defense. Approximately 55 percent were employed in industry and transport; 23 percent in science, education, and culture, 3 percent in finance and trade; and less than 1 percent in agriculture and meteorology.[37] Of the 1,800 or so mainframe and minicomputers installed in Chinese industrial enterprises, the great majority were used in process control; relatively few were dedicated to administrative tasks such as accounting, inventory, planning, or marketing.[38]

China's High-Tech Geography

Because computers are both highly capital-intensive and extremely knowledge-intensive, China's microelectronics industry has evolved in a geographically uneven fashion. The vast bulk of microprocessing technologies are being designed, developed, manufactured, and consumed in the relatively modern cities and newly created special economic zones (SEZs) along China's eastern seaboard. In 1985, the cities of Beijing and Shanghai together accounted for almost 40 percent of the nation's total inventory of 100,000 computers; the same two cities also claimed over half the country's 84,000 trained computer operators.[39]

At the other extreme, a survey published in 1984 revealed that a number of provinces had very few mainframe and minicomputers on hand. Of the 3,329 installed machines inventoried in the survey, less than two dozen were located in certain poor interior provinces such as Guangxi and Inner Mongolia; Ningxia and Qinghai could claim less than 10 machines each; Gansu and Tibet were not even represented in the survey.[40] Overall, the 7 provincial and municipal areas with the highest concentrations of mainframe and minicomputers accounted for over 71 percent of all installed machines, while the 7 areas with the lowest concentrations accounted for less than 4 percent. Such unbalanced geographic distribution reflects—and may well, in the short run, further exacerbate—existing economic, technical, and educational disparities between coast and interior, between China's industrial-commercial rimland and its rustic-agrarian heartland and border regions. The 1982 regional distribution of China's installed mainframe and minicomputers is shown in Table 3.

It is unclear whether a further increase in such regional disparities, and the uneven pattern of economic development they suggest, would ultimately promote or retard China's overall modernization. On the one hand, from a purely economic and technical point of view, it is clearly beneficial to have a high concentration of research institutes, skilled technicians, production facilities, and commercial entrepreneurs clustered in close proximity to a few major urban places—à la California's Silicon Valley. On the other hand, the existence of a geographically impacted high-tech culture, characterized by a few scattered enclaves of microelectronic modernization at the edge of a vast sea of interior underdevelopment, could arguably contribute to the perpetuation—and exacerbation—of China's historical pattern of unbalanced growth.

Local Area Networks

Despite the dense clustering of computing power and resources in a few key Chinese cities, very few Chinese computers—or end-users—are electronically interconnected in local networks. Indeed, the vast majority of Chinese computers of all types are currently employed as stand-alone, autonomous data processors. For the most part, these machines are unlinked, unshared, uncommunicative, and inaccessible to all but a single operator working at a single work station in a single location. Very few distributed, multi-terminal computer systems have been installed in China; local area networking (LAN)—that is, a number of small computers hard-wired to a central file server which permits electronic data transfers and the sharing of software or peripheral devices such as hard disks, printers, and modems—remains virtually in its infancy.[41] In 1985, China had only 0.5 telephones for every 100 people. Poor telecommunications facilities and low-quality phone circuits further hamper the ability of computers to "speak" to each other in China, thus grossly limiting the use of modems in electronic data transfers.[42]

Table 3 Regional Distribution of Installed Mainframe and Minicomputers, 1982

Beijing	681	Hebei	88	Xinjiang	29
Shanghai	617	Tianjin	88	Zhejiang	27
Liaoning	289	Shanxi	63	Guangxi	23
Shaan'xi	212	Jilin	63	Inner Mongolia	18
Jiangsu	207	Gueizhou	57	Fujian	11
Sichuan	176	Hunan	51	Qinghai	9
Hubei	171	Anhuei	46	Ningxia	6
Guangdong	121	Yunnan	36	(Gansu	0–5)
Heilongjiang	109	Jiangxi	35	(Tibet	0–5)
Henan	96	(Shandong	30–50)	*TOTAL*	3,329

Source: Gong Bingzheng, "Woguo jisuanji yenyong xianzhuang, juoguo ji yinyibu puji tuiguang de jianyi." Figures in parentheses were interpolated by the author from correlative data.

There are a few scattered exceptions to the rule of stand-alone, noncommunicating computers. One is the previously noted network of 21 IBM 4300-series computers installed by the State Statistical Bureau to record and analyze results of the 1982 national census;[43] another is the Univac 1100/70-based national airline reservation system, purchased by CAAC from the Sperry Corporation in 1985. This latter system reportedly will use satellite transmission rather than dedicated wires or telephone lines to link terminals in different cities.[44] A third exception is the multi-terminal LANs recently installed to process room reservations, guest registration, and billing at a few large tourist hotels in the Chinese capital.[45] Finally, a number of universities have installed multi-user LANs (either donated by foreign manufacturers or purchased with World Bank educational loan funds) in specially equipped computer labs for instructional purposes. Such networks are a clear exception, however, to the general rule of stand-alone, independent computing in the PRC.

The Language Barrier

One obvious reason for the lack of effective, interactive LANs in China is the extreme paucity—and awkwardness—of applications software capable of processing Chinese characters. Several proprietary Western software programs (for example, WordStar, SuperCalc, and

Dbase II) are readily available in (mostly unlicensed and often bug-ridden) Chinese versions in Hong Kong and the PRC. Some of these programs make use of abbreviated pinyin romanization to represent individual Chinese syllables (of which there are only a few hundred, exclusive of tonal variation); once the proper syllabic abbreviation is entered at the keyboard, a character-based menu containing a number of tonally grouped homophones "pops up" on the screen, requiring at least 2 further inputs from the operator to select the desired character. Other programs make use of 4-digit numerical codes, entered via the computer's keyboard, to represent individual Chinese characters (as opposed to syllables), which are then graphically displayed on the screen. This scheme, while obviously involving fewer discrete keyboard operations than the 3-step phonetic/syllabic technique, is not without drawbacks, since it requires the operator either to memorize several thousand 4-digit codes or to look up the code for each individual Chinese character before entering it at the keyboard.[46] Other schemes for processing Chinese characters have been developed both in China and abroad. Most of these systems require either a great deal of operator experience and expertise or a relatively large number of separate keyboard inputs.[47]

Because Chinese character processing is still in its relative infancy, the task of writing new software—or adapting existing software to Chinese requirements—continues to be slow and tedious. For this reason, English has remained the language of choice for Chinese computer programmers and end-users involved in word-processing, database management, and electronic communications. Only in such highly specialized fields as computer-assisted design (CAD), industrial process control, and quantitative data processing has the language barrier proved to be relatively insignificant.[48]

For all the above reasons the average Chinese computer user has yet to find the world of microcomputing particularly accessible or hospitable. Once Pac-Man and Space Invaders have been mastered (some computer programs are omni-lingual), the fact remains that precious few "user-friendly" software applications are currently available to the non-specialist in China.

Computerizing the Workplace

Despite the absence of strong consumer demand for expensive, software-starved computers in the private sector, a substantial market for multipurpose micros potentially exists at the level of the Chinese work unit, or *danwei*. In 1983, only around 1,500 Chinese *danwei*—less than half of 1 percent of the total—reportedly had computers in situ.[49] While the number of computer-owning *danwei* in China may have increased to as many as 10,000–20,000 by 1985—thereby possibly approaching current Soviet levels—this is nonetheless a very small fraction of the whole, leaving considerable room for future expansion.[50]

Over the next several years, the computerization of the Chinese workplace (including large and medium-sized urban enterprises) should receive a substantial boost as China's high-tech import-substitution policy begins to bear fruit and as domestically manufactured microelectronic products begin to rival foreign imports in quantity and quality.[51] If the gap between domestic and foreign technology begins to narrow, foreign-exchange restrictions and heavy import duties will no longer constitute prohibitive barriers to computerization. At the same time, lower per-unit production costs—made possible by economies of scale in domestic manufacturing, a growing pool of skilled (but still relatively low-paid) Chinese technicians and workers, improved production techniques, and anticipated advances in Chinese IC and LSIC technology—should enable low-cost domestic microcomputers to handle much of the work load previously performed by more expensive mainframes, minis, and foreign-made micros. Moreover, Chinese computer manufacturers, forced to vie for customers in an increasingly competitive high-tech marketplace (rather than having their market shares guaranteed, as in the past, by a unified state plan), may well be constrained to streamline operations, increase productivity, and raise quality control standards.[52] Finally, anticipated advances in Chinese-language software development,[53] coupled with increasing standardization of programming languages and operating systems, expanded local area networks, and improved telecommunications capabilities, will tend to make China's computers considerably more versatile, efficient, and user-friendly in the not-too-distant future.

For all these reasons, and despite the existence of numerous obstacles, it is not unrealistic to expect China's *danwei*-centered computer boom to continue for several more years, albeit at a reduced and probably uneven rate of growth. Certainly the possibility of a few hundred thousand microcomputers operating in Chinese work units by the year 1990—some standing alone, some linked to mainframes or minis through local area networks, and still others communicating by means of special, high-quality telephone lines or satellite relays—neither severely strains economic credulity nor excessively taxes the technological imagination.

But what of the *sociopolitical* imagination? How, for example, will the spread of microelectronic data processing and communications technology affect Chinese decision making? How will it affect traditional bureaucratic barriers to inter-organizational communication and coordination? Will it diminish or reinforce the party's traditional monopoly over the public dissemination of information? By the same token, what effect will the proliferation of computerized databases and LANs—and the nascent "information revolution" engendered thereby—have upon China's highly secretive, endemically factionalized political culture? How, in short, will the dawning of the computer era impact and feed back upon the ethos and institutions, the organizational and behavioral environments, of contemporary China?

THE COMPUTER AND THE POLITY

At the outset of the Great Leap Forward in the late 1950s, China's state statistical system was thrown into chaos, as rural cadres, under intense pressure to achieve success in agricultural communization, fed grossly inflated production figures to higher-level officials. As falsified data percolated upward toward Peking, reporting errors were progressively amplified until the nation's entire network of information-gathering services virtually collapsed under a mountain of misleading statistics.

The problem was not unrelated to high technology. In 1958—the first full year of the Great Leap—China's local grain-production figures were gathered not by means of direct on-site inspection, as in previous years, but indirectly via a newly completed telephone network linking each of the PRC's 2,000 counties with the central government in Beijing. Speaking into an impersonal handset, shielded from the prying eyes and ears of government inspectors, local cadres could engage in wholesale fabrication of the facts with relative ease.

Although this case was clearly unusual, serious distortions and impedances in the flow of information are routine, everyday occurrences in the PRC. To cite but one additional example, several years ago at an economic development seminar held in South China and attended by this author, one of the featured speakers—a cadre from the Guangdong provincial planning bureau—gave a series of highly detailed provincial labor-productivity statistics. Throughout his presentation one member of the audience (a "responsible comrade" from the State Planning Commission in Beijing) was observed taking copious notes. During a break in the proceedings, the Beijing cadre was asked why he bothered taking notes, since the information under discussion was presumably available on demand to higher-level planning authorities. Not so, averred the "responsible comrade," somewhat taken aback by the question; "We don't have that information in Beijing."

The Impedance of Information

This anecdote illustrates a type of communications blockage commonly encountered in the PRC. The underlying reasons for such blockage are manifold. For one thing, Chinese organizations tend to be rigidly bounded, tightly compartmentalized, and supervised by a plurality of administrative "grannies." In addition, systemic rigidities in national economic planning and administration powerfully constrain *danwei* at each level to stockpile resources and internalize transaction costs as a hedge against external uncertainty. With each unit thus confronting a highly demanding, uncertain, and potentially troublesome operating environment, a strong incentive exists to suppress and distort information that might adversely affect the unit's future claim to resources. In such a milieu, communications among different *danwei*, even within the same parent organization, tend to flow quite slowly and selectively—if at all.

The Chinese have a name for this phenomenon. They call it "unit-firstism" (*benwei zhuyi*). "Unit-firstism" is a generic game of organizational survival and self-aggrandizement played by myriad insecure, cellularized *danwei* concerned primarily with maintaining or improving their relative status and resource base. It is a game frequently characterized by adoption of such self-protective strategies as concealment, bargaining, and "back-door" barter.[54] At the best of times, "unit-firstism"–along with its territorial analog, "localism" (*difang zhuyi*)–leads to routine distortion or outright suppression of inter-organizational communication and feedback (*vide* the previous example of the Guangdong planning bureau). At the worst of times–such as the outset of the Great Leap–it may have far more catastrophic consequences.

"Guanxi" Networks

The information impedances engendered by *danwei* cellularity and uncertainty may be offset, at least in part, by the existence of informal, non-institutionalized channels of social communication. Such informal conduits–known generically as *guanxi* (personal-connection) networks–are composed of clusters of mutually dependent, mutually supportive individuals whose organizational affiliations may cut across conventional *danwei* boundaries. Although such extra-bureaucratic pathways ostensibly serve to ameliorate and soften the rigid horizontal compartmentalization of the system, the existence of significant status differentials between "insiders" and "outsiders," leaders and followers, tends to foster a high degree of intra-network secretiveness and exclusivity, sometimes taking the form of esoteric, allegorical communications. As a result, Chinese *guanxi* networks, far from uniformly facilitating a more free and open flow of social communications, may actually contribute to their further impedance.[55]

Buttressed by the existence of exclusive *guanxi* networks, the Chinese system of cellular sub-national administration thus erects powerful barriers against the free flow of information–microelectronic or otherwise. Because of such systemic impedance, it is extremely difficult to foresee in China a wholesale replication of Japan's spectacular, microelectronic-based "information revolution" of the past two decades. In Japan, information is generally viewed as a universal public good, to be freely disseminated; in China it is much more likely to be regarded as a factional, organizational, or unit resource, to be jealously hoarded.[56]

Closely related to the generic problem of information impedance is the glaring lack of technical coordination and standardization–the virtual technological anarchy—that has characterized the computerization process in China. The proliferation of dozens of different, often mutually incompatible computer systems, programming languages, software, and LANs in work units throughout China, together with a chronic shortage of technical support, makes it virtually impossible for individual *danwei* to coordinate and integrate their own internal electronic information systems—let alone to communicate effectively with other units.

Computer Security

A further obstacle to effective microelectronic communication is the need for computer security. Given the restrictive, highly stratified nature of organizational information flows in China, an inordinate amount of attention must be paid to devising procedural safeguards against unauthorized access to *danwei* computer terminals and databases. Since access to information is an important measure of personal status (as well as a vital perquisite of *danwei* leadership), the maintenance of secure, restricted channels of communication is a major requirement of any unit-installed computer system. Various devices have been employed toward this end, including hard-wired, closed-circuit LANs, guarded computer rooms and terminals, elaborate user passwords, and the encoding of sensitive data. In each case, the effect is to safeguard the flow of privileged communications by excluding unauthorized outsiders.[57]

Alongside these systemic impedances, there exist other, more subtle but equally compelling constraints limiting the potential impact of a nascent Chinese microelectronic "information revolution." As noted above, China is a land of myriad intersecting cells and compartments, separated by formal bureaucratic barriers which are cross-cut, in turn, by fluid networks fashioned of interdependent human relationships. These networks are flexible; and they provide the resiliency—and subtlety—needed to meliorate endemic Chinese organizational tendencies toward excessive formalism and rigidity. Computers, however, are no great respecters of subtle human relationships; not do they function well in fluid, ambiguous, or indeterminate environments. They are hard-wired automatons; when properly programmed, they are capable of performing a wide variety of clearly defined, repetitive tasks. They can perform these tasks rapidly, routinely, and virtually without fail. They do not get tired, bored, cold, or hungry. They have no feelings, no politics. They do not belong to factions and are not subject to "spiritual pollution." They are logical and predictable to the nth degree.

"Garbage In; Garbage Out"

In the boringly repetitive, mechanical world of the computer, the program—and the programmer—are all-important. Unlike the carpenter, who can blame his materials, his tools, or even his wife's infidelity for his failings, the computer will normally have but one source of error—the procedure writer. In this respect, computers are virtually immune to the entire repertory of post-programming denials, rationalizations, and deceptions used by human beings to mislead each other and thereby avoid responsibility for mistakes and shortcomings. Computers cannot—without programmer complicity—cover up failure or scapegoat others; they cannot—unless so instructed—dissemble or otherwise engage in ambiguous, esoteric communications. Nor are they readily fooled by human tricks and illusions.

In China, failed policies have sometimes been attributed to uncontrollable exogenous forces, such as bad weather (the collapse of the Great Leap) or lingering historical traces (the "remnant feudal autocracy" of the late Maoist era). Most often, however, policy failures have been blamed on the deviant behavior of heretical individuals or conspiratorial groups—the "Li Li-san line," the "renegade, traitor and scab" Liu Shaoqi, the "Lin Biao anti-Party clique," the "Gang of Four," and so forth. The computer changes all of that, making it possible more or less precisely to calculate the correlation between an event, its antecedents, and its aftermath. By hard-wiring—and thereby removing from the realm of random, discretionary, or subjective selection and interpretation—the effects that flow from any particular, programmable cause, the computer can virtually eliminate exogenous sources of variability and error from the routines performed under its command/control program. In so doing, it can also locate the source of a problem—and thereby reduce the likelihood of a successful avoidance of responsibility.

Few of contemporary China's 10 million state administrative cadres are likely to respond with great enthusiasm to the programmable certitude of the computer. Buck passing, paper pushing, and other garden-variety forms of responsibility avoidance have become a virtual way of life under China's "bureaucratic mode of production." And, while there are clear indications that Deng Xiaoping's reform efforts have succeeded in bringing a certain measure of dynamism, vitality, and innovative energy to this sclerotized monolith, a steady stream of Chinese media reportage concerning persistence of bureaucratic inertia and irresponsibility cautions us against expectations of quick or easy societal transformation.[58]

CONCLUSION: COMMUNISM AND THE COMPUTER

We turn, finally, to what many observers regard as the most important—and most enigmatic—political variable affecting China's computerization drive: the CCP's monopolistic control over the media of mass communication. Precisely because computers make it technically feasible to circumvent official, approved channels of communication in the dissemination of ideas and information, they are sometimes seen as posing an innate threat to the power of Communist elites and institutions. Christopher Evans thus argues that the spread of cheap, universal computer power will erode traditional restraints on the flow of information in Leninist systems by "encouraging . . . the spread of information across the base of the social pyramid." Such a development, he argues, "favors the kind of open society which most of us in the West enjoy today, and has just the opposite effect on autocracies." And he predicts that, by the end of the 1980s, "even the most ardent Marxist will probably have to bow to the overwhelming [democratizing] testimony of the microprocessor."[59]

Not all analysts, however, are so certain of the corrosive effects of the information revolution on Communist Party control. Erik Hoffmann argues that the process of computerization will almost surely *not* produce major changes in the Soviet polity, since "modern information technology is likely to be integrated or absorbed into existing bureaucratic value systems and behavioral patterns." By carefully controlling both access to and end-uses of computers, he argues, Soviet elites can ensure that control over electronic data processing will remain in the hands of "generalist" politicians and loyal party technocrats, thereby serving to "reinforce and intensify centralized control over strategic decisions." With computers remaining a narrowly confined producer good rather than a freely available consumer good, there is relatively little danger, Hoffmann avers, of a significant, computer-driven erosion of Soviet political power.[60]

To date, the Soviet experience has proven inconclusive. Few knowledgeable observers of Soviet affairs have claimed to detect any substantial diminution in the degree of bureaucratic control or Communist Party domination in recent years. Yet, the much-publicized *glasnost* (openness) campaign of Mikhail Gorbachev has raised profound questions about the causal relationship between an ever-rising flow of secular communications in Soviet society (facilitated, *inter alia*, by the spread of telephones, television, computers, and other technological carriers of the "information revolution") and the generation of internal pressures for political democratization.[61]

If the Soviet case seems ambiguous, the Chinese case is, in some ways, even more perplexing. On the one hand, China is clearly at a more primitive stage of technological development than the USSR, its economy less information-dependent than the latter. With only about 6 telephones per 1,000 population, China lags well behind both the Soviet Union (92 telephones per 1,000) and Taiwan (210 per 1,000).[62] In this connection, a leading Western expert on the "information revolution" has argued that the critical point beyond which an autocratic regime finds it difficult to maintain its power monopoly is when 20 percent of its population have telephones.[63] Measured against this rather impressionistic yardstick, China is at least one or two decades away from the threshold of political criticality.

Television, on the other hand, has become an immensely popular and readily available medium of mass communications in the PRC. With almost 50 percent of China's urban households reportedly owning TV sets, it has been estimated that the 1985 Superbowl may have had a larger potential viewing audience in China (where it was nationally broadcast several months after the event via microwave relay) than in the United States.[64] Other things being equal, it would appear that television has far greater potential as an electronic disseminator of new ideas, information, and values than computers, which are, after all, both exceedingly rare in China today (0.1 per 1,000 population, compared with 50 TV sets per 1,000) and almost exclusively *danwei*-owned and operated. Moreover, an ordinary xerox machine (of which there were an estimated 10,000 in China in

1985) is capable of making more copies of a given document—whether authorized text or underground *samizdat*—in a single hour than a high-speed dot-matrix printer attached to a computer can print out in an entire day. And, finally, while it is certainly true that the technology embodied in the microcomputer is revolutionary indeed, the nature and magnitude of its effects upon the state/Party apparatus and political culture of Communist systems has yet to be demonstrated with any clarity.

For all these reasons, it seems advisable to resist the temptation to overstate the probable political impact of the computer in China. In the long term, the ongoing global revolution in information technology (of which computerization is an important part) may well be subversive of Communist monocracy, and perhaps even fatally so; but the near-term effects of the computer per se upon the organization and operation of the Leninist polity remain both highly problematical and extremely difficult to document.

Conclusion

CONCLUSION

Science, Technology, and China's Political Future—A Framework for Analysis

RICHARD P. SUTTMEIER

The relationships among science, technology, and political systems (S-T-P) are inherently complex. Regrettably, we have no adequate theories for explaining and predicting them; yet, whether our concerns are with China, or with other countries, the cardinal issues facing mankind are rooted in S-T-P relationships.

There are certain identifiable *realms* where we can see effects of the S-T-P interactions. Harvey Brooks, for instance, has called our attention to issues of scale, centralization, and standardization in S-T-P interactions. There is also the issue of changed valuations of goods, services, and factors of production—what comes to be "economized" as a result of technological change.[1] We also know that S&T advances are likely to have political and economic distributive effects, effects on policymaking, and effects on the mechanisms of social choice. Finally, new developments in S&T influence our underlying cultural values and, through these, the bases of our perceptions.

Efforts at theorizing about S-T-P relations usually fall into two main categories. One focuses on the "impacts" of science and technology on political life, and on society generally; S&T are seen as the independent variables. The other approach assumes the opposite pattern of causation; politics and society are taken as the independent variables. While the inadequacies of both approaches are known, they have yet to be overcome. Due recognition of both causal perspectives will help to avoid misleading oversimplifications of the causation problem. Thus, in the analysis that follows, I consider first the influence of the political system in Chinese S&T development.

THE POLITICAL DETERMINATION OF CHINA'S S&T DEVELOPMENT

Of the various influences on the development of Chinese S&T since 1949, and, indeed, in the twentieth century, none has been greater than the political environment. Political instability and political interventions into science have been the main impediments to S&T development. Yet, major government programs and expenditures have also been the main spur to development since 1949.[2] Examples of how social and political factors have shaped Chinese S&T include the following:

China has developed nuclear weapons and reasonably sophisticated rockets and space satellites largely on its own as a result of high-priority government programs. China has made relatively little progress in developing an integrated national electric-power network, an integrated national telecommunications network, or integrated national railway or highway networks. Reportedly, barges on the Changjiang must put ashore to register as they enter new provinces.

Chinese researchers have often made impressive laboratory achievements. The record of bringing discoveries and inventions made in the laboratory into serial production has been poor in spite of thirty years of official lament about the separation of research from production.

In recent years, China has entered into negotiations with foreign companies for the procurement of large, complex technological systems. Two such cases are particularly notable. One is the nuclear power plant being built in Guangdong province. The other is a communications satellite. In both cases, the negotiation dragged on for an extended period of time. The foreign vendors expended large amounts of money, and incidentally transferred a fair amount of technology as part of the negotiations. In spite of what seemed like a firm commitment from the Chinese, in the end, the foreigners were disappointed with the result. While a contract for the nuclear power plant has now been signed, there has been speculation that its construction will be a financially losing proposition for the French, who are to be the suppliers. In the satellite case, China canceled the bidding in the face of indecision over what type of satellite to procure, foreign-exchange shortages, and successes in its own space program. Some foreign company may yet supply a satellite, but foreign vendors will be exceedingly cautious in how they approach the submission of bids in the future.

In efforts to transfer the technology needed to service and maintain modern commercial aircraft, one American company found a persistent Chinese inclination to avoid following established international practices in favor of earnest attempts to improvise with local materials and practices. In one example, the Chinese would approach the cleaning of hydraulic cylinders used in landing gears by using gasoline instead of the special cleaning and lubricating oil specified in the manuals. In the process they would shorten the life of crucial seals and gaskets, and thereby accelerate the need for costly overhauls of this component.[3]

In another case, the American manager of one of the most ambitious, complex, and challenging Sino-foreign cooperative enterprises lives with fears of what he calls "time bombs." Should these "go off" in the future, they would certainly set back the enterprise, and could eventually lead to its abandonment. When asked about these "time bombs," he points to future political instability born of an underlying lack of mutual trust in Chinese organizations and a deep-seated normlessness in their activities.[4]

These various observations, ranging from the subjective to the factual, point to factors of political import, exogenous to S&T themselves, that shape the directions of S&T. Indeed, these cases reveal some important aspects of the Chinese polity. At least six general propositions can be suggested.

(1) The nuclear weapons, space, and rocketry cases have demonstrated that China has a core of scientific and engineering talent that is quite respectable by international standards, and that *China has the institutional capacity to mobilize the talents and the material resources required to achieve high-priority national-security objectives.* Yet, the general inability to bring R&D results into commercial application *is symptomatic of a Chinese political economy ill-suited to technological innovation.*

(2) The electric power, telecommunications, and railroad cases sug-

gest an underdevelopment of those technologies that are usually assumed to be of fundamental importance for the achievement of political integration and the development of a prosperous modern society. The apparent neglect of these technologies is seemingly contrary to what Chinese leaders have been saying since 1949 concerning their interest in providing for the conditions necessary for a modern society of "wealth and power." *The fragmentation we see in these (infrastructure) technologies suggests an underlying institutional or political fragmentation, revealing a disjunction between the pretensions of a strong centralized state and the institutional requisites for technologies needed to make such pretensions credible.*

(3) The nuclear-power-plant and communication-satellite cases reflect *a Chinese tendency toward indecision which seems to be a function of unresolved bureaucratic rivalries.* Bureaucratic entities develop routines and vested interests built around technologies that bias later technological choices. When bureaucracies, with their distinct technological preferences, must work together in support of a large technological project, they must either be willing to hold in check their "techno-bureaucratic" interests (which they rarely do) or submit to a higher authority, if one exists. Seemingly, in China, these *higher authorities or central coordinating bodies have had severe difficulties coping with entrenched ministerial power.* Indeed, the attention of such powerful central coordinating bodies is among the scarcest of resources in a country like China.[5]

(4) These two cases of negotiations over the import of complex technological systems also suggest a powerful unresolved tension about the procurement of foreign technology. The tension is between the perception of the technical superiority of the foreign technology and the belief that China should be able to provide all or most of these systems itself. These cases point to *the existence of a "technological nationalism" in which "choice-of-technology" decisions are expanded beyond the bounds of strict economic, or even more general instrumental, rationality to include considerations that seem to have both political and psychological overtones.* This tension is clearly related to the goal of self-reliance, but it has had pernicious effects. "Technological nationalism" has not solved the problem of reliance on foreign science and technology; it has probably made it worse.

(5) The aircraft-maintenance case and the general experience of failure to bring laboratory results into production point to a paradox in

China's S&T which might be called *the maladjustment of theory and practice.* It is frequently said that Chinese science has a capacity for theoretical understanding, but an incapacity for linking theory to practice. Yet, in technological practice, such as aircraft maintenance, there is a tendency to be oblivious to the underlying theoretical principles from which the preferred, and often "right," technological practice is deduced. Instead, simple empiricism and inductivism characterized by ad hoc trial-and-error approaches are used.

(6) Finally, the case of the American manager concerned about political instability points to *the existence of widespread anxieties and insecurities, social mistrust, and normlessness in the underlying relationships that form the building blocks of collective life.* Political stability at the societal level in the face of these behavioral characteristics at the level of the enterprise may thus be difficult to achieve. These characteristics would, of course, also work against the successful operation of the technologically and organizationally complex enterprise the manager is trying to establish.

Thus, the record in China points to strong causal forces emanating from the political system shaping the course of science and technology. Less evident are signs that political change has been dependent on science and technology. Most observers of China, however, would almost reflexively argue that China's modernization drive will, if successful, certainly create forces from a modernized S&T that will profoundly alter the political system. These changes are likely to become especially interesting in the context of some of the recurring questions about Chinese politics that constitute the *problematique* of the political system.

ELEMENTS IN THE *PROBLEMATIQUE* OF CHINESE POLITICS

The Issue of Centralization

China's political system has yet to find the "right formula" for the relationship between centralization and decentralization of its political and administrative institutions.[6] The reforms of the 1980s in the economy, in S&T, and in the political system more generally, while portrayed as decentralizing, also have the objective of increasing effective central power. This is often stated in terms of achieving more effective macroeconomic control over the economy, but what is really meant is to enable the central authorities to do more effectively what they want to do. The current logic is that the accomplishment of this end requires that the central authorities try to do less by decentralizing. Many prerogatives that the center probably could not exercise in any case are therefore being given up to lower units.

Thus, decentralization is in evidence in the expanded decision-making powers of rural households, enterprises, and local governments. Yet the roles of such central coordinating bodies as the State Economic and Planning Commission, as well as the State Science and Technology Commission and the Science and Technology Leading Group in the S&T area, have clearly become more important in recent years. There is a logic of simultaneous centralization and decentralization at work.

The frequent adjustments in centralized and decentralized prerogatives, however, occasioned by such issues as capital construction, foreign-exchange expenditures, dissent, and student unrest, point to the fact that the issues of centralization/decentralization are far from being solved. It remains to be seen what the effects of an S&T development program will be on the centralization/decentralization issue, and whether the former can succeed without significant resolution of the centralization/decentralization issue.

The "National-Security State"

A second element of the Chinese political *problematique* is the influence of national-security concerns on national priorities and the influence of the military in politics. Given the central political role of the military in the past, and the fact that so much of the nation's investment and R&D had been committed to the military, it is not unreasonable to describe China at the end of the Maoist era as a "national-security state." A central issue facing the post-Mao leadership has been the extent to which the national-security state should be disestablished.

It is clear that the current leadership regards this question as cardinal. There is considerable evidence of changed priorities, and the Deng Xiaoping leadership has attempted to weaken the political influence of the military. Yet, although the modernization of national defense is often said to be the lowest priority of the Four

Modernizations, as Frieman rightly points out, the interpretation of this No. 4 position requires caution.[7]

There is considerable evidence that the Deng leadership is succeeding in shifting the resources of the defense industries, including defense R&D, to the service of the civilian economy. Defense plants are marketing motorcycles and refrigerators; ministries of the old military machine-building establishment are building nuclear power plants and marketing satellite launch services. China's recent efforts to export its *own* technology have featured the technologies of the defense industries.

While the economic significance of the defense sector's entry into the civilian economy remains to be calculated, the breadth of competence and versatility demonstrated supports the judgment that it had been in a privileged position in past investment decisions. Furthermore, the achievements of the defense sector both support, and are supported by, the ideology of "technological nationalism."

The reorientation of the defense industrial and R&D sectors raises questions of enormous potential importance for China's future. Is China undergoing a true conversion from military production and defense R&D? Is the national-security state being disestablished? Or is it only in a temporary transition which will allow the defense industry to become more technologically innovative as it faces commercial competition and gains access to advanced civilian or dual-use technologies? In short, is China creating a *modern* military industrial complex to replace the *pre-modern* and often irrational national-security state of the Mao era? What sort of S-T-P relations are likely to result from these changes?

The New Elite

In answer to the foreigners' question of whether the reform movement in China will survive Deng Xiaoping, the Chinese, in their strategy for political succession, seem to be providing an affirmative answer. Although the ouster of Hu Yaobang in January 1987 added a new uncertainty to the succession question, the selections made in the 13th Party Congress in October 1987 indicate that a more educated and technocratically inclined elite is emerging. Many questions remain about the predispositions of these new elite cohorts. Radical ideological Maoists they seem not to be. That they are of one mind as to where the country should go, however, is less certain.

There seems to be a bifurcation in the leadership between those who have backgrounds in economic work and those with engineering backgrounds. We would expect the former to approach the evaluation of policy options with criteria of economic efficiency and cost-effectiveness; for the latter—with more of a technocratic orientation—orderliness, control, and technical efficiency would be the relevant criteria. The former, for instance, would be expected to be more favorably disposed toward marketization, while the latter are more likely to have a deep-seated bias for an administered economy. On "choice-of-technology" decisions, the former would be guided by factor endowments (and thus biased toward labor intensity), while the latter would more likely choose technologies that provide enhancements to national technical capabilities.

Among the engineers, most have backgrounds in heavy industry, and quite a few who come out of the defense industries are in charge of creating China's high-tech future. Song Jian is a prominent example. One is thus left with the question of whether the national-security-state orientation of the past may not find a new expression in the political economy of the future.

The attitude-formation experiences of these new elites suggest that there may be very different attitudes between cohorts who are only a few years apart in chronological age. We would expect that the major disruptions these people have lived through would have had different consequences, depending upon the particular developmental phase they were in at the time. Furthermore, when we reflect upon the frequent societal and personal disruptions experienced by all these people, it is difficult to see how stable sets of values and attitude structures could emerge.

Thus, while their educational and occupational backgrounds would suggest pragmatic, problem-solving orientations, in terms of what we think we know about personality development, these are people whose life experiences have been characterized by great cognitive dissonance and the kinds of discontinuities that engender psychological insecurities. A capacity for unpredictability may therefore be mixed in bizarre ways with a commitment to the values of instrumental rationality. There is, therefore, reason to maintain a healthy scepticism about the capacity of new elites to carry through the reform program and set the kind of new tone to Chinese politics that accords with our wishful thinking.

State-Society Relations

The evolution of state-society relations is the most difficult to analyze, yet it may also be the most important. If there are to be signs of change in state-society relations, we would expect them in the following areas.

First, we would expect to see changes in the formal/legal proscriptions on the powers of the state. Despite evidence of a major new commitment to the role of law in society, the interest in law seems more directed to the creation of the proper infrastructure for economic development, and to the prevention of future arbitrary exercises of power by willful individuals, as during the Cultural Revolution, than to upholding civil rights. The evidence is not compelling that the development of the legal system will in some direct sense constrain the powers of the state.

The new interest in law could, however, indirectly lead to some eventual limitations on the state. We see some evidence of this in an evolving assignment of property rights. Private property now has a standing in Chinese law that is different from before, and the acknowledgment that there are rights to the ownership of intellectual property is a new development.

As economic growth proceeds and the economy becomes more complex, it will become increasingly difficult for the state to control it without expanding its own control apparatus. The Deng leadership seemingly does not intend to follow this path, as seen in their inclinations to control less but to control what they wish to control more effectively. This orientation is not inconsistent with the move toward greater societal autonomy, and indeed, this may require effective government.

The use of the Party as a control apparatus may be a different matter. The Party as an institution, and its control rôle in a reformist China, has become unclear. Reforms call for the formal role of the Party in the operation of the economy (especially at the level of the production unit) to be reduced. Party membership and active service as a Party cadre have lost appeal. On the other hand, the Party has stepped up its efforts to recruit new members from those occupational groups that will become strategic as the Four Modernizations program progresses. In short, the future role of the Party is a major uncertainty in assessing the extent to which greater societal autonomy will result from economic growth and development.

The role of the work unit, or *danwei*, in its appropriation of technical knowledge and control over the movement of personnel, is one of the most serious obstacles to the modernization of S&T. From the viewpoint of central authorities, it is also an obstacle to policy implementation. Yet, the power of the work unit lies in the fact that it supplies housing, medical care, and retirement benefits, which are not readily available to the individual through other means. Efforts to erode the power of the work unit by the establishment of national social-security schemes, for instance, could have important consequences for state-society relations.

Another key indicator of change in state-society relations would be the regime's tolerance of new forms of association and group formation. As yet, there is little significant change from the past pattern of officially sponsored and controlled associational life. There are a few emerging differences, however. In some areas, officially sponsored groups are more active in pressing the interests of their members.[8] Also, new types of organization are appearing which range from units like CITIC (China International Trade and Investment Corporation), to consulting and venture-capital firms, to new models for the organization of R&D, to new, high-technology "*min-ban*" (non-governmental) enterprises, such as the Stone Corporation. Change in state-society relations is likely to be advanced by further bounding and reducing the realm of the "political" (a move which has characterized the post-1979 reform period) and by the growth of social complexity resulting from further economic development. These forces in combination are likely to produce significant political change, but the forms that change will take remain unclear.

The Problem of Ideological Salience

A legacy of the Cultural Revolution is the disillusionment of large segments of the population with Marxism-Leninism. The problem is not solely one of the faith in a doctrine legitimizing Party rule; the more basic issue is one of the loss of faith in any common set of social values around which the society can cohere.

The various efforts to promote "socialist spiritual civilization" are intended to rediscover these common values. In many ways, however, the main thrust of the economic reforms runs counter to the value revitalization. The emphasis on material incentives, and the evident appeals of material gain to the population, are difficult to square with the norm of selfless service to the people so central to the socialist ethic of the past. Moreover, China's new and extensive interaction with the outside world, which in itself is an element of the political *problematique,* exacerbates the concern over ideology as seen with the regime's attacks on "spiritual pollution," venality resulting from interactions with the foreigner, and, in early 1987, in the attacks on "bourgeois liberalization." The deeper problem, of course, is the inevitable exposure to heterodox ideas which the new "open door" brings to the elites and general population alike. While many of these ideas need not be inconsistent with China's development objectives, they are inevitably a pluralizing and a relativizing force vis-à-vis an official ideology.

CHINESE POLITICS AND THE "REALMS" OF CHANGE IN S-T-P RELATIONS

To what extent will these elements of the *problematique* be ameliorated or exacerbated by the further development of Chinese science and technology? To what extent will they affect the prospects for success of the S&T development program? To answer these crucial questions, it will be helpful to return to the earlier discussion of the realms in which change in S-T-P relations occur.

Scale

One of the more important properties of modern technologies that have social and political effects is scale.[9] Scale changes are implicit in the kind of S&T development the Chinese envision. Much of this will be "upscaling" through the introduction of technologies that have economies of scale (for example, the shift in the technologies of fertilizer and steel production), or require networks of scale (telecommunications, transport, electric power). In addition, the scale of the domestic market will expand, which is likely to affect state-society relations. But the effects of scale are complex. S&T development also has a "downscaling" side as a result of the introduction of small-scale technologies (for example, household appliances, microcomputers), which will enhance the quality of life for individuals, small groups, and households, quite independent of services provided by the state. Such technologies can be empowering for individuals, households, and small groups.

The complexities of the challenge of scale are illustrated by the automobile.[10] As an individual artifact, it is a relatively small-scale technology that provides individuals with choices concerning transport and mobility (as well as status). As a technology of a transport system, however, automobile technology is of enormous scale and is capable of generating large-scale problems.

The impacts of scale involve the interactions of technologies and their institutional environments and the interrelatedness of scale and other factors, such as distribution. How microcomputers are distributed, for instance, may have much to do with the impact of their scale. As Baum points out, the microcomputer has not yet become the personal computer.[11] Nevertheless, there is a clear "soft determinism" or "telic inclination"[12] to the scale of the micro in the direction of empowering small social units.

Centralization

Issues of scale are also linked to the question of whether technologies are centralizing or decentralizing. As we have seen, problems of centralization and decentralization are cardinal to Chinese politics; yet answers to the questions of when a technology is centralizing or decentralizing are not always obvious. China has experienced a particularly pernicious blend of centralization and decentralization; some things have been inappropriately centralized, while others have been inappropriately decentralized. China's economic and institutional reform program is attempting to remedy the problem. The question is whether S&T development will be consistent with these aspirations. Clearly, some areas of S&T development will reinforce tendencies toward centralization. The development of infrastructure and information-management technologies will enhance the penetrative and integrative potential of the state. But the society to be penetrated will be much more complex and less reliant on the state for integration, as a result of the diffusion of smaller-scale technologies.

Centralization and decentralization are likely to take new forms, and the application of old categories may become very misleading. For instance, in cases where large-scale technologies seem to be centralizing, they may not be centralizing at the state level. Rather, they may foster the powers of large corporate entities. If these newly empowered entities elude close control by planning agencies of ministries, would that be centralization or decentralization?

Similarly, the same infrastructure technologies that are potentially supportive of the growth of central power may also be available to individuals and groups, should they wish to carve out zones of autonomy from the state. The very unpredictability of the scale and centralization effects of technology is likely to make their impact particularly difficult for a policy already troubled by confused distributions of discretionary authority. S&T development, therefore, will require increased selectivity in the exercise of central power, a reinforcement of trends already begun.

Standardization

The growth of modern science and technology both requires and is a force for standardization. On the other hand, many of today's advanced technologies, such as those for design and manufacturing, offer remarkable opportunities to escape the worst features of standardization.

China's experience with standardization is contradictory. In this highly bureaucratic society, in which, until recently, an enormous population was clothed in a standard uniform, one would assume that standardization as a norm would be second nature. Yet, there is a serious lack of industrial and technological standardization. China's more notable technological achievements have been customized technologies. An inability to serialize production according to some standard has been one of its prime technological weaknesses. At a behavioral level, we see efforts to standardize the implementation of policies, often when there should be greater room for discretion and customized application. Yet, actual behavior in response to standardized policy is often highly variant.

The interactions between technology and institutions in the area of standardization are seemingly perverse. Again, the microcomputer is an interesting case. Although China has an elaborate policy for the development of the computer industry, the industry suffers from an absence of standardization.[13] In the United States, by contrast, we have (for better or worse) an industry standard set epiphenomenally, not by policy intervention. Further evidence that centralized policy intervention is not always required for standardization is found in the standard-setting role of professional societies and industrial associations in the advanced capitalist societies. It may indeed be the case that societies with effectively operating markets and with traditions of private association are better positioned for providing this public good. These societies may provide the conditions for generating detailed product information necessary for workable standards, and offer incentives for the cooperative action required.

There are strong forces for standardization in Chinese society and very good technological and industrial reasons for standardization to proceed. But there are also social and cultural forces pushing against standardization which will find in new technologies the tools for diversification. Effective technological and industrial standardization should facilitate higher orders of national economic integration; ironically, industrial standardization may be a force for greater economic diversity.

Economizing

Choices of technologies and technological systems always involve judgments about those factors of production where there should be economies. Inanimate energy can be substituted for labor, information can be substituted for energy, and so on. Although these decisions are made in the context of an economic calculus, they also represent statements of societal and political values. China faces many challenges in its economizing decisions, but these challenges cannot be met effectively without a rational price structure. What would be the effects of rational prices on China's technological decisions, and on industrial structure more generally?

As Brooks and others have pointed out, modern technological progress has been characterized by economizing on labor, not on materials and energy.[14] However, because of concerns about the availability of materials and energy, technologies are increasingly chosen that also economize on these factors. But these technologies do not simply resubstitute labor. Instead, they involve the substitution of information for energy and materials; but the production of information is, in fact, labor-intensive. Therefore, societies making these economizing decisions place considerable value on those able to produce, process, and apply certain kinds of information.

Although China has a relative abundance of labor, it has a shortage of skilled labor (especially the kind that can produce, process, and apply technologically useful information). It also has shortages of energy and some kinds of materials, which would seem to argue for choices of technologies that economize on energy and materials by substituting information, not labor (as conventionally understood). The Chinese, as a matter of policy, intend to make some choices along these lines, whether narrowly "efficient" or not, on the basis that, in the long run, a reliance on labor-intensive technologies will not advance the quality of production and the level of technology employed. China's choices of technology are difficult because they will have diverse effects. The S&T development strategy requires a much higher valuing of skilled people by the society (this has not yet been unambiguously reflected in policy or behavior), and is likely to be (perhaps perversely) labor-releasing. Yet the expected shift in employment from agriculture to industry will increase the supply of labor for industrial work, with profound implications (such as the increased size of the urban proletariat) for the economy and the polity. Finally, as in other industrializing societies, there will be an expansion of the service or tertiary sector of the economy. Thus, economizing decisions molded by technology will have important effects on the occupational structure.

Distribution

There is little doubt that a dynamic program of S&T developments will give rise to questions about distribution. Given that S&T have been assigned such a prominent role as determinants of China's future, it is only natural that more and more people will want to have a piece of that future. That resources for S&T, especially manpower, are scarce means that competition for them has become intense. Issues of distribution may therefore become a factor contributing to increased political participation over demands for fair shares. Debates over the fairness of regional distribution have already surfaced.

There is also an underlying issue of whether the criteria for technically and economically rational decisions will have a central role in distributive, central, resource-allocation policies. If these distributive decisions are dominated by conservative, non-transformative political considerations, less can be expected from the S&T development. Of course, political considerations will come into play with regard to distribution, and especially to redistribution. Successful S&T development will distribute benefits unequally—by occupation, academic discipline, industry and ministry, and, as noted, by region. The management of dissatisfactions resulting from such distributive outcomes will require some redistribution.

A successful S&T development effort in the context of a more general modernization drive, as Bhopal and Chernobyl dramatically remind us, involves risks as well as benefits. How China "distributes" these risks, and whether such "distributions" are perceived to be just and equitable, are likely to become important political questions.

In the final analysis, China's ability to handle these questions depends on institutional innovation. It is highly unlikely that these distributive issues can be managed by a central planning system alone. Their volume and complexity defy synoptic solution. Innovations in marketization (especially the development of factor markets), new developments in the legal system, and the working out of insurance schemes are required. If China shows such capacity for institutional innovation, it will be testimony to the creativeness of its leadership. It may also lead to a more decentralized policy, which tolerates a society possessing more autonomy from the state than previously. What is certain is that S&T development will challenge the polity's distributive capabilities.

Mechanisms of Social Choice

China's promotion of market mechanisms should have a stimulating effect on the development of S&T, but it also may bring into sharper focus the problems of "market failure" resulting from technological development. Increasingly complex "intermediate goods" which affect the safety and reliability of final products[15] are already a problem in China. The absence of well-developed markets at present makes this problem worse than it has to be. In a market economy, competition and the fear of losing customers provide some incentives for attending to safety and reliability. As technology develops, however, it becomes more difficult to rely completely on market-oriented, *caveat emptor* principles. In the advanced industrial societies, governments have established elaborate regulatory mechanisms for assessing the properties of intermediate goods and the safety and reliability of final products. It is likely that China will also have to expand, modernize, and strengthen its regulatory apparatus. This produces a new kind of politics, which, by its very nature, juxtaposes particularism against the application of universal standards.

Another example of market failure is the problem of externalities, such as environmental pollution. While this problem can be approached by altering incentives in the context of market exchanges, it too requires regulatory and standard-setting administrations. The challenge is to increase the effective presence of such administrative bodies without slowing economic development unduly. In addition, such bodies must be kept relatively free from lapsing into a tradition where particularistic exceptions to rules become the norm of policy implementation. These issues surrounding China's mechanisms of choice are related to the challenges of allocating risks of standardization noted above. They are also related to policymaking.

Policymaking

As Halpern demonstrates, the central government increasingly seeks technical advice in the making of policy in order to assure more reliable empirical premises for its decisions.[16] The further development of S&T will continue this trend.

Expert involvement in policymaking can be a force for pluralism in the polity. Yet, the lines between advice and advocacy, advisor and adversary, are often fuzzy.[17] There is the possibility, therefore, that the growth of the role of expert advisors, unaccompanied by other changes in the polity, may simply give a very limited section of the population access to high-policy circles.

There is also the danger of the illusion that all policy problems are susceptible to technical solutions if only enough expert advice is available. In many situations, such "technical fixes" are attractive possibilities, but invariably they are only one part of the policy decision. The experiences of the advanced technological societies suggest that the determination of the value or normative premises of the decision, rather than the empirical, is often the more problematic part of the decision and the more in need of institutional innovation. Some members of the Chinese political elite may be particularly susceptible to the illusion of the technical fix. Their celebration of S&T's role in China's modernization is itself a manifestation of their susceptibility, as is the belief that learning modern management "science" will solve the problems of the Chinese enterprise. Yet, other Chinese are taking the problems of managing normative issues in policymaking more seriously. The growing role of the National People's Congress, interest in the further development of a legal system, and the emergence of a livelier, more investigative press are some indications of this.

Nevertheless, they point to the possibility of a bifurcated policymaking system in the future. The bifurcation would occur between those responsible for supplying the empirical premises of the policy debate (bureaucrats, planners, policy analysts, scientists, and engineers) and those concerned that the decision makers get the normative issues right (the Party, the NPC, the press, non-technical intellectuals, perhaps some scientists). Such a bifurcation would be consistent with trends in other countries.

Impacts on Underlying Values and Perceptions

Distorted policymaking, resulting from the primacy of technical advice and the illusion of the technical fix, are one manifestation of a larger problem referred to by Manfred Stanley as the "cognitive conquest" of "technicism."[18] The concept of technicism may be more appropriate for the post-Mao era than that of "scientism" introduced by Danny Kwok to describe an intellectual orientation toward the role of science in society in the pre-1949 period.[19] According to the "technicist" view, the power of instrumental reasoning, which has enjoyed such success in man's dealings with nature, should be applied to problems in all other realms of existence. Technical efficiency becomes the basis for problem solving and evaluation, driving out other values in the process. When "technicism" triumphs, instrumental rationality structures and dominates our cognitions.

Most observers would agree that the Chinese political culture could benefit from the infusion of instrumental rationality, an important characteristic of modernity. China's S&T development can be expected to foster the diffusion of more instrumental approaches to political life.

But technicism involves more than the infusion of instrumental rationality. It is rather a totalistic myth of the triumph of instrumental rationality—with its methodology, values, and standards of adequacy—over other modes of responding to experience. While technicism is evidently a false doctrine, it is nevertheless a seductive one. It may be particularly appealing to adherents of Marxism-Leninism who do not regard totalistic, "scientific" doctrines as uncongenial. The danger of the cognitive conquest of technicism in China may be to reinforce a static Marxism-Leninism, and thereby retard needed ideological innovation.

Among the fallacies of technicism are its denial that noninstrumental values matter and its insistence that there are simple solutions to dissent over values. But technicism does not provide answers in situations where one person's instrumental rationality is another person's source of annoyance. Yet, the disruptions and dislocations of modernization evoke such dissensus. Since technicism is mute on this subject, it does not offer a plausible formula for social meaning to which a regime could appeal in managing the inevitable conflicts generated by modernization.

It is not inevitable, however, that China will be susceptible to the cognitive conquest of technicism. Technicism is not a necessary concomitant to S&T development. Indeed, it is antithetical to the tentativeness of science and the "open-systems" thinking of many forms of modern engineering practice. One could imagine these latter habits of experiencing the world taking root in China, most likely in "pockets" of research institutes, universities, and production enterprises. For the influence of these pockets to spread, however, there would have to be more tolerance of cultural diversity and organizational autonomy than heretofore. But, again, there is nothing necessary or inevitable about the diffusion of these values of science and technology.

SCIENCE, TECHNOLOGY, AND THE POLITICAL FUTURE

As in the past, the political system continues to have a strong influence on S&T. Yet, on the basis of comparative experience, it is reasonable to expect that progress in S&T will influence the evolution of the political system. The impacts are often likely to engender unexpected consequences, and ultimately, the outcomes will be a function of interactions of S&T, policy, and politics. While there is some evidence of Chinese efforts to anticipate and plan for these impacts, they have only begun. The hard problems of political and institutional analysis of the impacts of S&T have not yet been tackled.

The unpredictability of the impacts of S&T is also a function of the fluidity and lack of institutionalization in the polity. The various challenges, discussed above, are not likely to be forces for stability; they will be upsetting and problematic. But this does not mean that they are necessarily forces for political instability. As suggested, they could also be forces for the reorganization of power which could turn out to be quite congenial with modernity.

The consequences of the interactions between S&T development and the political system depend to a great extent on the timing and sequences of changes. S&T could be important forces for the modernization of the polity, the enhancement of its capabilities, and its evolution toward a more liberal and open order. Yet, there is nothing necessary or automatic about the actualization of this possibility. In fact, should it prove threatening, S&T development could be sacrificed by political leaders wishing to protect positions of power. From the discussion above of the eight realms of S-T-P relations, it is difficult to see how the political system, regardless of political will, could avoid some exceedingly disruptive consequences from the success of the S&T development program.

For example, successful S&T development may hinder the search for the right formula for political and administrative centralization. As we have seen, the scale, centralization, and standardization effects of modern technologies are complex, at times contradictory, and defy the simple prescription of institutional remedies. While some centralization is necessary for dealing with distributive problems and S&T may improve central policymaking, centralization will not solve all of the problems involved with deciding on mechanisms of social choice.

The eight realms of change in S-T-P relations have interesting implications for the future of the national-security state. The defense sector may be China's one institutional system capable of managing the effects we have described. While organizational changes will be necessary, the bureaucratic mission and extant organization of the defense sector can not only accommodate these changes, but can profit from them. This would suggest that the military industry and R&D system are likely to play an important role in the shaping of the direction of the S&T development program, and that aspects of the "modern military-industrial complex" version of the future of the national security state are likely to be prominent features of the polity.

The impact of S&T on intra-elite stability is not encouraging. The challenges from successful S&T development will require a high degree of elite consensus. Yet, when we consider how some of the challenges discussed above—those pertaining to mechanisms of social choice, the economizing question, principles of distribution—will impact the elite, increased intra-elite conflict seems likely. The bifurcation of elites based on professional orientation is likely to be transformed into one based on political interests and constituencies by the choices that will have to be made. Problems of elite consensus will not be helped by the increasing number of (and increasingly complex) distributive, redistributive, and regulatory policy challenges that are likely to appear.

S&T development will also affect state-society relations, because successful S&T development is likely to contribute to more autonomous social action. A society that is more autonomous of the state may in principle seem more desirable. However, in the context of China's problems of governance and policy implementation, this may actually create greater political instability.

The problems of governance will not be helped by the confusion over cultural values which accompanies successful S&T development. Technicism may be appealing to a China experiencing S&T modernization, but, as previously stated, it is ultimately a false doctrine for providing meaning to social life and articulating shared social values necessary for the pursuit of common objectives.

These rather pessimistic conclusions about the Chinese political future could be said to be overstated. They do not take into consideration the unique adaptive behaviors that any society exhibits and, relatedly, the inevitable shaping of the effects of S&T development by the political system. As noted above, the question of timing will also be important.

Yet, it is difficult to escape the conclusion that S&T development will generate a range of new challenges for political action, which, in volume and complexity, will be difficult for the political system to assimilate. If the past is a guide to the future, the active political intervention into S&T development necessary to ameliorate the more disruptive effects of S&T will seriously constrain the success of the development effort, which now enjoys high priority. Thus, China's choices are difficult indeed, and it may face a considerably more uncertain political future than is commonly supposed.

Notes
Index

NOTES

Introduction, by Merle Goldman and Denis Fred Simon

1. Qian Xuesen, "Scientific-social Revolution and Reform," *BR,* 23 March 1987, p. 22.
2. Nathan Sivin, "Why the Scientific Revolution did not take place in China or didn't it?" *Chinese Science* 5:45–66 (1982).
3. Richard Baum, "Science and Culture in Contemporary China," *Asian Survey,* December 1982, p. 1170.
4. Ibid.
5. Richard P. Suttmeier, "New Directions in Chinese Science and Technology", in John Major, ed., *China Briefing* (Boulder: Westview Press, 1986), pp. 91–102.
6. Genevieve Dean, *Technology Policy and Industrialization in the People's Republic of China* (Ottawa: International Development Research Center, 1979).
7. Han Baocheng, "On U.S. Technology Transfer," *BR,* 20 April 1987, p. 4.
8. David Michael Lampton, ed., *Policy Implementation in Post-Mao China* (Berkeley: University of California Press, 1987).
9. Ibid.
10. Li Honglin, "Socialism and Opening to the Outside World" *People's Daily,* 15 October 1984, p. 5.
11. Martin Fransman, *Technology and Economic Development* (Boulder: Westview Press, 1986).

12. Edward Roberts, ed., *Generating Technological Innovation* (New York: Oxford University Press, 1987), especially pp. 222–242.
13. Jan Prybyla, *Market and Plan Under Socialism: The Bird in the Cage* (Stanford: Hoover Institution Press, 1987). See also Joseph Berliner, *The Innovation Decision in Soviet Industry* (Cambridge: M.I.T. Press, 1976).
14. Charles E. Lindblom, *Politics and Markets* (New York: Basic Books, 1977).
15. Peter Hall and Ann Markusen, eds., *Silicon Landscapes* (Boston: Allen & Unwin, 1985).
16. Song Jian quoting Deng Xiaoping in "Science Reforms Vital," *Science*, 9 August 1985, p. 526.
17. Morris Crawford, "Programming the Invisible Hand: The Computerization of Korea and Taiwan" (Cambridge: Program on Information Resources Policy, Harvard University, 1986).
18. See the papers on Japan by L. Lynn and T. Ozawa in Nathan Rosenberg and C. Frischtak, eds., *International Technology Transfer* (New York: Praeger, 1986).

1. Technocratic Organization and Technological Development in China: The Nationalist Experience and Legacy, 1928–1953, by William C. Kirby

1. See Deng's statements on "talent" in the chapter by Leo A. Orleans in this volume.
2. Sun Yat-sen, *The International Development of China* (New York and London, 1922 [Taipei: China Cultural Service, 1953]), p. 208.
3. See William C. Kirby, *Germany and Republican China* (Stanford: Stanford University Press, 1984), pp. 30–32, 58–61, 78–81; Arthur N. Young, *China's Nation-Building Effort, 1927–1937: The Financial and Economic Record* (Stanford: Hoover Institution, 1971), pp. 292–294.
4. Douglas S. Paauw, "The Kuomintang and Economic Stagnation, 1928–1937," *Far Eastern Quarterly* 16. 2:213–214 (February 1957).
5. Bodo Wiethoff, *Luftverkehr in China, 1928–1949* (Wiesbaden: Otto Harrassowitz, 1975), p. 318.
6. Instructive in this regard is the experience of Ding Wenjiang as Director General of Shanghai. See Charlotte Furth, *Ting Wen-chiang: Science and China's New Culture* (Cambridge: Harvard University Press, 1970), pp. 166–191.
7. Hung-mao Tien, *Government and Politics in Kuomintang China, 1927–1937* (Stanford: Stanford University Press, 1972), pp. 21–22. Lloyd E. Eastman, *China Under Nationalist Rule: Two Essays* (Urbana: Center for Asian Studies, University of Illinois, 1980), p. 11.
8. See Kirby, *Germany and Republican China*, pp. 81–85.
9. See Richard Suttmeier's chapter in this volume.
10. See Furth, *Ting Wen-chiang*, pp. 216–219.
11. See the chapter by Nina Halpern in this volume.
12. SHA, 47(4), Weng Wenhao, "Guofang sheji weiyuanhui zhi mudi ji shuoming" (Goals of the National Defense Planning Commission; December

1932); SHA, 47(32), "Canmo benbu guofang sheji weiyuanhui mishuting gongzuo baogao" (Report of Secretariat of the National Defense Planning Council of the General Staff; 1934); CAS, *Ziyuan weiyuanhui yange* (Successive changes of the National Resources Commission; Nanjing, 1949), p. 1b; SHA 28(5965), "Zhonggongye jianshe jihua shuomingshu" (Explanation of the plan for heavy industries, 1936); Cheng Linsun, "Lun kang Ri zhanqian ziyuan weiyuanhui de zhonggongye jianshe jihua" (The NRC's heavy industrial plan in period prior to the anti-Japanese war), *Jindaishi yanjiu* (Research in modern history), 1986, no. 2; and William Kirby, "Kuomintang China's 'Great Leap Outward': The 1936 Three Year Plan for Industrial Development," in *Essays in the History of the Chinese Republic* (Urbana: Center for Asian Studies, University of Illinois, 1983).

13. On individual enterprises, see, *Ziyuan weiyuanhui yange; ZYYK* 1.2:85–100, 158–159, 337 (June 1939); 3:163–166 (July 1939); 2.1:37ff. (January 1940); *Guofang gongye ji wuqi fazhan* (National-defense industries and armaments development), ed. Combined Services Forces (Taibei, n.d.). On varieties of Sino-foreign joint ventures of the period, see Kirby, "Joint Ventures, Technology Transfer and Technocratic Organization in Nationalist China, 1928–1949," *Republican China* 12.2 (April 1987).
14. CAS, *Ziyuan weiyuanhui heban shiye heyue huibian* (Assembled NRC agreements for cooperative ventures; April 1941).
15. These included tungsten, antimony, and tin. See SHA, 28(2), "Ziyuan weiyuanhui wuye guanlichu zuzhi dagang" (Outline organization of the NRC tungsten-administration office; July 1937); and Kirby, "National Policy and Local Politics: The Kiangsi Tungsten Trade during the Nanking Decade," paper delivered to the Midwest Conference on Asian Affairs (1981).
16. This is still the dominant interpretation in the PRC: see Wu Taichang, "Guomindang zhengfu de yihuo changzhai zhengce he ziyuan weiyuanhui de kuangchan guanli" (The Guomindang Government's barter and debt-repayment policies and the NRC's mining administration), *Jindaishi yanjiu* (Research in modern history) 17:83–102 (July 1983).
17. Weng Wenhao, "Wode yijian buguo ruoci" (My humble opinion), *Duli pinglun* (Independent critic) 15 (28 August 1932). See correspondence between Weng and Hu Shi in *Hu Shi laiwang shuxin xuan* (Selected correspondence of Hu Shi; Beijing, 1979–1980), Vols. II and III.
18. *ZYYK* 1.3:163 (July 1939); ibid. 1.2:85–100 (June 1939); *GYTL* III, 839, 903–904; *Guofang gongyue ji wuqi fazhan*, p. 225.
19. CAS, *Ziyuan weiyuanhui guowai maoyi shiwusuo baogao* (Reports of the NRC Foreign Trade Office), monthly and annually, 1939–1942, 1946; AH, NRC, file, "Dui Su maoyi jiaoshi" (Trade negotiations with the USSR), April 1947; AH, NRC, file, "Niuyue gongzuo baogao, 1940–1944" (New York Foreign Trade Office Report, 1940–1944); Arthur N. Young, *China's Wartime Finance and Inflation, 1937–1945* (Cambridge: Harvard University Press, 1965), pp. 98–99, 104–105.
20. *Jingjibu gongbao* 1.13:608 (16 August 1938). Details on the practice of control and regulatory powers may be found in Ch'ao-ting Chi (Ji Chaoding), *War-*

time Economic Development of China (New York: Institute of Pacific Relations, 1939), Ch. 3, pp. 6–8, 12–17.

21. AH, NRC, file, "Ziyuan weiyuanhui jianjie" (Brief introduction to the NRC), 1942, pp. 1–7; *China Handbook 1937–45* (New York: Macmillan, 1947), p. 364; Chi-ming Hou, "Economic Development and Public Finance in China, 1937–45," in Paul K. T. Sih, ed., *Nationalist China During the Sino-Japanese War, 1937–45* (Hicksville, NY: Exposition Press, 1977), 211.
22. AH, NRC, 1941–1946 files entitled "Ge jiaoyu jiguan jieshao xuesheng" (Introductions of students by various educational agencies); "Zhongyang jianjiaoyu hezuo weiyuanhui" (Central Educational Cooperative Committee); "Kaoshiyuan kaoxuan weiyuanhui" (Examination Committee of the Examination Yuan); "Xuanyong gedaxue biye xuesheng laihui" (Selection of college graduates to enter the commission); "Zijianan" (Cases of self-recommendation).
23. AH, NRC, "Ziyuan weiyuanhui jianjie," p. 3; FDR, AWPM, box 40, Cragg manuscript, p. 133. See the production index of NRC-operated plants in *China Handbook 1937–45*, p. 366.
24. FDR, AWPM, Cragg, p. 97.
25. Ibid.
26. FDR, AWPM, box 36, notes of a conference with NRC representatives, 18–19 September 1944; *China Handbook 1937–45,* p. 366.
27. See, for example FDR, AWPM, box 6, report of Edwin K. Smith, 1 June 1945; ibid. box 7, memorandum to H. LeRoy Whitney, 16 June 1945; ibid., box 10, NRC report to Jacobsen, September 1944, pp. 7ff. On the "great prestige" enjoyed by technicians and engineers in Republican China, see Y. C. Wang, *Chinese Intellectuals and the West, 1872–1949* (Chapel Hill: University of North Carolina, 1966), p. 470.
28. AH, NRC, WWH, letter, Weng Wenhao to F. Pan, 4 March 1948.
29. AH, NRC, file, "Ziyuan weiyuanhui jianjie," speech by Qian Changzhao of 21 January 1942. See also Qian Changzhao, "Zhonggongye jianshe xianzai ji jianglai" (Present and future of heavy industrial development), *ZYGB* 3.3:49–54 (September 1942). Even T. V. Soong discussed following a Soviet model of industrialization in a BBC speech of 8 August 1943; see Yuan-li Wu, *China's Economic Policy: Planning or Free Enterprise?* Sino-International Economic Pamphlets, no. 4 (New York, 1946). See also "Sulian gongye jianshe zhi yejin" (Evolution of Soviet industrial development), *ZYGB* 3.6:43–47 (December 1942).
30. SHA, 28(2)934, file, "Zhanhou gongye jianshe chubu shishi jihua."
31. PAC, file, "Gongye jianshe jihua huiyi" (Conference on industrial development plan); Wu Cheng-lo, "China's International Industrial Capital," *China at War* 15.5–6:50 (November–December 1945).
32. AH, NRC, "Ziyuan weiyuanhui jianjie"; ibid., file, "Xunlian renyuan gedan ziliao" (Materials on the training of various units); ibid., files "Jishu weiyuanhui xinhan" (Correspondence of the technical committee, 1944–1945). SHA, 28(18971), files, "Zhu Mei jishutuan" (Technical Office in the United States).
33. See SHA, 28(2)50, file, "Zhu Mei daibiaotuan banshichu sanshiwu niandu

baogao" (1946 annual report of the NRC office in the United States); AH, NRC, file, "Waiji zhuanjia" (Foreign experts), list of 16 August 1945; ibid., "Gexiang jishu hetong" (Various consulting agreements), 1945–1946.

34. SHA, 52(2)1517, passim; HST, papers of Harry S. Truman, files of Edwin A. Locke, Jr., box 8, "Précis of Dr. J. L. Savage's Preliminary Report on the Yangtze Gorge Project"; NA, records of the Bureau of Reclamation of the Department of the Interior, file 090.09, "Foreign Activities, China, 1944–45."
35. See Kenneth Lieberthal and Michel Oksenberg, "Waiting for the Three Gorges Dam," *China Business Review* 13.5:7–9 (1986).
36. AH, NRC, Qian Changzhao papers, telegram, Weng Wenhao to Yun Zhen, 16 May 1947.
37. PAC, "Jingji jianshe wunian jihua biaojie" (Chart of the Five-Year Plan for economic construction), 1946.
38. See *UNRRA in China, 1945–47* (Washington: United Nations Relief and Rehabilitation Administration, 1948), pp. 287–307; George Woodbridge, ed., *UNRRA: The History of the United Nations Relief and Rehabilitation Administration* (New York: Columbia University Press, 1950), II, 443–447.
39. AH, NRC, WWH, correspondence, Weng Wenhao with Xingzhengyuan peichang weiyuanhui (Reparations Committee of the Executive Yuan), 1946–1947; ibid., Qian Changzhao papers, Yun Zhen to Yu Dawei, 12 September 1946.
40. *FRUS,* 1942, China, p. 228, Gauss to Hull, 8 September 1942.
41. Generally, see Kirby, "Planning Postwar China: China, the United States and Postwar Economic Strategies," *Proceedings of the Conference on Sun Yat-sen and Modern China* (Taibei, 1986).
42. CIA research report SR-8 (March 1948), III-23.
43. *ZYGB* 13.4:40 (October 1947).
44. See SHA, 28(2)831, for Sun Yueqi's reports of takeover difficulties in the northeast, 1945–1946; see also AH, NRC, WWH, correspondence, Weng Wenhao and Zhang Cikai, 1945–1946.
45. Shen Yi, "Ziyuan weiyuanhui yu wo" (The NRC and me), *Zhuanji wenxue* (Biographical literature) 44.2:73 (1984).
46. NA, Dept. of State, 893.60/10–1646, Bayne (Shanghai) to Penfield (Washington), 16 October 1946.
47. At least this was the case until the NRC lost its ability to place university graduates in jobs in the final year of the war; see Suzanne Pepper, *Civil War in China: The Political Struggle, 1946–1949* (Berkeley: University of California Press, 1978), p. 88.
48. Weng Wenhao, "Huigu wangshi" (Looking back on past events), *Wenshi ziliao* (Literary and historical materials) 80:13 (February 1982).
49. See Wu Taichang, p. 86; Guan Demao, "Weng Wenhao qi ren yu shi" (Weng Wenhao, the man and his work), *Zhuanji wenxue* 36.4:77 (1980); "Weng Wenhao, 1889–1971," ibid. 42.6 (1982).
50. On the relationship of the Political Study faction to Chiang, see Hung-mao Tien, *Government and Politics in Kuomintang China, 1927–37* (Stanford: Stanford University Press, 1972), pp. 67–68.

51. For a different view from the standpoint of the Nanjing decade, see Parks M. Coble, Jr., *The Shanghai Capitalists and the Nationalist Government 1927–1937* (Cambridge: Council on East Asian Studies, 1980), pp. 233–240.
52. Lloyd E. Eastman, *Seeds of Destruction: Nationalist China in War and Revolution, 1937–1949* (Stanford: Stanford University Press, 1984), p. 112.
53. Although Weng apparently formally joined the GMD during the war, his Party membership appears to have been understood as the minimum gesture needed from a cabinet minister. One source claims that he never did join the Party; see Guan Demao.
54. Eastman, *Seeds,* pp. 109–113; Xiao Zheng, "Kangzhan houqi shi tudi gaige yu dangzheng gexin yundong" (Land reform and Party and government renovation in the latter stages of the war of resistance), *Zhuanji wenxue* 34.3:104 (March 1979). On Weng's attitude toward the Youth Corps, see *FRUS* 1945, China, pp. 410–411.
55. Eastman, *Seeds,* pp. 170–202; *FRUS,* 1948, China, p. 720.
56. These consisted almost exclusively of NRC personnel already on Taiwan in charge of NRC industrial facilities there.
57. *Lun gongshangye zhengce* (Policy toward industry and commerce), ed. Jiefangshe (n.p.: Xinhua, 1949), pp. 23ff; *Gongshangye zhengce yu zhigong zhengce* (Policy toward industry and commerce and policy toward staff and workers; n.p.: Zhonggong Huabei zhongyuan ju, January 1949).
58. See Pepper, pp. 390–391. On the particularly smooth transition in Tianjin, see *Jiefang hou de Tianjin gongye* (Tianjin industry after liberation; Tianjin, 1950), pp. 11ff.
59. Perhaps too much so by Pepper, pp. 392–393.
60. Interview, Sun Yueqi, Beijing, 16 July 1985; Sun Yueqi, "Guomindang zhengfu ziyuan weiyuanhui liu zai dalu zhi jingguo" (The circumstances of the NRC's remaining on the mainland), *Wenshi ziliao* (Literary and historical materials) 69:191–192 (1979); Xie Peihe, "Huiyi Shanghai gongye de jieguan" (Recalling the takeover of Shanghai industry), *Shanghai wenshi ziliao* (Shanghai literary and historical materials) 46:272 (1985).
61. Sun Yueqi, pp. 193ff.
62. CAS, "Ziyuan weiyuanhui yewu weiyuanhui huiyi jilu" (Records of meetings of the NRC's industrial committee), meeting of 1 June 1949.
63. Ibid., meetings of 30 May to 15 October 1949.
64. S. Kuznets, "Growth and Structural Shifts," in W. Galenson, ed., *Economic Growth and Structural Change in Taiwan: The Postwar Experience of the ROC* (Ithaca: Cornell University Press, 1979).
65. Alice H. Amsden, "The State in Taiwan's Economic Development," in Peter B. Evans, et al., *Bringing the State Back In* (Cambridge: Cambridge University Press, 1985), p. 91. For personnel information regarding NRC corporations on Taiwan, see Guoshi guan (Academia historica), ed., *Ziyuan weiyuanhui dangan shiliao chubian* (Documentary materials on the NRC, first compilation; Taibei, 1985), II, 715ff. The Premier was Sun Yunxuan. Generally, on the strong state managerial role in Taiwan's economic development, see Amsden, pp. 78–106; Thomas B. Gold, *State and Society in the Taiwan Miracle* (Armonk:

M. E. Sharpe, 1986); and Mohuan Hsing, *Taiwan: Industrialization and Trade Policies* (Oxford: Oxford University Press, 1971), esp. pp. 169–170, 201–202.

66. Chu-yuan Cheng, *Communist China's Economy, 1949–1962* (South Orange: Seton Hall University Press, 1963), 9.
67. Ibid., p. 8; Cheng, *China's Economic Development: Growth and Structural Change* (Boulder: Westview, 1982), p. 138; Hsueh Mu-ch'iao (Xue Muqiao), *The Socialist Transformation of the National Economy in China* (Beijing: Foreign Language Press, 1960), p. 20; Thomas G. Rawski, *China's Transition to Industrialism* (Ann Arbor: University of Michigan Press, 1980), p. 30.
68. Interviews, Xie Peihe, Wang Danxiang, Lin Aiyuan, in Shanghai, 22 August 1985.
69. For recent foreign complaints see "Why Investors are Sour on China," *New York Times*, 8 June 1986, p. F-7; and "Great Wall," *Wall Street Journal*, 17 July 1986, p. 1.
70. Max Weber, *The Theory of Social and Economic Organization* (Glencoe: Free Press, 1957), p. 337; Hans Rosenberg, *Bureaucracy, Aristocracy and Autocracy: The Prussian Experience, 1660–1815* (Cambridge: Harvard University Press, 1958); Chalmers Johnson, *MITI and the Japanese Miracle: The Growth of Industrial Policy, 1925–1975* (Stanford: Stanford University Press, 1982), pp. 36–37; Guillermo A. O'Donnell, *Modernization and Bureaucratic-Authoritarianism* (Berkeley: Institute of International Studies, 1973); Karen L. Remmer and Gilbert W. Merkx, "Bureaucratic-Authoritarianism Revisited," *Latin American Research Review* 17.2:3–50 (1982); John B. Sheahan, "Experience with Public Enterprise in France and Italy, in William G. Shepherd, ed., *Public Enterprise: Economic Analysis of Theory and Practice* (Lexington, MA: Lexington Books, 1976), pp. 123–183. V. V. Ramanadham, ed., *The Nature of Public Enterprise* (New York: St. Martin's Press, 1984), pp. 227–238; LeRoy P. Jones, ed., *Public Enterprise in Less-developed Countries* (Cambridge: Cambridge University Press, 1982); Pranab Bardhan, *The Political Economy of Development in India* (Oxford: Basil Blackwell, 1984), pp. 23–39; Jagdish N. Bhagwati and Padma Desai, *India: Planning for Industrialization* (London: Oxford University Press, 1970), pp. 111–122, 135–167.
71. See Kendell E. Bailes, *Technology and Society under Lenin and Stalin: Origins of the Soviet Technical Intelligentsia, 1917–1941* (Princeton: Princeton University Press, 1978).
72. But on the eventual dominance of state over party bureaucracies even in the Soviet Union, see John Lukacs, "The Soviet State at 65," *Foreign Affairs* 65.1: 21–36 (Fall 1986).
73. A situation best described by Lloyd Eastman, *The Abortive Revolution: China Under Nationalist Rule, 1927–1937* (Cambridge: Harvard University Press, 1974), pp. 278–286.
74. There is, of course, no shortage of examples in Chinese or Western history of bureaucratic bodies exchanging one set of rulers for another, even under "revolutionary" conditions. On the position of technocrats after three distinct "revolutionary" transitions, see Bailes, *Technology and Society under Lenin and Stalin*, pp. 44–62; Karl-Heinz Ludwig, *Technik und Ingenieure im Dritten Reich*

(Düsseldorf: Droste, 1974), pp. 103–160; and A. James Gregor, *Italian Fascism and Developmental Dictatorship* (Princeton: Princeton University Press, 1979), p. 323.

75. Richard Suttmeier, *Science, Technology and China's Drive for Modernization* (Stanford: Hoover Institute, 1980), p. 34.
76. See Harry Harding, *Organizing China: The Problem of Bureaucracy, 1949–1976* (Stanford: Stanford University Press, 1981), esp. 70–78.
77. The renewed eminence of former NRC chairmen Qian Changzhao (Vice-Chairman of the National People's Political Consultative Conference) and Sun Yueqi (who honorifically supervises work on the Three Gorges dam project, initially planned by the NRC), is public evidence of this, as are several major scholarly projects on the history and organization of the NRC.

2. *Learning from Russia: Lysenkoism and the Fate of Genetics in China, 1950–1986, by Laurence Schneider*

1. See Vice-Premier Fang Yi, "Zai quanguo kexue dahui shang de baogao" (Report to the National Science Conference), 18 March 1978, section 2, which sets a priority for "genetic engineering." *GMRB* 29 March 1978.
2. The phrase *Hundred Schools* is derived from the slogan made famous in the mid-1950s: "Let a Hundred Flowers Bloom; Let a Hundred Schools of Thought Contend." (The "flowers" are the arts and literature; the "schools" are the natural sciences and other areas of scholarship.) For a succinct outline of the "professional" model of science associated with the policy and its counter "mobilization model" in China, see Richard P. Suttmeier, *Research and Revolution* (Lexington, MA: Lexington Books, 1974).
3. For the background of PRC geneticists, see my "Genetics in Republican China," in John Z. Bowers, et al., eds., *Science and Medicine in Twentieth Century China* (Ann Arbor, University of Michigan Press, forthcoming).
4. These judgments are based, *inter alia,* on these sources: S. D. Richardson, *Forestry in Communist China* (Baltimore: Johns Hopkins University Press, 1966), pp. 143–149; Y. C. Ting, "Genetics in the People's Republic of China," *BioScience* 28.8:506–510 (August 1978); Jack R. Harlan, "Plant Breeding and Genetics," in Leo Orleans, ed., *Science in Contemporary China* (Stanford: Stanford University Press, 1980), pp. 295–312. Also my interviews with Y. C. Ting, Professor of Biology, Boston College and Visiting Researcher at the CAS Genetics Institute (Beijing, 1984); C. C. Tan, Director of Fudan University Genetics Institute (Shanghai, 1984); C. C. Li, former Chair, Biostatistics Department, University of Pittsburgh, and Chair, Agronomy Department, Beijing Agricultural University, 1949–1950 (Pittsburgh, 1985).
5. For this summary, I have relied on the standard sources on Lysenkoism by Mark Adams, Loren Graham, David Joravsky, and Zhores Medvedev. Of special recent interest are two papers: Nils Roll-Hansen, "A New Perspective on Lysenko?" (unpub. paper, Oslo, Norway, 1984); and Douglas R. Weiner, "The Roots of 'Michurinism': Transformist Biology and Acclimatization as Currents in the Russian Life Sciences," *Annals of Science* 42:243–260 (1985).

6. *Jeifang ribao* (Liberation), 10 April 1940, p. 21; November 1941, p. 4; 23 October 1942, p. 2; 30 January 1943, p. 4; 27 March 1946, p.2. And Michael Lindsay, ed. *Notes on Educational Problems in Communist China 1941–1947* (New York: Institute of Pacific Relations, 1950). Also see *Kangri zhanzheng shiji jiefanggu koxue jishu fazhan shi ziliao* (Historical materials on the development of S&T in the Liberated Areas during the anti-Japanese war period, 4 vols.; Beijing: Zhongguo xueshu, 1983–1985), esp. II, 85–97.
7. Formally, T. D. Lysenko came to full power in 1948, at the July-August meeting of the Lenin Academy of Agricultural Sciences. He used the presidency of this academy for his headquarters. For a chronicle of these Soviet tours, see *KXTB* 3 (1950); 1:17–24 (1953); 5:100 (1953). Also see *BNDB* 2.1:2 (April 1956). Also interviews with C. C. Li and C. C. Tan and with Professor Mi Jingjiu of Beijing Agricultural University (Beijing 1986). Mi was an interpreter for some of the Russian advisors and he became a major publicist for Lysenkoism in China.
8. For examples, see *KXTB:* 1:32 (1950) and 3:194 (1950); and *Kexue dazhong* (Mass science) 12:445 (1954).
9. See "Genetics Dies in China," *The Journal of Heredity* 41.4:90 (April 1950).
10. Interviews with Li Peishan, Deputy Director of CAS History of Natural Science Institute (Beijing 1986), Tang Peisong, CAS Plant Physiology Institute (Beijing 1986), and with C. C. Tan. And see Tan's "Pipan wo dui Michulin shengwu kexue de zuowu kanfa," (Criticism of my mistaken attitudes toward Michurinism) *KXTB* 8:562–563 (1958). Also see Liang Xi, "Wo duiyu Luo Tianyu tongzhi suo wei zuowu de ganxiang" (The political errors of Comrade Luo Tianyu), *People's Daily*, 29 June 1952; and anon., "Wei jianchi shengwuxue di Michulin fangxiang er naozheng" (Carry on the struggle in biology for the Michurin line), *People's Daily*, 29 June 1952. At this time, CAS was by default responsible for most science planning.
11. "Wei jianchi," *People's Daily*, 29 June 1952.
12. Ibid.
13. C. C. Tan interview, and self-criticism (see note 10 above).
14. For further discussion and a list of key secondary works on the structure of Chinese S&T and the Soviet influence, see my bibliographical essay "Science, Technology, and China's Four Modernizations," *Technology In Society* 3:291–303 (Autumn 1981). Also see O. B. Borisov and B. T. Koloskov, *Soviet Chinese Relations* (Bloomington: University of Indiana Press, 1975), and Mikhail Klochko, *Soviet Scientist in Red China*, tr. A. MacAndrew (New York: Praeger, 1964).
15. For Lepeshenskaya, see, for example, *Ziran kexue* (Natural science) 2:49–54 (January 1952) and 2:217–221 (March 1952). Also see A. V. Dubrovina, *Darwin zhuyi* (Darwinism; Shanghai, 1953). This is a translation of her lectures. And for Glushchenko, see *KXTB* 11:44–58 (1955); and *BNDB* 2.1:2 (April 1956).
16. Interview with Professor Bao Wenkui of Beijing Agricultural University (Beijing, 1986). *People's Daily*, 25 August 1956 and 18 December 1956; and Yu Youlin's biography of Bao Wenkui in *ZBFT* 3:85–93 (1979).

17. C. C. Tan interview and interviews with Liu Zutong, Professor of Human Genetics, Fudan Institute of Genetics (Shanghai, 1984, 1986).
18. For example, see Glushchenko's harassment of Zhu Xi (1908–1962) the eminent French-trained embryologist, in *KXTB* 1:80–81 (1953) and 4:70–73 (1953).
19. For the October 1955 CAS Michurin Centenary celebration, see *KXTB* 10, 11, 12 (1955). Also see *Michulin xuanji* (Michurin's works) compiled by the North China Agriculture Research Institute (Beijing, 1956); and Liu Fulin, tr., *Michulin yibai-nian jinian wenji* (Centenary collection of Michurin's works; Beijing, 1957).
20. The reconstruction of events from January through April is based on: Yu Guangyuan, "Zai 1956 Qingdao yichuanxue hui shang de jianghua," (Talks at the Qingdao Symposium) *ZBFT* 5:5–13 (1980), Gong Yuzhi, "Fazhan kexue de biyou zhi lu," (Develop the road that must be followed in science), *GMRB*, 28 December 1983, p. 2; C. C. Tan interview; Li Peishan, et al., "Qingdao . . . lishi beijing ho jiben jingxian" (Qingdao's historical background and basic events), *ZBFT* 4:41–49 (1985); Interviews with Yu Guangyuan (Beijing 1986), and Li Peishan.
21. Discussion of the symposium is based on interviews with C. C. Tan and Liu Zutong, and on the *Qingdao yichuanxue zuotanhui fayen jilu–neibu* (Proceedings of the Qingdao Genetics Symposium–for restricted circulation; Shanghai, 1957). I have edited and translated this document as *Lysenkoism in China: Proceedings of the 1956 Qingdao Genetics Symposium* (Armonk: M. E. Sharpe, 1986).
22. The director, Wang Shou, taught classical genetics at the Nanjing-Cornell program in the 1930s. See his *Zhongguo zuowu yuzhongxue* (Crop-breeding in China; Nanjing, 1936). The Deputy Director, Tai Songen, received his BS from Nanjing in 1931 and his PhD from Cornell in 1936. Zu Deming, leader of the Michurinists, was also a Deputy Director.
23. See CAS Genetics Institute research reports in *Yichuan xue jikan* (Genetica sinica) 4 (1959) and thereafter.
24. Li Ruqi, "Yichuanxue zhangzuo" (Lectures on genetics), a 13-part serial in *SWXT* 3 (1957)–3 (1958). Also see, Wu Zhongjian, tr., *Zhiwu zajiao de shixian* (Mendel's works; Beijing, 1957); Lu Huilin, tr., *Jiyin lun* (Morgan's theory of the gene; Shanghai, 1959); Sheng Zujia, *Yichuan mima* (The genetic code; Shanghai, 1960); *SWXT* 1:11–17 (1957) for translation of Dubinin's work on hybrid corn; Fudan Genetics Institute, tr. *Fushe yichuanxue wenti* (Dubinin's papers on radiation genetics; Shanghai, 1964).
25. Li Jingxiong, *BNDB* 3.1:1–22 (1957); *Hong Qi* (Red flag) 2, 3 (1962). *People's Daily*, 8 January 1963; Wu Shaoma, *Nongyeh xuebao* (Acta agricultura sinica) 10.5:360–367 (1959); Bao Wenkui, *GMRB*, 24 October 1962 and 27 October 1962.
26. Tong Dizhou, *GMRB*, 17 March 1961; and interviews with C. C. Tan, Liu Zutong, and Li Peishan.
27. Liu Zutong interviews.
28. See reports of CAS Biochemistry Institute in *Acta Biologica Sinica* 1961–1965;

and Tan's radiation work and collaboration with Dubinin's student Arseneva in *Scientia Sinica* throughout 1964.

29. This renewal of the Hundred Schools policy began formally with a *Hongqi* editorial, 5 March 1961. Eighty-one of these articles, in 653 pages, are collected in Fudan University institute of Genetics, ed., *Yichuanxue wenti taolun ji* (Collected materials on the genetics question); 3 vols. (Shanghai, 1961–1963).
30. C. C. Tan interview. I learned about Tong Dizhou from his former student and colleague Dr. Shi Yingxian, Deputy Director, CAS Developmental Biology Institute (interviews, Beijing 1984, 1986). For the experience of the CAS Genetics Institute see: *China: Science Walks on Two Legs* (New York: Avon 1974), pp. 122–130; Zhou Bao, "Achievements of Open Door Research," *Scientia Sinica* 19.2:171–174 (March 1976); "Open Door Research," *BR*, 2 February 1976, pp. 4–5.
31. For the 1972 symposium, see *ZBFT* 6:16 (1980). For a summary of Tong's 1972–1976 collaborative work with M. C. Niu, see "Genetic Manipulation in Higher Organisms" (in English), *Scientia Sinica* 20.4 (1977).
32. *Yichuanxue bao* (Acta genetica sinica) began publication in 1974. Its management was merged with that of the National Genetics Society in December 1978.
33. Interview with Shao Qiquan, Deputy Director of the CAS Genetics Institute, and Secretary of the National Genetics Society (Beijing, 1984). For a report of this organizational effort and the first conference, see *Yichuanxue bao* 6.1 (1979). *Yichuan* (Hereditas) began in January 1979.
34. Shi Yingxian interview.
35. Interviews with Dr. Wang Bin, Director CAS Genetic Engineering Laboratory (Beijing 1986), and Shao Qiquan (Beijing 1986).
36. Interview with Dr. Xue Jinglun, Fudan Genetics Institute (Shanghai 1986); and with Wang Zhiya, Director, and Luo Deng, Vice-Director, Shanghai Center of Biotechnology (Shanghai 1986).
37. Ibid., and 1986 interviews with Shi Yingxian and Shao Qiquan. For summaries of high-level policy relevant to these issues, see Premier Zhao Ziyang "Revamping China's Research System," *BR*, 8 April 1955, pp. 15–21; and Zhou Chengkui, "Revamping the Science and Technology System," ibid., 16 June 1986, pp. 21–26.
38. For example, see Fan Dainian, "Guan yu Zhongguo Koxue Yuan . . . ," (About the CAS), *ZBFT* 2:21–30 (1986).
39. This is the *ZBFT.* Its editor and informing intelligence is Mr. Fan Dainian.
40. Shi Xiyuan, "Lysenko qi ren" (This man Lysenko), *ZBFT* 1:65–81 (1979); Li Jingyen, "Lepeshenskaya xin xipaoxue shuo de xing-wang" (Rise and fall of the new cytology), *ZBFT* 1:59–63 (1981); Yu Youlin, "Bao Wenkui," *ZBFT* 3:85–93 (1979); Li Peishan, "Qie qie lao-ji lishi shang de jiangxian jiaoxun" (Never forget the lessons of historical experiences), *ZBFT* 1:13–16 (1981); Li Peishan et al., "Qingdao . . . de lishi beijing ho jiben jingxian," *ZBFT* 4:41–49 (1985).
41. Li Peishan et al., eds., *Baijia zhengming–fazhan kexue de biyou zhi lu: 1956 Qingdao yichuanxue zuotanhui jishi* (A Hundred Schools contending–Develop the road that must be followed in science; Beijing: Commercial Press, 1985).

42. For example, see Gong Yuzhi, *Wenhui Bao*, 29 April 1986; Lu Dingyi, *GMRB*, 7 May 1986; Yu Guangyuan, *People's Daily*, 16 May 1986; and various authors in *People's Daily*, 6 and 13 June 1986.
43. Li Peishan and Huang Shun'e, *People's Daily*, 2 May 1986. In the *New York Times*, 16 June 1986, John Burns reviews an English language version of this article which appeared in *BR*, 26 May 1986.
44. The symposium was held in Beijing and sponsored by many scholarly organizations, including the CAS. See *People's Daily*, 16 June 1986, for the symposium speeches. The description of Xu Liangying's remarks is from an eyewitness who wishes to remain anonymous.

3. Reform of China's Science and Technology Organizational System, by Tony Saich

1. For further discussion of this point, see R. Volti, "Science and Technology in Communist Systems: Introduction," *Studies in Comparative Communism* 15.1 and 2:1–5.
2. This definition refuted the idea developed by Mao Zedong and promoted by the "Gang of Four" that S&T were part of the superstructure of society and thus a legitimate arena for political struggle. Deng Xiaoping, "Speech at Opening Ceremony of National Science Conference," *Peking Review* 21.12:11 (24 March 1978).
3. In the first year after the arrest of the "Gang of Four," some 16 academic conferences were held for specialists in the field of S&T.
4. During the CR, the number of institutes under CAS was reduced from around 100 to less than half that number.
5. Zhao Ziyang, "Gaige keji tizhi, tuidong keji he jingji, shehui xietiao" (Reform the science and technology system and promote its coordination with the economy and society), *People's Daily*, 21 March 1985, pp. 1, 3.
6. According to Mr. Luo Wei, Deputy Director of the Science Policy Study Office of CAS, while both sides had to accept blame, he felt that industry should bear greater responsibility. Interview 16 January 1985.
7. See Chen Xianhua, "Zhuajin jianli hangye jishu kaifa zhongxin" (Speed up the establishment of technology development centers for industry), *Keyan guanli* 1:35–36 (1984).
8. Xia Yulong and Liu Ji, "It is also Necessary to Eliminate Erroneous 'Leftist' Influence on the Science and Technology Front," *Jiefang ribao*, 2 June 1981, p. 4, translated in *China*-81-112, 11 June 1981, K8-K12.
9. Xie Shaoming, "Shixing youchang hetongzhi" (Implement the compensation contract system), *GMRB*, 22 October 1984, p. 2.
10. "Combine Research With Production," *BR* 28.11:4 (18 March 1985).
11. Zeng Decong, "Gaodeng xuexiao keyan guanlizhong de jige lilun wenti" (Several theoretical problems in the management of scientific research among institutes of higher education), *Keyan guanli* 1:56–60 (1984).
12. Zhao Ziyang, "Gaige keji tizhi, tuidong keji he jingji, shehui xietiao," pp. 1, 3.
13. The appointment of Song Jian to head the SSTC was a further indication of

the need to integrate the work of the two sectors. Song had previously worked within the military S&T sector.

14. The SSTC of the PRC, *Statistical Data on Science and Technology of 1985* (Beijing, 1985), p. 13.
15. About 14% of China's researchers work in the 122 institutes of CAS, consuming about 10% of the nation's research funds. Interview with Mr. Luo Wei, 16 January 1985 and *Zhongguo kexue jishu zhengce zhinan* (A guide to scientific and technological policies in China; Beijing, 1986), p. 231.
16. Zhou Guangzhao, "Develop Science and Technology, Promote the National Economy—Some Thoughts on the Work at the Chinese Academy of Sciences," *Zhongguo keji luntan* 1:8–10,34 (1985) translated in *JPRS - CST* - 86 001 27 January 1986, p. 41.
17. Ibid., p. 44. In terms of financial allocations, around 15% was given to basic research, a little under 50% to applied research topics, and 35% to developmental research. Interview with Mr. Luo Wei, 16 January 1985.
18. "Zhonggong zhongyang guanyu kexue jishu tizhi gaige de jueding" (Decision of the Central Committee of the CCP on reform of the scientific and technological system), *People's Daily*, 20 March 1985, pp. 1, 3.
19. Interview with Mr. Luo Wei, 16 January 1985.
20. These institutes also conduct basic research. In terms of funding, basic research accounted for 20% of the total funding in 1985 (the SSTC of the PRC, *Statistical Data*, p. 13). Most production ministries have their own research and training institutes. Indeed, the Ministry of Agriculture, Animal Husbandry and Fishery, and the Ministry of Public Health, among others, have their own academies.
21. The impact of contract research is discussed in the following section.
22. The World Bank calculates that, in China, government spending on education as a percentage of its total expenditures is 7.1%. The median for other developing countries is 15.5%. WB, *China: Long-term Issues and Options. Annex A: Issues and Prospects in Education* (1985), p. 45.
23. Interview at the Ministry of Education, 17 January 1985; and Wang Yihang, "Tasks Set for Educational Reform," *BR* 28.51:22 (1985).
24. D. F. Simon, "Rethinking R and D," *The China Business Review*, July-August 1984, p. 27.
25. *China Exchange News* 13.2:37 (1985). These universities are Nanjing, Shandong, Fudan, Jilin, and Zhongshan.
26. These were the "Decisions Relating to the Transformation of the Economic Accounting System in Research Units and Design Institutes."
27. See Du Shunxing, "Kexue yanjiu danwei de jingji guanli" (Economics management in institutions of scientific research), *Keyan guanli* 4:73–76 (1980); and Li Guoguang and Fan Qiongying, "Shixing youchang hetongzhi dashi suoqu" (Compensation contract system is an irresistible trend), *People's Daily*, 13 December 1984, p. 3.
28. This did not include the research institutes of CAS.
29. Leaders of China's S&T system have likened China's labor force to a "pool of stagnant water" and policies have been introduced to stir up the water and get

it flowing. Increased movement of research personnel is seen as a good way of speeding up the dissemination of knowledge. Measures have included sending scientists on short-term contracts to work in remote areas and bringing technicians and engineers out of retirement. The most important measure has been curtailing the principle of lifelong tenure. For further details on personnel policy, see T. Saich, *China's Science Policy in the 1980s* (Manchester University Press, 1988), Chapter 5.

30. Zhao Ziyang, quoted in Yu Ren, "Yitiao xinluzi" (A new road), *Kexue yu kexue jishu guanli* (Studies of science and management of science and technology) 2:7–9 (1984).
31. "Zhonggong zhongyang guanyu kexue jishu tizhi gaige de jeuding."
32. For a discussion of the different types of research and production alliances and their advantages, see Chen Chuansheng and Ma Aiye, "Shilun woguo de kexue shengchan lianheti" (A discussion of our nation's scientific and production alliances), *Kexue yu kexue jishu guanli* 3:2–5 (1984).
33. See, for example, Wu Shumin, "Lüelun hangye jishu kaifa zhongxin" (A brief discussion on developing centers for industrial techniques), *Keyan guanli* 1:40–41, 39 (1984) and Li Minquan, "Shilun hangye yanjiusuo tiaozheng chongshi wei hangye jishu kaifa zhongxin" (A discussion of the readjustment and revitalization of industrial research institutions and their transformation into industrial technology development centers), *Keyan guanli* 1:42–46, 50 (1984).
34. Xinhua, 24 March 1985, translated in *JRPS - CST* 85 -010, p. 1 and Xinhua, 7 May 1985, translated in *Summary of World Broadcasts: the Far East* 7946.
35. T. P. Bernstein, "China in 1984: the Year of Hong Kong," *Asian Survey* 25.1:38 (1985).
36. As the American economist Jan Prybyla notes, the function of prices is to act as "backup 'value' instruments to quantitative orders." Some sectors (such as coal) can thus face unfavorable price structures and have become accustomed to operating with large subsidies. By contrast, industries that use coal have been receiving hidden subsidies as a result of the low price of coal. Changes in pricing and subsidies can also affect individuals who have been used to overt price stability. China's attempts to raise prices in 1980 led to unrest and thus, when price reform was returned to the policy agenda in 1984, it was stressed that price rises should not result in a fall in the real income of urban and rural inhabitants. Thus, price reform can upset interests deeply rooted in the system. Quote from Prybyla, *The Chinese Economy* (New York: Columbia University Press: 1981), p. 129.
37. BBC World Service, 7 November 1987.
38. Mu Gongqian, Wang Lin, and Huang Shengli, "Gaige guojia dui yanjiusuo guanli de sanxiang cuoshi" (Three measures to reform the state's management of research institutions), *GMRB*, 21 May 1984, p. 2.
39. Zheng Ruizeng and Zheng Fuming, "Yao 'yang ji chan dan' bu yao 'sha ji qu luan'" (Raise the chicken in order to lay eggs; don't kill the chicken to get the eggs), *GMRB*, 21 May 1984, p. 2.
40. The SSTC of the PRC, *Statistical Data*, p. 5. Of the 317 that were financially self-supporting, 39 came under the State Council and 278 under provincial-

level administration. Of the 648 receiving less financial support from the government, the totals were 87 and 553 respectively. There were no CAS institutes said to be self-supporting nor financially independent. Ibid. p. 6.

41. Xinhua, 4 December 1985.
42. P. Kneen, *Soviet Scientists and the State* (Basingstoke: Macmillan, 1984), pp. 103–104.
43. For an interesting account of Chinese reactions to the "new technological revolution," see W. L. van Woerkom-Chong, "The 'New Technological Revolution,' China's Modernisation and World Economy: Some Chinese Discussion Themes," *China Information* 1.2:33–44 (1986).
44. In addition, in October 1984, the Central Committee issued its decision on the reform of the economic system and, in May 1985, the decision on the reform of the education system.
45. For details on the process of the drafting of the 1985 decision, see T. Saich, *China's Science Policy in the 1980s,* Chapter 1.
46. See Gong Jinxing, "Tantan suozhang zerenzhi" (A talk on the responsibility system for institute directors), *Keyan guanli* 1:40–43 (1986). In the decentralization of powers, institutes are being given increased autonomy and have powers over questions of adjustment of orientation tasks, arranging the planning of scientific research, utilization of research topic expenditures, and appointment and dismissal of middle-level cadres.
47. Sun Jian and Zhu Weiqun, "Dangqian luoshi zhishi fenzi zhengce qingkuang jiujing ruhe? Jiangsusheng de diaocha biaoming: 'zuo' de yingxiang wangu, wenti yuanwei jiejue" (How is the current situation of implementing the policies on intellectuals? An investigation in Jiangsu province shows "Leftist" influence is powerful and problems are far from being solved), *People's Daily,* 8 July 1984, p. 3.

4. Reforms and Innovations in the Utilization of China's Scientific and Engineering Talent, by Leo A. Orleans

1. Xie Shaoming, "Paid Contract System is Key to Reform of Research System," *GMRB,* 22 October 1984; translated in *JPRS*-CST-85-004, 5 February 1985.
2. Merle Goldman, "The Zigs and Zags in the Treatment of Intellectuals," *China Quarterly,* December 1985, pp. 709–715.
3. New China News Agency, 1 June 1957, as cited in Leo A. Orleans, *Professional Manpower and Education in Communist China* (Washington: National Science Foundation, 1961), p. 95.
4. "Serious Misemployment of University Graduates," *Jingji ribao,* 1 February 1983; translated in *JPRS,* No. 83, 105, 21 March 1983, p. 65.
5. "Suitable Jobs for Graduates," *China Daily,* 8 September 1984, p. 3.
6. "Promotion Process Wastes Talent," *BR* 37:29 (16 September 1985).
7. Henan Provincial Service, 28 November 1985; translated in *FBIS*-PRC, 6 December 1985, p. P3.
8. Xinhua, 13 March 1986; translated in *JPRS*-CPS-86-029, 11 April 1986, p. 38.
9. Xinhua, 15 July 1983; *FBIS*-PRC, 19 July 1983, p. K4.

10. "State Council Approves 1984 Plan for Assignment of Postgraduate Students," *People's Daily*, 3 October 1984; translated in *JPRS*-CPS-84-074, 30 October 1984, p. 22.
11. "Colleges Outline Job Placement Reform," *China Daily*, 13 March 1985, p. 3.
12. Ibid.
13. Wang Yibing, "Tasks Set for Educational Reform," *BR*, No. 51, 23 December 1985, p. 21. In March 1985, Vice-Minister of Education Huang Xinbai announced that institutions of higher education would be allowed to enroll 33,000 employees from Chinese enterprises (5.9% of the new entrants) on a contract basis. *China Exchange News* 13.2:36 (June 1985).
14. Ibid.
15. "Students Want Jobs to Fit Their Talents," *China Daily*, 26 September 1985, p. 3.
16. "Majority of Students Assigned by State," *China Daily*, 23 July 1986, p. 1.
17. Ibid.
18. Yang Xuewen, "Important New Policy To Reform the College Graduate Placement System," *GMRB*, 10 July 1985; translated in *JPRS*-CPS-85-090, 11 September 1985, p. 24.
19. "Companies Poach Graduates," *China Daily*, 10 September 1985, p. 1.
20. "Suitable Jobs for Graduates," *China Daily*, 8 September 1984, p. 3.
21. "Serious Misemployment of University Graduates," *Jingji ribao*, 1 February 1983; translated in *JPRS*, No. 83, 105, 21 March 1983, p. 65.
22. Xu Ying, "Ministry of Education Tries to Improve Graduate Assignments," *GMRB*, 21 January 1983; translated in *JPRS*, No. 83, 059, 11 March 1983, p. 184.
23. *China Daily*, 25 February 1986, p. 3.
24. "Students Want Jobs to Fit Their Talents," *China Daily*, 26 September 1985, p. 3.
25. Xinhua, 28 May 1984; translated in *JPRS*-CST-033, 24 October 1984, p. 17.
26. Xinhua, December 4, 1984; translated in *FBIS*-PRC, December 5, 1984, p. K7.
27. "'Floating Talents,'" *China Daily*, 30 November 1985, p. 3.
28. Ding Shiyi, "Break Through the Obstacles Standing in the Way of Reasonable Talent Exchange," *Banyue tan* (Semi-monthly talks), No. 23, 10 December 1984; translated in *JPRS*-CEA-85-040, 25 April, 1985, p. 113.
29. "Comment on Recruitment," *Gongcandang yuan* (*Communist party member*), No. 1, 8 January 1985; translated in *JPRS*-CPS-85-038, 22 April 1985, p. 62.
30. "Flexible Personnel Policy 'Could Curb Waste of Skills,'" *China Daily*, 1 May 1985, p. 4.
31. "We Should Have a More Flexible Management System of Able Persons," *People's Daily*, 22 April 1985; translated in *FBIS*-PRC, 26 April, 1985, p. K9.
32. Xu Fan, "Inviting Applications for Jobs Is a Reform of the Personnel System," *People's Daily*, 26 August 1983; translated in *JPRS*, No. 84, 561, 19 October 1983, p. 64.
33. Xin Xiangrong, "Rational Flow of Skilled Personnel," *BR* No. 32:4 (6 August 1984).

34. Chengdu Sichuan Provincial Service, 30 August 1985; translated in *JPRS*-CPS-098, 23 September 1985, p. 59.
35. Kunming Yunnan Provincial Service, 1 August 1984; translated in *FBIS*-PRC, 3 August 1984, p. Q2.
36. Guo Shuyan, "Three Problems in the Management of Scientific and Technical Personnel," *People's Daily*, 10 April 1986; translated in *JPRS*-CST-86-028, 18 July 1986, p. 6.
37. *China Daily*, 17 August 1984, p. 4.
38. Ding Shiyi, "Break Through the Obstacles Standing in the Way of Reasonable Talent Exchange," *Banyue tan* (Semi-monthly talks), No. 23, 10 December 1984; translated in *JPRS*-CEA-85-040, 25 April 1985, p. 113.
39. *China Daily*, 8 June 1984, p. 1; in *FBIS*-PRC, 8 June 1984, p. 15.
40. Xinhua, 19 June 1984; translated in *JPRS*-CEA-84-055, 9 July 1984, p. 70.
41. Xinhua, 4 December 1984; *FBIS*-PRC, 5 December 1984, p. K8.
42. *Banyue tan*, p. 114.
43. "Guangdong Province Allows S&T Personnel To Resign and Seek Jobs Elsewhere," *GMRB*, 22 August 1984; translated in *JPRS*-CST-84-034, 29 October 1984, p. 13.
44. Changsha Hunan Provincial Service, 18 December 1984; translated in *JPRS*-CPS-85-006, 14 January 1985, p. 45.
45. Chen Xian and Xu Wenpao, "The Situation of 'Unit Ownership' of Talented People Should be Broken Up," *Xinhua ribao*, 15 January 1985; translated in *JPRS*-CPS-85-028, March 22, 1985, p. 72.
46. Quan Feng, "Why Can't Certain Places in Northern Jiangsu Keep Their Talented People?" *Xinhua ribao*, 7 January 1985; translated in *JPRS*-CPS-85-028, 22 March 1985, p. 61.
47. *China Daily*, 17 August 1984, p. 4.
48. Jing Shi and Liu Zhengcai, "Strengthen Horizontal Ties to Enhance the Mobility of Personnel," *Wen Hui bao*, 29 January 1985; translated in *JPRS*-CPS-85-064, 28 June 1985, p. 20.
49. Xiao Guan, "Shanghai Talent Bank Helps in the Rational Mobility of More Than 500 Scientists and Technicians," *China Daily*, 19 January 1986.
50. "State Plan to Improve Flow of Scientists," *China Daily*, 23 July 1984, p. 1.
51. *China Daily*, 17 August 1984, p. 4.
52. Wang Kang, "Some Problems with the Hiring System," *Zhongguo keji luntan* (Forum on science and technology in China), No. 1, September 1985; translated in *JPRS*-CST-86-008, 1 March 1986, p. 20.
53. "Students Sing for Their Supper," *China Daily*, 14 July 1986, p. 1.
54. Jin Luzhong, "In the Spirit of Reform Resolve the Management and Employment Problems of Scientific and Technical Personnel," *GMRB*, 17 October 1984; translated in *JPRS*-CST-85-004, 5 February 1984, p. 13.
55. Xinhua, 11 March 1986; translated in *JPRS*-CST-86-012, 8 April 1986, p. 65.
56. *China Daily*, 17 May 1986, p. 4.
57. *State Council Bulletin*, No. 15, 20 July 1984; translated in *JPRS*-CPS-85-022, 8 March 1985, p. 9.

58. Yang Xiaoping, "Research Bodies to Tender Bids for Key Science Work," *China Daily*, 5 February 1986, p. 1.
59. Jin Juzhong, "In the Spirit of Reform Resolve the Management and Employment Problems of Scientific and Technical Personnel," *GMRB*, 17 October 1984; translated in *JPRS*-CST-85-004, 5 February 1984, p. 13.
60. Xinhua, 4 February 1986; translated in *FBIS*-PRC, 5 February 1986, p. K16.
61. Zheng Haining, "Adjusting the Internal Relations of the Scientific and Technical Contingent," *GMRB*, 10 December 1985; translated in *JPRS*-CST-86-010, 19 March 1986, p. 39.
62. Xinhua, 7 February 1986; translated in *FBIS*-PRC, 5 February 1986, p. K15.
63. Xinhua, 13 April 1986; translated in *FBIS*-PRC, 17 April 1986, p. K12.
64. "Views on Setting Up Experimental Points in R&D Units To Put Operational Expenditure Under a Contract System," *State Council Bulletin*, No. 15, 20 July 1984; translated in *JPRS*-CST-85-022, 8 March 1985, p. 9.
65. Xinhua, 22 June 1985; translated in *FBIS*-PRC, 26 June 1985, p. K5.
66. Xie Shaoming, "Paid Contract System is Key to Reform of Research System," *GMRB*, 22 October 1984; translated in *JPRS*-CST-85-004, 5 February 1984, p. 35.
67. "Funding Scheme to Make Full Use of Scientific Skills," *China Daily*, 14 September 1984, p. 3.
68. Wang Kang and Wang Tongxun, "Report on an Investigation Into the So-Called 'Economic Crimes' Committed by Quite a Few Scientists and Technicians," *GMRB*, 1 August 1984; translated in *FBIS*-PRC, 8 August 1984, p. K2.
69. "The State Science and Technology Commission Issues a Circular on Eight Lines of Distinction for Determining the Direction of Reform and Against Unhealthy Tendencies," *Jishu shichang* (Technology market), 25 June 1985; translated in *JPRS*-CST-85-034, 1 October 1985, p. 10.
70. Xiao Guangen, "We Cannot Be Too Restrictive of Spare Time Jobs for Scientists and Technicians," *People's Daily*, 9 April 1986; translated in *JPRS*-CST-86-028, 18 July 1986, p. 16.
71. "Opinions of Some Questions in Current After-Work Hours Technical Service of Scientists and Technicians," *Hebei ribao*, 9 April 1984; translated in *JPRS*-CST-85-003, 28 January 1985, p. 19.
72. For a further discussion of this subject, see Leo A. Orleans, "Soviet Influence on China's Higher Education," in Ruth Hayhoe and Marianne Bastid, eds., *China's Education and the Industrialised World* (Armonk: M. E. Sharpe, 1987).
73. Xinhua, 1 February 1986; *JPRS*-CST-86-008, 1 March 1986, p. 57.
74. Xinhua, 21 June 1984; *JPRS*-CEA-84-055, 9 July 1984, p. 44.
75. Xinhua, 8 May 1985; translated in *FBIS*-PRC, 30 May 1985, p. K1.
76. It was recently reported that "almost all of China's colleges of engineering have established cooperative contracts with enterprises and businesses in the science and technology fields." (Xinhua, 1 February 1986; *JPRS*-CST-86-008, 1 March 1986, p. 57.)
77. Xinhua, 21 June 1984; *JPRS*-CEA-84-005, 9 July 1984, p. 44.
78. "University Faculty, Staff Receive Bonus for Consulting Work," *China Daily*, 7 February 1984, p. 3.

79. Tao Xinliang, "Some Viewpoints on Science and Technology Development in Higher Education," *Kexue yu jishu kexue guanli* (Science of science and management of S&T), No. 11, 15 November 1985; translated in *JPRS*-CST-86-008, 1 March 1986, p. 30.
80. Ibid., p. 31.
81. Ibid., p. 30.
82. Xia Zuoyun, "More Scientists Should Combine Research with Business Management," *Tianjin ribao*, 12 November 1985; translated in *JPRS*-CST-86-009, 13 March 1986, p. 29.
83. *China Daily*, 11 April 1986, p. 2.
84. Wang Yougong, "A New Contingent of Scientific and Technical Troops—Survey on Science and Technology Service Companies Run by the People of Nanjing," *People's Daily*, 4 October 1985; translated in *JPRS*-CST-86-005, 11 February 1986, p. 39.
85. Xinhua, 26 July 1984; *JPRS*-CST-84-033, 24 October 1984, p. 16.
86. Xinhua, 22 October 1985; *JPRS*-CST-85-039, 13 November 1985, p. 12.
87. Zhao Yingchun, "Scientific and Technological Consulting in the New Situation," *Keyan guanli* (Science research management), No. 4, October 1985; translated in *JPRS*-CST-86-012, 8 April, 1986.
88. Tao Weijiang, "Some Reflections on the Function of Higher Level Institutions in the Technology Markets," *Kexue yu jishu guanli* (Science of science and management of S&T), No. 11, 15 November 1985; translated in *JPRS*-CST-86-008, 1 March 1986, p. 25.
89. "Persist in Reform and Promote Rational Transfers of Qualified Personnel," *People's Daily*, 23 July 1983; translated in *FBIS*, 30 July 1986, p. K25.
90. "Education Fuels Modernization Drive," *China Daily*, 16 July, p. 4.
91. In part, this is probably due to the significant number of students graduating from 2- or 3-year specialized technical colleges who fill (at least temporarily) the middle-level technical positions. (See Leo A. Orleans, "Graduates of Chinese Universities: Adjusting the Total," *The China Quarterly*, September 1987, pp. 444–449.)
92. "Education Fuels Modernization Drive," *China Daily*, 16 July 1986, p. K25.

5. *China's Industrial Innovation: The Influence of Market Forces, by William A. Fischer*

1. Richard P. Suttmeier, *Research and Revolution: Science Policy and Societal Change in China* (Lexington, MA: Lexington Books, 1974).
2. Richard P. Suttmeier, "Science and Technology in China's Socialist Economic Development," paper prepared for the World Bank Science and Technology Unit, Project Advisory Staff, January 1982.
3. Suttmeier, "Science and Technology."
4. William A. Fischer, "The Structure and Organization of Chinese Industrial R&D Activities," *R&D Management* 13:63–81 (April 1983).
5. William A. Fischer, "Scientific and Technical Planning in the People's Republic of China," *Technological Forecasting and Social Change* 25:189–207 (May 1984).

6. Hu Yongchang, Jiang Chengcheng, Chen Changqing, Luo Deng, and Huang Aizhu, "The Policy Adopted for the Total Synthesis of Insulin and Yeast Alanine tRNA and the Manner in Which the Work Was Organized," paper presented to the National Academy of Sciences US-China Science Policy Conference, Washington, D.C., January 1983.
7. Tony Saich, *The Evolution of Science and Technology Policy in the People's Republic of China Since the Death of Mao Zedong* (Sinologisch Instituut, Leiden University, December 1984).
8. Denis Fred Simon, "Implementation of China's S&T Modernization Program," in David Michael Lampton, ed., *Policy Implementation in Post-Mao China* (Berkeley: University of California Press, 1985).
9. Marshall I. Goldman, *USSR in Crisis: The Failure of an Economic System* (New York: Norton, 1983).
10. Harley D. Balzer, "Soviet Science in the Gorbachev Era," *Issues in Science and Technology* 1:29–54 (Summer 1985).
11. Suttmeier, "Science and Technology," pp. 60–61.
12. Ibid., p. 74.
13. Saich, p. 15.
14. Liu Guoguang and Zhao Renwei, "Relationship Between Planning and the Market Under Socialism," in George C. Wang, ed., *Economic Reform in the PRC* (Boulder: Westview Press, 1982).
15. Denis Fred Simon, "Science and Technology Reform," *The China Business Review,* March-April 1985.
16. Roy Rothwell and Walter Zegveld, *Industrial Innovation and Public Policy* (London: Francis Pinter, ltd., 1981); Harvey A. Averch, *A Strategic Analysis of Science & Technology Policy* (Baltimore: The Johns Hopkins University Press, 1985).
17. William A. Fischer, "Update on Enterprise Reforms," *China Business Review,* September-October 1986, pp. 42–45.
18. William A. Fischer, "Chinese Industrial Management: Outlook for the Eighties," in the Joint Economic Committee of the US Congress, *China's Economy Looks Toward the Year 200,* Vol. I (Washington: USGPO, 1986).
19. Chen Jiyuan, Xu Lu, Tang Zongkun, and Chen Lanton, "Management Reforms in the Qingdao Forging Machinery Plant," in William Byrd, Gene Tidrick, Chen Jiyuan, Xu Lu, Tang Zongkun, and Chen Lantong, *Recent Chinese Economic Reforms,* World Bank Staff Working Papers, Number 652 (Washington: The World Bank, 1984).
20. Denis Fred Simon and Detlef Rehn, "Innovation in China's Semiconductor Components Industry: The Case of Shanghai," *Research Policy* 16.5:259–277 (October 1987).
21. Zhou Rixin, "Xian Aircraft Corp. Flying High," *BR,* 27 January 1986, p. 26.
22. Wu Fumin and Xia Ruge, "Hao Jianxiu Speaks," Xinhua Domestic Service, 24 December 1985, translated in *FBIS*-CHI-85-248, Daily Report–China, 26 December 1985, p. K4.
23. "State Gives Firms New Trade Rights," *China Daily,* 18 January 1986, p. 2.
24. Joseph Alutto, personal communication.
25. William A. Fischer, "The Transfer of Western Managerial Know-how to

China," report prepared for The Office of Technology Assessment, May 1986.
26. Zhao Jinming, "Profits Rise as Link-up Pays Off," *China Daily*, 18 April 1986, p. 2.
27. Fischer, "Chinese Industrial Management."
28. Ji Dong, "My Humble View of the Key to and Breakthrough Point in the Reform of Scientific and Technological Research Systems," *Scientology and Management of Science and Technology*, March 1985, pp. 19–22, translated in *JPRS*-CST-85-019, China Report-Science and Technology, 18 June, pp. 27–36.
29. Meng Xiangjie, "Technical Markets Are Flourishing; Bridges and Links—A General Survey of Technical Markets Nationwide," *Outlook Weekly*, 21 January 1985, pp. 32–33, translated in *JPRS*-CST-85-012, China Report-Science and Technology, 23 April 1985, pp. 15–17.
30. Yang Yi, "China's Need Cited to Link Research With Market," *China Daily*, 7 April 1986, p. 4.
31. William A. Fischer and Michael Farr, "Dimensions of Innovative Climate in Chinese R&D Units," *R&D Management* 15:183–190 (July 1985).
32. William A. Fischer, "The Commitment to R&D Among Chinese Enterprises," School of Business Administration, University of North Carolina at Chapel Hill, Working Paper, Summer 1986.
33. William A. Fischer, "The Management Center at Dalian," *China Exchange News* 11:12–14 (June 1983).
34. William A. Fischer and William Woodward, "Technical Decision Making in Chinese Enterprises," School of Business Administration, University of North Carolina at Chapel Hill, Working Paper, Summer 1986.
35. This is a refrain the author has repeatedly heard from legions of Chinese managers from 1980 up to the present.

6. *Organizational Reforms and Technology Change in the Electronics Industry: The Case of Shanghai, by Detlef Rehn*

The research contained in this paper was conducted in Shanghai in July 1985 and in Beijing in January 1986 under a grant from the Stiftung Volkswagenwerk in Hannover, FRG. The views expressed by the author are solely his own and do not necessarily represent the views of Stiftung Volkswagenwerk.

1. See, for example, Roger Miller and Marcel Cote, "Growing the Next Silicon Valley," *Harvard Business Review*, July/August 1985, pp. 114–123.
2. Nancy Dorfman, "Route 128: The Development of a Regional High Technology Economy," *Research Policy* 12:299–315 (1983).
3. Everett Rogers and Judith Larsen, *Silicon Valley Fever* (New York: Basic Books, 1984).
4. The lectures were published in Zhonggong Zhongyang Zuzhi Bu (Department of Organization of the CCP Central Committee) et al., eds., *Yingjie xin de jishu geming* (Welcome to the new technological revolution; Changsha, 1984), 2 vols.
5. Ma Hong, "Zhuazhu jihui, yingjie shijie xin de jishu geming de tiaozhan" (Let

us take the opportunity and welcome the challenges of the new technological revolution in the world), Zhonggong Zhongyan Zuzhi Bu (see note 4 above), "Jishu geming", pp. 27–75.

6. Li Tieying, "Jianli jicheng dianlu gongye jidi wo guo gongye fazhan de yi ge zhanlüexing wenti" (The establishment of IC industrial bases is a strategic question of our industrial development), *China Computerworld* 14:3 (20 July 1983).
7. Zhang Jinquan, "Guanyu jianshe kexue gongye qu ruogan wenti chutan" (Discussion of some questions concerning science and industrial parks), *Keyan guanli* 4:11–14 (1986).
8. See, for example, Fang Yi's speech at the National Science Conference in March 1978. Fang Yi, "Zai quan guo kexue jishu da hui shang de baogao (Zhaiyao)" (Selections from The National Conference of Science and Technology), *People's Daily*, 29 March 1978.
9. "Zhonggong Zhongyang guanyu jingji tizhi gaige de jueding" (Resolution of the CCP Central Committee on the reform of the economic system), *People's Daily*, 21 October 1984; "Zhonggong Zhongyang guanyu kexue jishu tizhi gaige de jueding" (CCP Central Committee's decision on the reform of the science and technology management system), *People's Daily*, 20 March 1985.
10. See "Dianzi Gongye Bu 'Guanyu tuijin dianzi gongye jingji guanli tizji gaige de baogao' (Zhaiyao)" (Ministry of Electronics, Report on the implementation of the reform of the management system of the electronics industry [Extracts]), *Zhongguo dianzi bao*, 21 March 1986, p. 2. China's electronics industry is characterized by a multitude of institutions belonging to different "systems." Engaged in electronics are institutions run by industrial ministries of the central government, the military and CAS, factories and research institutes at the provincial level, and others.
11. See Li Peng, "Dianzi he xinxi chanye yao wei si hua jianshe fuwu" (The electronics and information industries have to serve the construction of the Four Modernizations), *Jingji ribao*, 14 January 1985, and Xinhua, 11 January 1985, *FBIS*–China, 15 January 1985, pp. K25–27.
12. Li Peng, "Dianzi he xinxi."
13. Xinhua, 11 January 1985.
14. "Dianzi Gongye 'Qi Wu-jihua gangyao" (Outline of the 7th Five-Year Plan for the electronics industry), *Zhongguo dianzi bao*, 2 May 1986.
15. Jiang Zemin, "Wo guo dianzi gongye de fazhan zhanlüe wenti" (Problems of the development strategy of the electronics industry in our country), *Zhongguo jixie dianzi gongye nianjian 1985* (Yearbook of China's machine-building and electronics industry; Beijing, 1985), I, 4–8.
16. Li Tieying, "Jianchi gaige, jiasu fazhan, nuli zhenxing dianzi gongye" (Continue the reform, speed up the development, actively invigorate the electronics industry), *Zhongguo dianzi bao*, 21 January 1986. Also "Nuli fazhan xiaofei lei dianzi chanpin, jianli dianzi gongye jingji liangxing xunhuan" (Actively develop consumer electronics, establish a healthy economic cycle in the electronics industry), *Zhongguo dianzi bao*, 31 December 1985, p. 2.
17. Li Zhaoji, "Yao zhongshi touzi lei chanpin de kaifa" (Pay attention to the

development of investment-type products), *Zhongguo dianzi bao,* 11 March 1986, p. 1.

18. Chen Jiyuan, "Diqu jingji jiegou de duice" (Measures regarding the regional economic structure), in Sun Shangqing, ed., *Lun jingji jiegou duice* (On the economic structure; Beijing, CASS Publishing House, 1984), pp. 318–360.
19. Shanghai Shi Tongji Ju, ed., *Shanghai tongji nianjian 1983* (Statistical yearbook of Shanghai; Shanghai: People's Publishing House, 1984), p. 93; Guojia Tongji Ju, ed., *Zhongguo tongji nianjian 1983* (Statistical yearbook of China, Beijing: Statistics Publishing House, 1983), p. 253.
20. A summary of the different standpoints in the debate is presented in Shen Junbo et al., "Guanyu Shanghai changyuan guihua de ge zhong jianjie" (Views on Shanghai's long-term development), *Shehui kexue* 3:70–75 (1981).
21. Nicholas Lardy, *Economic Growth and Distribution in China* (Cambridge: Cambridge University Press, 1978).
22. Xu Zhihe and Li Douhan, "Shanghai gongye jiegou helihua wenti chutan" (A preliminary survey of the rationalization of Shanghai's industrial structure), *Shehui kexue* 1:25–29 (1982).
23. "Hu Yaobang Hopes that Shanghai Will Play a Vanguard Role in the Reform," *Ming pao,* 21 October 1984, translated in *FBIS*-PRC, 24 October 1984, p. W3.
24. Rui Xingwen, "Tuanjie qi lai, wei ba Shanghai jianshe cheng wei kaifangxing, duo gongneng he gaodu wenming de shehuizhuyi xiandaihus chengshi er fendou" (Join together and fight for the transformation of Shanghai into an open, multi-faceted and culturally advanced modern socialist city), Report at the 5th delegation conference of the Shanghai Communist Party, *Wenhui bao,* 10 March 1986.
25. "Guowuyuan guanyu Shanghai shi chengshi zongti guihua fang'an de pifu" (State Council on the general program for the urban development of the City of Shanghai), *Wenhui bao,* 6 November 1986.
26. *Shanghai jingji fazhan zhanlüe wenji* (A selection of articles on strategy for Shanghai's economic development; Shanghai: Academy of Social Sciences, 1984), pp. 49–71.
27. "Jiao ta shidi, yingjie xin tiaozhan" (Firmly welcome the new challenge—Interview with Liu Zhenyuan, Vice-Mayor of Shanghai), *Kexuexue yu kexue jishu guanli* 4:8–12 (1984).
28. The Jiangsu Leading Group has been jointly established by the MEI and Jiangsu province; in the (local) Beijing Leading Group, Zhang Xuedong, Vice-Minister of the MEI, is a deputy chairman. *Zhongguo dianzi bao,* 4 June 1985 and 8 July 1986.
29. *Zhongguo dianzi bao,* 30 July 1985, p. 1.
30. "Wei fazhan Zhongguo dianzi gongye zou chu yi tiao xin luzi" (Go a new way in the development of China's electronics industry), *Zhongguo dianzi bao,* 24 January 1986, p. 1.
31. *Guoji dianzi bao,* 11:1 (11 June 1986).
32. One group, especially the MEI, argued in favor of Wuxi because of its excellent environment (clean water, low air pollution) and the good conditions for research and production owing to the location of a number of China's most

advanced electronics facilities there. Another group preferred Shanghai because of its large R&D potential, the high educational level of its labor force, and the existence of strong support industries.

33. *People's Daily,* 17 July 1986. Also Zhang Feng, "Li Tieying yi du tan" (Discussion with Li Tieying), *Shijie jingji daobao* 300:15 (4 August 1986).
34. Zhang Feng, "Li Tieying yi du tan."
35. Linear ICs are of an analog type, i.e. the relationship between inputs and outputs varies continuously over a specific range. In digital ICs, the relationship between inputs and outputs is a binary (on/off) one. Franco Malerba, *The Semiconductor Business* (Madison: University of Wisconsin Press, 1985), p. 14.
36. "Zai gaige maichu xin de yi bu" (A new step in the reform), *Zhongguo dianzi bao* 142:1 (7 November 1986).
37. *Wenhui bao,* 11 March, 1986; *GMRB,* 15 March 1986.
38. Discussions in Shanghai, July 1985.
39. Denis Simon and Detlef Rehn, "Innovation in China's Semiconductor Components Industry: The Case of Shanghai," unpublished manuscript, February 1986.
40. *Zhongguo dianzi bao,* No. 157, 30 December 1986 and No. 160, 6 January 1987.
41. *Zhongguo dianzi bao,* 25 February 1986, p. 1.
42. Madelyn Ross, "Shanghai's Push into High-Technology," *China Business Review,* March/April 1985, pp. 36–39.
43. Faculty of Industry and Foreign Trade, Shanghai Jiaotong University, "Xin jishu geming yu ruanjian chukou kaifa" (The new technological revolution and the software export and development), *Shanghai jingji fazhan zhanlüe wenji,* pp. 133–142.
44. Ross.
45. The negative impact of the low level of Shanghai's computer industry has already been felt, as may be seen from the decision of the State Council Electronics Leading Group to prefer Beijing as the main site for the production of large and medium-sized computer systems. Discussions in Beijing, January 1986.
46. Gene Gregory, *Japanese Electronics Technology: Enterprise and Innovation* (Tokyo: Japan Times Press, 1985), pp. 419–424.

7 Scientific Decision Making: The Organization of Expert Advice in Post-Mao China, by Nina P. Halpern

In addition to the participants in the conference on "China's New Technological Revolution," Harvard University, May 1986, especially the organizers, Denis Fred Simon and Merle Goldman, I should like to thank the following individuals for helpful comments on an earlier draft of this chapter: Alexander George, Carol Hamrin, Kenneth Lieberthal, Denis Sullivan.

1. Deng Xiaoping, "Respect Knowledge, Respect Trained Personnel," in *Selected Works of Deng Xiaoping (1975–1982)* (Beijing: Foreign Languages Press, 1983), p. 54.

2. Wan Li, "Making Decisions With a Democratic and Scientific Approach is an Important Aspect in Restructuring the Political System," *FBIS,* 19 August 1986, p. K28.
3. Harry Harding, *Organizing China* (Stanford: Stanford University Press, 1981), p. 17.
4. I have argued elsewhere that the reform group, at least, had little impact on policymaking. See my "Making Economic Policy: The Influence of Economists," in US Congress Joint Economic Committee, ed., *China's Economy Looks Toward the Year 2000* (Washington: US GPO, 1986), pp. 137–140.
5. See Kenneth Lieberthal, "The Three Gorges Dam Project," in Kenneth Lieberthal and Michel Oksenberg, *Bureaucratic Politics and Chinese Energy Development* (Washington: US GPO, 1986), Chapter 6.
6. In my discussion of the various advisory bodies, in addition to documentary sources, I draw on interviews conducted in China and the United States between October 1985 and March 1986 with three officials of the SSTC's NRCSTD; a member of SSTC's Department of Comprehensive Management; an official of the State Council's Economic, Technical, and Social Development Research Center; and a former researcher at the Institute of Industrial Economics of CASS, who also consulted for TERC and other advisory bodies.
7. Ma Hong, "Jiaqiang jishu jingji yanjiu, wei 'sihua' jianshe fuwu" (Strengthen research on technical economics, serve Four Modernizations construction), in Ma Hong, *Kaichuang shehui kexuede xin jumain* (Initiate a new situation in social science; Beijing: Shehui kexue chubanshi, 1984), pp. 207–208.
8. See also the discussion of the Center's purpose in *Jingjixue dongtai* (Trends in economics), July 1981, pp. 5–6.
9. These are by now well documented. See, e.g., Michel Oksenberg, "Economic Policy Making in China: Summer 1981," *China Quarterly* 90:165–194; Denis Fred Simon, "Understanding the Electronics Industry," *The China Business Review,* March-April 1986, pp. 10–15; and Susan Shirk, "The Domestic Political Dimensions of China's Foreign Economic Relations," in Samuel S. Kim, ed., *China and the World* (Boulder: Westview Press, 1984), pp. 57–81.
10. *Jingjixue dongtai,* July 1981, pp. 5–6.
11. Diplomatic sources reported this, based on an interview with a high-level figure in the Center.
12. According to Ma Hong, before TERC was set up, Zhao Ziyang had seriously considered a proposal to establish a "Technical Advisory Commission" under the State Council. Zhao had ultimately rejected this idea because he felt that the specific types of questions likely to arise in policymaking were not appropriately addressed by any fixed group of experts; rather, different experts should be drawn into the process depending on the type of question under study. See Ma Hong, *Kaichuang shehui kexuede xin jumian,* p. 207.
13. I would like to thank Professor Kenneth Lieberthal for pointing this out to me.
14. Chen Yun has been particularly outspoken in criticizing the SPC; see his harsh indictment of its performance in his 1982 speech, "Jianquiang jihua jingji"

(Strengthen the planned economy), in *Sanzhong quanhui yilai zhongyao wengao xuanbin* (Collection of important documents since the 3rd Plenum), Vol. II (Beijing: Renmin chubanshe, 1982; reprinted Taipei: Zhongyang yanjiu zazhishe, 1983), p. 1134.

15. For a history of the Baoshan project, see Martin Weil, "The Baoshan Steel Mill: A Symbol of Change in China's Industrial Strategy," in US Congress Joint Economic Committee, ed., *China Under the Four Modernizations,* Vol. I, (Washington: US GPO, 1982), pp. 365–393. At the end of 1980, scientists, perhaps motivated by the desire for an equivalent to the recently established ERC, lobbied for the establishment of scientific advisory groups under the State Council and Party Secretariat, arguing that the need for such groups was demonstrated by the fact that neglect of scientists' advice in the past had led to such erroneous decisions as Baoshan. (These demands were reported in *People's Daily* on 10 November 1980; see Xinhua, 10 November 1980, in *FBIS*, 12 November 1980, p. L33.) TERC may have been a response to this lobbying, as well as to the leadership's own awareness of faulty decision making over the prior few years.
16. Chen Yun's Finance and Economics Commission appeared to concentrate on the readjustment issue; its only known meeting was a September 1979 one at which Chen presented his views on economic readjustment (see *Sanzhong quanhui yilai zhongyao wengao xuanbian* I, 171–176), and Chen is known to have argued the priority of this issue from 1978 on.
17. The following type of procedure appears to be fairly typical. When the State Council was presented with a proposal to build a pipeline to transport coal, it asked TERC for advice. The Center organized relevant departments to perform feasibility studies and afterwards organized a meeting at which the results of these studies were debated. On issues that were unresolved after much discussion, Ma Hong stated that TERC would report to the State Council the areas of disagreement and the differing view of the participants. See Ma Hong, *Kaichuang shehui kexuede xin jumian,* p. 208.
18. Students of US presidential decision making have suggested two alternative models of how an advisory body can address the problem of structuring the flow of advice from the bureaucracy to the executive. One is a system of "centralized management," in which the advisory body's staff is relied upon "to distill, analyze, and present options to the president for his choice. . . . the president . . . is shielded from too much direct exposure to undigested, 'raw' disagreements over policy within the executive branch." (Alexander L. George, "The Case For Multiple Advocacy in Making Foreign Policy," *American Political Science Review,* September 1972, p. 752.) Such a system elevates the advisory body, which presumably shares the president's perspective more than do the bureaucratic agencies, to a position *above* the latter. The alternative model is one labeled "multiple advocacy," which "is designed to expose the President to competing arguments and viewpoints made by the advocates themselves rather than having viewpoints filtered through a staff to the President." (Roger B. Porter, *Presidential Decisionmaking: The Economic Policy Board* [Cambridge: Cambridge University Press, 1980], p. 241.) The key

difference is whether or not disagreements among different lower-level actors are presented to the executive for decision, or whether he is shielded from them. TERC appears to fit the model of "multiple advocacy" much better than that of "centralized management."

19. Lin Xi, "Important Capital Construction and Technological Transformation Projects to be Examined by Consulting Organs Before Finalization," *People's Daily*, 8 February 1986, p. 1, in *FBIS*, 21 February 1986, p. K11.
20. Commentator's Article, "An Important Reform in Managing Construction Projects," *People's Daily*, 8 February 1986, p. 1, in *FBIS*, 21 February 1986, p. K12.
21. A smaller study of this type, inspired by World Bank visitors to China in 1981, was done by the State Science and Technology commission in 1981–1982. I should like to thank Carol Hamrin for this information.
22. In addition to interviews, information on the study is found in Ma Hong, "Kaizhan '2000 niande Zhongguo' de yanjiu" (Develop research on "China in the Year 2000"), in *Kaichuang shehui kexuede xin jumian*, pp. 103–123; *JPRS* #84524 (13 October 1983), S&T 210: 165–166; *JPRS* #84374 (21 September 1983), Economic Affairs 384: 71–72; *FBIS*, 4 November 1985, p. K17; and *BR* 44:18–20 (1985).
23. CAST Constitution, published in *People's Daily*, 28 March 1980, p. 2, in *JPRS* #75797, 30 May 1980, p. 1.
24. In addition to interviews with officials of the NRCSTD, this discussion is based upon a Chinese-English pamphlet issued by the Center, titled "National Research Center for S&T for Development," published in Beijing.
25. Ibid., and *Kexue guanli yanjiu* (Science research management) 4:23 (August 1983).
26. Only brief summaries of TERC's report have so far been published, so it is difficult to make this comparison.
27. *FBIS*, 3 April 1986, p. K2.
28. Most NRCSTD members who are currently called "economists" actually have a mathematical or scientific-engineering background; their exposure to economics is said to consist largely of attending occasional lectures on the subject rather than of any systematic training.
29. Strictly speaking, the Commission is considered a decision-making rather than an advisory organ, but it organizes much research of an advisory nature.
30. *Jingjixue dongtai*, August 1982, pp. 8–10.
31. In a personal communication, Professor Kenneth Lieberthal suggests that one purpose of the groups may be to increase the personal staff resources available to the leaders.
32. My information about the S&T leading group comes largely from an extremely useful unpublished report by Martha Caldwell Harris of the US Congress Office of Technology Assessment, "Trip Report: Chinese Decision-making on Technology Importation."
33. Denis Fred Simon, "Understanding the Electronics Industry," p. 12.
34. This is very clear, e.g., in the ongoing debate over the Three Gorges dam project and in the attempt to produce a coordinated development strategy for the

electronics industry. On the latter, see Denis Fred Simon, "Understanding the Electronics Industry."

35. Wan Li, "Making Decisions With a Democratic and Scientific Approach Is an Important Aspect in Restructuring the Political System," p. K26.
36. Cao Jiarui, "China's Technological Imports–Present Status and Problems (Part III)," *Liaowang Overseas Edition* 20:12–13 (19 May 1986), in *FBIS,* 30 May 1986, p. K16.
37. Lin Xi, "Important Capital Construction and Technological Transformation Projects to be Examined by Consulting Organs Before Finalization," p. K11.
38. I should like to thank Carol Hamrin for clarifying this point for me.
39. Alexander L. George, "Limits of Rationality in Designing Public Policy and Developing Policy-Relevant Theory: A Discussion Paper," unpublished paper presented at the Dartmouth College Forum on the Limits of Rationality in Designing Public Policy, October 1985.
40. For an argument that Communist elites in Czechoslovakia have been able to make use of expert advice without compromising their power over the system, see Sharon L. Wolchik, "The Scientific-Technological Revolution and the Role of Specialist Elites in Policy-Making in Czechoslovakia," in Michael J. Sodaro and Sharon L. Wolchik, eds., *Foreign and Domestic Policy in Eastern Europe in the 1980s* (New York: St. Martin's Press, 1983), pp. 111–132.

8. The Impact of Returning Scholars on Chinese Science and Technology, by O. Schnepp

This work was supported in part by a grant from the National Science Foundation.

1. *China Exchange News* 15.3 and 4:17 (1987).
2. Ibid.
3. Xinhua News Agency release of 23 April 1984 in English; reproduced in Daily Reports, China, *FBIS,* 24 April 1984, page B 12. Consultations in Beijing, May 1984.
4. *Beijing banyuetan* (Semi-monthly talks), 25 August 1985, p. 39; translated in *JPRS,* China Report, S&T, 29 October 1985, p. 4.
5. David M. Lampton and CSCPRC staff, *A Relationship Restored* (Washington: National Academy Press, 1986).
6. Consultations in Beijing, May 1984.
7. Leo A. Orleans, "Chinese Students and Technology Transfer," *Journal of Northeast Asia Studies* 4.4:3 (Winter 1985).
8. O. Schnepp, "The State of China's Science and Engineering Manpower," paper presented at the conference on China in a New Era: Continuity and Change, Manila, August 1987.
9. Lampton.
10. Consultations in Beijing, May 1984.
11. *China Exchange News* 15.3 and 4:17 (1987).
12. Orleans.

13. O. Schnepp, "The Chinese Visiting Scholar Program in Science and Technology," unpublished paper.
14. Shen Xiaodan, *Kexuexue yanjiu* (Studies in science of science) 3.1:80 (1985). Shen Xiaodan, *Kexuexue yu kexue jishu guanli* (Scientiology and management of science and technology) 11:23 (12 November 1984); translated in China Report, S&T, *JPRS*, 22 August 1985, p. 24.
15. *GMRB*, 25 November 1984, p. 1, translated in China Report, S&T, *JPRS*, 20 February 1985, p. 10.
16. Li Tianhou and He Zhou, *Kexuexue yu kexue jishu guanli* 11:28 (12 November 1985), translated in China Report, S&T, *JPRS*, 1 March 1985. p. 1.
17. Xinhua News Agency release in English, 27 June 1985; reprinted in China Report, S&T, *JPRS*, 18 July 1985, p. 29. *Jiefang ribao* (Liberation daily), Shanghai, 18 November 1985, p. 1, translated in China Report, S&T, *JPRS*, 11 February 1986, p. 11.
18. Shen Xiaodan, *Kexuexue yanjiu.*
19. Orleans.
20. See Schnepp, "The Chinese Visiting Scholar Program."
21. The statistical information included in Table 1 was obtained during the survey visits from responsible administrative personnel, in most cases themselves scientists, concerning the number of faculty sent abroad by each institution.
22. Lampton.
23. Comparison with the data presented in Table 3-10 of Lampton's book indicates a discrepancy in the percentage of scholars sent by CAS, presumed to be mostly researchers. It is concluded that the percentage given in this article is high and is not representative of the whole population of visiting scholars.
24. *GMRB*, 25 November 1984, p. 1.
25. Shen Xiaodan, *Kexuexue yanjiu.*
26. Li Tianhou and He Zhou.
27. If, however, the scholar had obtained a PhD during his sojourn abroad, some attempt was made to place him in an environment where he could function best, according to government policy announcements of July 1984. More recently, "Postdoctoral mobile stations" have been established at a number of universities and research institutes for the purpose of facilitating the effective absorption of returning young PhDs from abroad. The returnee can spend up to four years in such laboratories before moving on to a suitable permanent position.
28. *GMRB*, 29 August 1984, p. 1, translated in China Report, S&T, *JPRS*, 28 January 1985, p. 23.
29. *People's Daily*, Overseas Edition, 15 March 1985, p. 1. Many recent promotions to professorial rank were reported in this reference, indicating an end to the freeze.
30. Xinhua News Agency release in English, 27 June 1985, reprinted in China Report, S&T, *JPRS* 18 July 1985, p. 29.
31. *Jiefang ribao*, Shanghai, 18 November 1985.
32. Xinhua News Agency report in English, 2 December 1984, reprinted in China Report, S&T, *JPRS*, 3 January 1985, p. 15.

33. *Jiefang ribao*, 18 November 1985, p. 1, translated in China Report, S&T, *JPRS*, 11 February 1986, p. 9.
34. *GMRB*, 25 November 1984, p. 1; Shen Xiaodan, *Kexuexue yanjiu.*
35. *People's Daily*, Overseas Edition, 23 December 1987, p. 1.
36. Xinhua News Agency report of 2 December 1984.
37. *GMRB*, 29 November 1981, p. 1; ibid. p. 2. *BR*, 4 January 1982, p. 9. *People's Daily*, 28 December 1985, p. 1, translated in China Report, S&T, *JPRS*, 20 February 1986, p. 2.
38. Shen Xiaodan, *Kexuexue yu kexue jishu guanli.*
39. *People's Daily*, Overseas Edition, 12 November 1985, p. 4, translated in China Report, S&T, *JPRS* 17 January 1986, p. 30.
40. *People's Daily*, 27 April 1985, p. 3.
41. *Jiefang ribao*, Shanghai, 18 November 1985.
42. Shen Xiaodan, *Kexuexue yanjiu.*
43. *Jiefang ribao*, Shanghai; 18 November 1985.
44. *GMRB*, 17 September 1985, p. 3, translated in China Report, S&T, *JPRS*, 5 November 1985, p. 1.
45. *Zhong'guo keji luntan* (Forum on science and technology in China) 1:54–56; September 1985, translated in *JPRS*, S&T, 13 March 1986, p. 15.
46. Ibid.
47. Ibid.
48. Xinhua News Agency release, 15 July 1987; reproduced in *FBIS*, 21 July 1987, p. K2. *GMRB*, 18 July 1987. p. 1, translated in *FBIS*, 27 July 1987, p. K15.
49. This was the personal experience of the author when he worked in Israel during the years 1960–1965.
50. O. Schnepp, "The State of China's Science and Engineering Manpower."
51. *Science Indicators*, The 1985 Report, National Science Board, Washington, 1985; Appendix Table 3–10, p. 144–145.

9. Issues in the Modernization of Medicine in China, by Gail Henderson

Research for this paper was supported by a grant from the Committee on Scholarly Communication with the People's Republic of China. Comments on earlier drafts from the following colleagues are gratefully acknowledged: Chen Xiangming, Myron S. Cohen, Kenneth DeWoskin, Renée C. Fox, William C. Hsiao, William Fischer, Arthur Kleinman, Li Hanzao, Liu Yuanli, Martin K. Whyte, and Glenn Wilson. Special acknowledgement of the contributions of Elizabeth Murphy and Sam Sockwell should be noted. They are collaborators on a study of resource allocation in medical care in China, and the impact that expensive technology has had upon the cost of care and the urban work-unit insurance systems. Our work together over the past two years has had a major impact upon my conceptualization of those issues.

*Note: The 1986 dollar-yuan conversion rate of $1.00 = 3.7 yuan was used.

1. Paul Starr, *The Social Transformation of American Medicine* (New York: Basic Books, 1982).

2. Renée C. Fox, "Medicine, Science, and Technology," in Linda Aiken and David Mechanic, eds., *Applications of Social Science to Clinical Medicine and Health Policy* (New Brunswick: Rutgers University Press, 1986).
3. Larry Churchill, *Rationing Health Care in America: Perceptions and Principles of Justice* (South Bend: University of Notre Dame Press, 1987).
4. Joshua Horn, *Away With All Pests: An English Surgeon in People's China* (New York: Monthly Review Press, 1969).
5. Victor Sidel and Ruth Sidel, *Serve the People: Observations on Medicine in the People's Republic of China* (New York: Josiah Macy Foundation, 1973).
6. Robert J. Blendon, "Can China's Health Care be Transplanted Without China's Economic Policies?" *New England Journal of Medicine* 300:1453–1458 (1979).
7. *Zhongguo weisheng nianjian* (China Health Yearbook; Beijing: Renmin Weisheng Chubanshe, 1985), p. 55.
8. AnElissa Lucas, "Changing Medical Models in China: Organizational Options or Obstacles?" *China Quarterly* 83:461–488 (1980).
9. Golladay and Liese, *Health Problems and Policies in the Developing Counties* (Washington: The World Bank, 1980); Sidel and Sidel.
10. *Zhongguo weisheng nianjian,* p. 54.
11. Dean Jamison et al., *China: The Health Sector* (Washington: The World Bank, 1984); Blendon.
12. "Public Health Facilities on the Rise" *BR* 9:34 (3 March 1986).
13. *Zhongguo weisheng nianjian,* p. 59.
14. Ibid., p. 40.
15. Liu Zongxiu and Yu Xiucheng, "Shilun weisheng jihua gongzuo gaige" (Discussion of the reform of health planning work) *Zhongguo weisheng jingji yanjiuhui,* December 1984.
16. "Wo yuan yu chengxiang jicheng yiyuan guagou xiezuo: jianli yiliao lianheti de zuofa he tihui" (Cooperative linkages between my hospital and basic level urban and rural hospitals: The method and conclusions regarding building up a medical network), Wuhan First Municipal Hospital, 1986.
17. *Zhongguo weisheng nianjian,* p. 45.
18. William C. Hsiao, "Transformation of Health Care in China," *New England Journal of Medicine* 310:932–936 (5 April 1984).
19. "Collective and Private Hospitals on the Increase," *China Daily,* 12 January 1987, p. 1.
20. Zhu Hongzhen, "Health Care System in Chinese Rural Areas Today" (unpublished manuscript).
21. Gail Henderson and Scott Stroup, "The Reconstruction of Cooperative Medical Insurance in Rural Guangdong Province: The Case of Shijing District" (unpublished manuscript).
22. Jamison.
23. In July 1985, reforms were instituted which introduced a new formula for salaries. They now consist of a basic living allowance, similar for all people; an occupational allowance, which depends upon the job one performs; a seniority allowance, which provides an allowance for each year worked; and regional

allowances, which take into account variations in the cost of living. Bonuses to reward performance are in addition to these inputs.

24. Gail Henderson, Liu Yuanli, Guan Xiaoming, and Liu Zongxiu, "The Rise of Technology in Chinese Hospitals" *International Journal of Technology Assessment in Health Care* 3:253–264 (1987).
25. Interview, H.15, 1986.
26. Interview, H.6, 1986.
27. Shanghai Municipal Health Bureau, "Shanghai shi yiyuan tuixing 'liangzhong shoufei' gongzuo de pingjia" (Evaluation on the work of pursuing the "two-price system" in Shanghai), *Weisheng jingji* (Health economics), March 1985, p. 15–17. See also: Wang Shaohua, et al., "Gaijin liangzhong shoufei zujin yiliao shiye fazhan" (Improve the two-price system in order to enhance the development of health business) *Weisheng jingji*, September 1985, pp. 38–39.
28. Ibid.
29. *Zhongguo weisheng nianjian*, p. 52.
30. Zhang Yifang, "Shoudu yiyuan gaige shidian gongzuo qingkuang" (A summary of pilot tests in the reform of capital hospital), *Zhongguo yiyuan guanli* (Chinese hospital management), February 1985, pp. 28–33.
31. Jamison; Interview H.15, 1986.
32. Interview, H.3, 1986.
33. "Woguo weisheng shiye de jiben qingkuang ji cunzai de wenti" (The basic conditions and current problems in China's health sector), Ministry of Health Document, 1986.
34. Deng Xiulan, "Anzhao ban disan chanye de yuanze banhao weisheng shiye" (Run health business well, according to the principles of managing the third industrial product) *Weisheng jingji*, August 1985, pp. 28–30.
35. Lu Dajing, "Gaige yusuan fenpei de tansuo" (Report on the reform of government subsidy), Shanghai Municipal Health Department, 1986.
36. "Woguo weisheng shiye. . . . "
37. John E. Wennberg, Benjamin A. Barnes, and Michael Zubkoff, "Professional Uncertainty and the Problem of Supplier-Induced Demand," *Social Science and Medicine*, XVI, 811–824 (1982). On the consumer's demand side, people often need a medical service too urgently to behave as rational purchasers; they often cannot obtain adequate information to make an informed purchase; and frequently, even given enough time, the technical nature of the product dictates a situation of imperfect information. On the supply side, the peculiar nature of medical care is such that the suppliers often create demand for their product; thus, an increase in the supply of medical beds and personnel usually results in an increase rather than decrease in demand for those services.
38. "Woguo weisheng shiye. . . ."
39. Gail Henderson and Myron S. Cohen, *The Chinese Hospital: A Socialist Work Unit* (New Haven: Yale University Press, 1984); Jamison.
40. Henderson and Cohen; Sidel and Sidel.
41. "Wo yuan yu Cheng Xiang . . ."

42. D. Michael Lampton, *The Politics of Medicine in China* (New York: Westview Press, 1977).
43. Terence J. Johnson, *Professions and Power* (London: Macmillan Press, 1972).
44. Sidel and Sidel.
45. Leo Orleans, "Chinese Students and Technology Transfer," *Journal of Northeast Asian Studies* 4.4:3–25.
46. Interview, H.19, 1986.
47. Henderson and Cohen.
48. Interview, H.19, 1986.
49. Interview, H.4, 1986.
50. Interview, H.8, 1986.
51. "Woguo weisheng shiye . . . "
52. "Gouzhi daxing zhenliao yiqi shebei ying zuo zhonghe kaolu" (Unified consideration should be made in the purchase of large diagnostic and treatment medical equipment), *Weisheng jingji,* 1985.
53. Henderson, Liu, Guan, and Liu.
54. "Zhongguo yikeyuan qicaichu guanli piluan langfei yanzhong" (Serious waste and mismanagement in the materials department of the Chinese Academy of Medical Sciences), *Jiankang bao* (Health daily), 22 February 1986.
55. Pan Rennian, "Analysis of the Cost-Benefit of the CT Scan at Shanghai First Medical College, Hua Shan Hospital," 1982.
56. He Zhibang, "Goujin CT yao jiangjiu jingji xiaoyi" (In importing a CT, economic profit must be considered) *Weisheng jingji,* August 1985, p. 35–36.
57. Gail Henderson, Elizabeth A. Murphy, Samuel T. Sockwell, Zhou Jiongliang, Shen Qingrui and Li Zhiming, "High-Technology Medicine in China: The Case of Chronic Renal Failure and Hemodialysis," *New England Journal of Medicine* 318:1000–1004 (14 April 1988).
58. "Cong liuxue huiguo keji renyuan shiyong qingkuang de faguan diaocha yanjiu woguo zhili yinjin de zhanlue" (Investigation on my country's strategy of importing intellectual capabilities, from the functional conditions of the returned technical scholars) *Kexuexue yanjiu* (Research on the science of science) 3.1 (1985).
59. According to a recent study conducted by the Committee on Scholarly Communication with the PRC, approximately 11% of the Chinese studying abroad are in the biomedical sciences. See also Orleans.
60. Stanley Reiser, *Medicine and the Reign of Technology* (Cambridge: Cambridge University Press, 1978).
61. China Rural Development Research Group, "Economic Growth and Rural Development," *BR* 29:10, 10 March, 1986, p. 20. The authors also state: "Under the former rigid planning system, the imbalance of social and economic development among different localities has actually not been corrected. On the contrary, the gap between regions has widened in recent years."
62. Chen Pichao and Tuan Chihsien, "Primary Health Care in Rural China: Post-1978 Development" *Social Science and Medicine* 17.19:1411–1416 (1983).
63. *Zhongguo weisheng manjian,* p. 45.

64. Zhu Hongzhan.
65. Hsiao.
66. Interview, G. 7, 1986.
67. Jamison.
68. Interview, H.11, 1986.
69. Interview, G.7, 1986.
70. *Zongguo weisheng nianjian.*
71. Judith Banister, *China's Changing Population* (Stanford: Stanford University Press, 1987).
72. Zhong Shi, "Yiyuan bu zheng zhi feng de qingkuan diaocha" (An investigation into irregularities in hospitals), *Shehui* (Society) 3:40–44 (August 1982), translated in *Chinese Sociology and Anthropology* 1984:36–48.

10. Science and Technology in the Chinese Countryside, by Arthar Hussain

I am very grateful for comments and guidance to Hans Binswanger, Merle Goldman, Dwight Perkins, Vernon Ruttan, Denis Simon, Edward Schuh, Robert Solow, and Thomas Wiens

1. The figures are from *Statistical Yearbook of China 1985* (hereafter *SYB/85*).
2. See FAO, *Production Yearbook 1983,* Tables 1 and 3; *World Development Report,* 1985, Table 21.
3. *SYB/85.*
4. For population projections, see WB, *China, Long-Term Development, Issues and Options* (1985), pp. 136–140; N. Keyfitz, "The Population of China," *Scientific American,* February 1984, pp. 38–47.
5. *SYB/85,* p. 213.
6. WB, *China: Long-Term Development,* p. 41.
7. Ren Mei'i, Yang Renzhang, and Bao Haoshang, *An Outline of China's Physical Geography* (Beijing: Foreign Languages Press, 1985), p. 47; F. Leeming, *Rural China Today* (London: Longman, 1985), Chapter 2.
8. M. Abramovitz, "Rapid Growth Potential and Its Realization," in E. Malinvaud, ed., *Economic Growth and Resources* (New York: St. Martins Press, 1978).
9. *SYB/85,* p. 263.
10. NAS, *Agricultural Production and Efficiency 1975,* pp. 137–138.
11. Zhan Wu, "The Development of Socialist Agriculture in China," in G. Johnson and A. Maunder, eds., *Rural Change: The Challenge for Agricultural Economists* (Totowa, NJ: Allanheld Osmun, 1981), p. 278.
12. *SYB/83,* pp. 204, 209; Luo Hanxian, *Economic Changes in Rural China* (Beijing: New World Press, 1985), p. 203.
13. WB, *China: Agriculture to the Year 2000,* Appendix 2 to *China: Long-Term Development,* pp. 46–50.
14. *SYB/85,* p. 580.

15. Y. Hayami and S. Yamada, "Agricultural Research Organization in Economic Development: A Review of the Japanese Experience," in L. G. Reynolds, *Agriculture in Development Theory* (New Haven: Yale University Press, 1975), p. 227.
16. NAS, *Wheat in the PRC,* 1977.
17. WB, *China: Agriculture to the Year 2000,* pp. 50–51.
18. Tong Dalin and Hu Ping, "Science and Technology," in Yu Guangyuan, *China's Socialist Modernization* (Beijing: Foreign Languages Press, 1984), pp. 630–631; Tony Saich's chapter in this volume.
19. For a general discussion, see Hu Ping, "On Restructuring the Management Systems in Scientific Research Institutions," in P. Lalkaka and Wu Mingyu, eds., *Managing Science Policy and Technology Acquisition Strategies for China and a Changing World* (New York: United Nations, 1984), and, for examples, see V. W. Ruttan, "Travel Notes on a Visit to Study Organization and Function of Chinese Agricultural Research System" (mimeographed, 1985).
20. On linkages, see WB, "China: Agriculture to the Year 2000," pp. 50–51; Ruttan, "Travel Notes."
21. NAS, *Wheat in the PRC;* IRRI, *Rice Research and Production in China: An IRRI Team's View,* 1978.
22. WB, *China: Agriculture to the Year 2000,* pp. 58–69.
23. Ruttan, "Travel Notes."
24. FAO, *Learning from China* (1978); FAO, *China: Agricultural Training System* (1980); NAS, *Plant Studies in the PRC* (1975); NAS, *Wheat in the PRC;* J. E. Nickum, *Hydraulic Engineering and Water Resources in the PRC* (Stanford University: US-China Relations Program, 1977); Dwight Perkins, "A Conference on Agriculture," *China Quarterly* 76:596–610 (September 1976); G. F. Sprague, "Agriculture in China," in P. Abelson, ed., *Food, Politics, Economics, Nutrition, and Research* (American Association for the Advancement of Science, 1975); S. Wortman, "Agriculture in China," *Scientific American,* June 1975, pp. 13–21.
25. See Benjamin Stavis, "Agricultural Research and Extension Services in China," *World Development* 5:631–645 (1978); NAS, *Wheat in the PRC,* p. 122.
26. For a comparative assessment, see WB, *Agricultural Research and Extension,* 1985.
27. Perkins, p. 603.
28. FAO, *Agricultural Training System,* Chapter 12.
29. See T. B. Wiens, "The Evolution of Policy and Capabilities in China's Agricultural Technology," in *Chinese Economy Post-Mao* (Washington: Joint Economics Committee of Congress of the United States), 1978; R. Barker, *The Philippine Rice Program: Lessons for Agricultural Development,* (Department of Agricultural Economics, NY State College of Agriculture and Life Sciences, 1984).
30. WB, *China: Agriculture to the Year 2000,* p. 53.
31. Ibid., pp. 51–52; Tong and Hu, pp. 645–647.
32. *SYB/85,* p. 591.

33. On the lack of career structure for researchers, see Tong and Hu, pp. 645–646.
34. C. Greer, *Water Management in the Yellow River Basin of China* (Austin: University of Texas Press, 1979), p. 55; Ren et al., Chapters 3 and 4.
35. J. E. Nickum, *Irrigation Management in China*, WB Staff Working Papers, No. 545 (1983), p. 58; Zhan Wu and Liu Wenpu, "Agriculture," in Yu Guangyuan, *China's Socialist Modernization* (Beijing: Foreign Languages Press, 1984), pp. 266–267.
36. FAO, *Production Yearbook 1983.*
37. WB, *China: Agriculture to the Year 2000*, p. 33.
38. See Fang Yi, *BR* 14:8 and 16:8–9 (1978).
39. Nickum, *Hydraulic Engineering*, p. 52; Nickum, *Irrigation Management*, passim; NAS, *Wheat in the PRC*, p. 3.
40. Luo Hanxian, *Economic Changes in Rural China* (Beijing: New World Press, 1985), p. 203.
41. *SYB/85*, p. 281.
42. FAO, *The Role of Seed Science and Technology in Agricultural Development* (1975), Section 3.
43. WB, *China: Agriculture to the Year 2000*, p. 40.
44. H. Hanson, N. E. Borlaug, and R. G. Anderson, *Wheat in the Third World* (Boulder: Westview Press, 1982), p. 74.
45. Ibid.
46. See the income and expenditure accounts of the rural communes in *SYB/83*, p. 209.
47. Nicholas Lardy, *Agricultural Prices in China*, WB Staff Working papers, No. 606 (1983); F. Leeming, "Progress Towards Triple Cropping in China," *Asian Survey* 19.5:50–67 (May 1979).
48. *SYB/85*, p. 281.
49. On HYVs of wheat, see D. G. Dalrymple, "Evaluating the Impact of International Research on Wheat and Rice Production in the Developing Nations," in M. Arndt et al., eds., *Resource Allocation and Productivity in National and International Agricultural Research* (Minneapolis: University of Minnesota Press, 1977), p. 178.
50. *SYB/85*, p. 253.
51. NAS, *Wheat in the PRC*, p. 61.
52. WB, *China: Agriculture to the Year 2000*, pp. 59–67; IRRI, *Rice Research and Production in China.*
53. WB, *China: Agriculture to the Year 2000*, p. 61; C. F. Zhou, *Production Constants and Research Priorities in Southern Winter Wheat Regions of China* (CIMMYT and UNDP, 1984).
54. NAS, *Plant Studies in the PRC.*
55. IRRI, *Rice Research and Production in China;* W. R. Coffman and S. S. Virmani, "Advances in Rice Technology in the PRC," in R. Barker and B. Rose, eds., *Agricultural and Rural Development in China Today* (New York State College of Agriculture and Life Sciences, 1984).
56. He Guiting et al., *The Economics of Hybrid Rice Production in China*, IRRI Research Paper Series, No. 101 (1984).

57. *FBIS* China. 9 September 1983, p. K20; the figure for the sown area is from *SYB/85*, p. 254.
58. Lin Shih-cheng and Yuan Loung-ping, "Hybrid Rice Breeding in China," in IRRI, *Innovative Approaches to Rice Breeding* (1980), pp. 35–52.
59. For an assessment, see K. Griffin, *The Political Economy of Agrarian Change*, 2nd ed. (London, Macmillan, 1979).
60. For a general discussion, see FAO, *China: Recycling of Organic Wastes in Agriculture* (1977); for a quantitative estimate, see Y. Y. Kueh, "Fertilizer Supplies and Food Grain Production in China, 1952–1982," *Food Policy* 9:206–218 (1984).
61. WB, *China: Agriculture to the Year 2000*, pp. 34–40; D. Richter, "The Facts about Fertilizer," *The China Business Review*, January/February 1985; Wiens, "The Evolution of Policy."
62. WB, *China: Agriculture to the Year 2000*, p. 37.
63. T. B. Wiens, "Microeconomic Study of Organic Fertilizer Use in Intensive Farming in Jiangsu Province, China," in IRRI, *Organic Matter and Rice* (1984), pp. 533–556.
64. WB, *China: Agriculture to the Year 2000*, p. 34.
65. *SYB/85*, p. 283.
66. For examples, see Shen Jin-hua, "Rice-Breeding in China," in CAS and IRRI, *Rice Improvements in China and Other Asian Countries* (1980), pp. 9–30.
67. For a case study, see T. B. Wiens, "The Limits to Agricultural Intensification," in R. Barker and B. Rose, eds., *Agricultural and Rural Development in China Today* (New York State College of Agriculture and Life Sciences, 1983).
68. For an assessment of triple-cropping, see Guo Yixian, "The Rice-Based Cropping System and Its Development in China," in IRRI, *Cropping System Research in Asia* (1982).
69. Wiens, "Limits."
70. *SYB/85*, pp. 252, 281; see also WB, *Socialist Economic Development* (1983), II, 73.
71. FAO, *Production Yearbook 1983.*
72. *SYB/85*, p. 267.
73. For case studies, see S. Ishikawa, "China's Food and Agriculture–A Turning Point," *Food Policy* 2.2:90–102 (May 1977); Leeming, *Rural China Today*, pp. 110–114; Wiens, "Limits."
74. IRRI, *Rice Reserach and Production.*
75. NAS, *Wheat in the PRC*, p. 19.
76. Ibid., pp. 21–24; Ruttan "Travel Notes"; D. L. Plucknett and H. O. Beemer, Jr., *Vegetable Farming System in China* (Boulder: Westview Press, 1981), Chapters 1 and 7.
77. WB, China: *Socialist Economic Development*, II, 74; Zhan and Liu, p. 267.
78. For a general discussion, see H. P. Binswanger, *Agricultural Mechanization: A Comparative Historical Perspective*, WB Staff Working Papers, No. 673 (1984).
79. Stavis, "Agricultural Research"; Hsu.
80. Xiang Nan (Hsiang Nan), *Certain Problems of Agricultural Mechanization*, Survey of Communist Mainland Press, No. 2900 (1962); *Stable and High Yields*

and Agricultural Mechanization, Survey of Communist Mainland Press, No. 3436 (1965).

81. Luo Hanxian, *Economic Changes in Rural China* (Beijing: New World Press, 1985), Chapter 7.

11. China's Military R&D System: Reform and Reorientation, by Wendy Frieman

1. David Holloway, "Innovation in the Defense Sector," in R. Amann and J. Cooper, eds., *Industrial Innovation in the Soviet Union* (New Haven: Yale University Press, 1982), pp. 288–294.
2. I am grateful to Ben Ostrov, Chinese University of Hong Kong, for providing me with the first four chapters of his dissertation, which discuss the history of these organizations in great detail.
3. Jerry Grey, "China Aims for Modern Wings," *Astronautics and Aeronautics* 19.2:56–64 (1981).
4. Ellis Joffe, *The Chinese Army After Mao* (Cambridge: Harvard University Press, 1987), p. 84.
5. Chu Yuan Cheng, *The Machine Building Industry in Communist China,* (Chicago: Aldine, 1971), pp. 9–16.
6. Joffe, p. 101.
7. CIA, "Chinese Defense Spending, 1965–1979" (Washington, July 1980), p. 1.
8. Yuan Li Wu, Robert B. Sheeks, *The Organization and Support of Scientific Research and Development in Mainland China* (New York: Praeger, 1970), p. 365.
9. Conversation with Peter Smith, NASA International Affairs Office, October 1985.
10. Office of Technology Assessment, "Energy Technology Transfer to China: A Technical Memorandum" (Washington, September 1985), p. 33.
11. See, for example, K. C. Yeh, "Industrial Innovation in China with Special Reference to the Metallurgical Industry," Rand Note N-2307 (Santa Monica, May 1985), p. 32; and Jonathan Pollack, "The Chinese Electronics Industry in Transition," Rand Note N-2306 (Santa Monica, May 1985), p. 64.
12. Robert Manning, "New Sellers in Arms Bazaar, *US News and World Report,* 3 February 1986, p. 37; "Arms Fall Off," *Far Eastern Economic Review* 9 (27 February 1986).
13. Manning, pp. 37–39.
14. "Defense Industry Urged to Produce Civilian Goods," *FBIS Daily Report: China,* 29 August 1983, pp. K15–18.
15. "Survey of Civilian Goods Produced by China's Ordnance Industry," *JPRS China Report: Economic Affairs. Almanac of China's Economy 1982,* 8 August 1983, pp. 318–325.
16. "Defense Industry Told to Help Boost Economy," Xinhua Domestic Service, 5 March 1985.
17. Zhonggou Xinwen She, 3 May 1985, translated in *FBIS Daily Report: China,* 7 May 1985, p. K11.

18. "Military Technology Transfer Increases," Xinhua Domestic Service, 12 December 1985.
19. "Defense Research Institutions to Use Contract System," Xinhua Domestic Service, 12 December 1985.
20. Sun Zhenhuan, "A Study on the Question of Integrated Military-Civilian Industrial System," *Jingji yanjiu* 5:68–73 (20 May 1985).
21. Joffe, p. 82.
22. Ibid., p. 85.
23. Eliot Cohen, "Distant Battles: Modern War in the Third World," *International Security* 10.4:159 (Spring 1986).
24. Ibid., pp. 158–163.
25. Joseph P. Gallagher, "China's Military Industrial Complex: Its Approach to the Acquisition of Modern Military Technology," *Asian Survey* 27.9:991–1002 (September 1987).
26. Harlan Jencks, *From Muskets to Missiles* (Boulder: Westview Press, 1982), p. 205.

12. Technology Transfer and China's Emerging Role in the World Economy, by Denis Fred Simon

1. Madelyn Ross, "China and the United States' Export Controls System," *The Columbia Journal of World Business,* Spring 1986, pp. 27–34.
2. Ernst Haas, "Why Collaborate: Issue-Linkage and International Regimes?" *World Politics,* May 1980, pp. 357–405.
3. Joseph Grunwald and Kenneth Flamm, *The Global Factory: Foreign Assembly in International Trade* (Washington: Brookings Institution Press, 1985).
4. Michael Yoshino, *Japan's Multinational Enterprises* (Cambridge: Harvard University Press, 1976).
5. Gerard O'Neill, *The Technology Edge: Opportunities for America in World Competition* (New York: Simon and Schuster, 1983).
6. Denis Fred Simon, *Taiwan, Technology Transfer and Transnationalism* (Boulder: Westview Press, forthcoming).
7. Alan Altshuler et al., *The Future of the Automobile* (Cambridge: MIT Press, 1984).
8. Even in those cases where labor-intensive tasks remain, technological change and advance have allowed those tasks to be specifically excerpted from the overall production scheme, thereby diminishing to some appreciable extent the scope and level of technology transfer. See Robert Ballance and Stuart Sinclair, *Collapse and Survival: Industry Strategies in a Changing World* (London: Allen and Unwin, 1983).
9. Reich argues that "it is far easier to move into flexible system production by upgrading manufacturing skills and know-how and elaborating networks of suppliers, distributors, and customers already in existence than by leaping into a totally unchartered sea of products and processes unrelated to an industrial base of the past." Robert Reich, *The Next American Frontier* (New York: Penguin Books, 1983), p. 130.

10. See C. C. Markides and N. Berg, "Manufacturing Offshore Is Bad Business," *Harvard Business Review,* September/October 1988.
11. Michael Porter, ed., *Competition in Global Industries* (Cambridge: Harvard Business School Press, 1986).
12. David Teece, *The Multinational Corporation and the Resource Cost of International Technology Transfer* (Cambridge: Ballinger, 1976).
13. William Abernathy et al., *Industrial Renaissance: Producing a Competitive Future for America* (New York: Basic Books, 1983).
14. William Rushing and Carol Brown, eds., *National Policies for High Technology Industries* (Boulder: Westview Press, 1986).
15. See Song Jian, "The Chinese Restructuring and Opening to the Outside," *Zhongguo keji luntan* 1:4–7 (September 1985). Song is the Minister-in-Charge of the State Science and Technology Commission.
16. Li Boxi et al., *Zhongguo jishu gaizao wenti yanjiu* (Research on the problems of technical transformation in China; Shanxi: Shanxi People's Publishing House, 1985).
17. See "Excerpts of the 7th FYP Released," Xinhua, 14 April 1986, translated in *FBIS-PRC,* 8 April 1986, pp. K29–30. Monies provided for this program in the 6th FYP were increased 66.6% over funds contained in the 5th FYP, which is further testimony to the importance attached to this effort.
18. Hao Qi, "Watch Out, Suppliers," *Intertrade,* April 1987, pp. 16–17.
19. R. P. Suttmeier, "New Directions in Chinese Science and Technology," in John Major, ed., *China Briefing 1985* (Boulder: Westview Press, 1986), pp. 91–102.
20. One of the principal uncertainties is that, under the decision to decentralize control over the operation of factories from within the former MEI and the SMBC, the authority for approving technology imports is still retained by the Commission. Thus, while the factory manager may have the prerogative to seek out new and more efficient modes of production, he does not have the necessary power to formally sign an agreement.
21. Ge Yuehua and Li Yuming, "Why Some Industrial Enterprises Are Not Enthusiastic About Purchasing New Technology," *Jishu shichang bao,* 18 February 1986, pp. 1–2.
22. Kazimierz Poznanski, "The Environment for Technological Change in Centrally Planned Economies," *World Bank Staff Working Papers,* No. 718 (Washington, 1985).
23. See "More Than 700 Items of Military Industrial Technology transferred to Civilian Uses in Shaanxi," *Jingji ribao,* 27 September 1986, p. 2.
24. "National Defense Requires All-Round Economic Growth," *China Daily,* 28 April 1986, p. 1.
25. SSTC, *Zhongguo kexue jishu zhengce zhinan* (Beijing: Science and Technology Publishing House, 1986).
26. Li Du, "China is Technologically Paving the Way for An Economic Takeoff," *Liaowang Overseas Edition,* 41:8–9 (13 October 1986), translated in *FBIS-PRC,* 28 October 1986, pp. K6–8.

27. Jiang Shaogao, "3,000 Pieces of Technology Imported in 3 Years," *People's Daily Overseas Edition,* 4 December 1986, p. 3.
28. Ibid.
29. *Business China,* 12 September 1985.
30. See Ministry of Foreign Economic Relations and Trade, *Technology Transfer to China: A Comprehensive Guide* (Hong Kong: Economic Information & Agency, 1986).
31. "Technology: a Boon to Industry," *China Daily,* 6 May 1987, p. 3.
32. Wang Xiaodong, "Results and Prospects for China's Technology Imports," *Liaowang* (Overseas Edition) 1:13–14 (5 January 1987), translated in *JPRS-CEA*-87-014, 26 February 1987, pp. 53–54.
34. Cao Jiarui, "The Present Condition of and Problems in China's Technological Imports," (Parts 1–3), *Liaowang* (Overseas Edition), 5 May 1987, pp. 14–15; 12 May 1987, pp. 18–19; and 19 May 1987, pp. 12–13.
35. Zhang Liang, "Redundant Technological Imports," *Intertrade,* March 1987, pp. 30–31.
36. Cao Jiarui, "China's Technological Imports: Present Status and Problems," Part II, *Liaowang* (Overseas Edition), 12 May 1986, p. 19.
37. Zhang Aiping, "Strengthen Leadership and Do a Better Job in Importing Technology," *Hongqi* 24:4–9 (16 December 1985).
38. See *China Daily,* 19 June 1985, p. 2.
39. The new regulations specifically limit the inclusion of 9 such clauses, including any restrictions on the buyers'/licensees' freedom to buy raw materials or components from sources other than the technology supplier; restrictions on the recipients' freedom to further develop and improve the acquired technology; unreasonable restrictions on markets that can be serviced by the recipients' products that employ the imported technology, etc.
40. See Denis Fred Simon and Christopher Engholm, *The China Venture: Corporate America Encounters the People's Republic of China* (Chicago: Scott Foresman, Inc., forthcoming).
41. Masaki Yabuuchi, "Japanese Technology Transfer and China's Technological Reform," *China Newsletter,* November-December 1986, pp. 10–12.
42. For a list of the provisions, see "Provisions of the State Council for the Encouragement of Foreign Investment, 11 October 1986," *China Trade Quarterly* 1.4:66–67 (December 1986).
43. Gail Bronson, "The Long March," *Forbes,* 15 December 1986, pp. 182–185.
44. "Textiles Earned $17.2 Billion in Foreign Exchange in the Past 5 Years," *People's Daily Overseas Edition,* 20 December 1985, p. 3.
45. It also manufactured a growing amount of its own textile machinery. Deng Zhongyuan, "More than 600 Types of New Textile Machines Produced in Five Years," *Jingji ribao,* 23 January 1986, p. 2.
46. Personal communication from Detlef Rehn.
47. Wang Jie and Fang Xing, "Developmental Institutes Should Integrate Science Research, Production, Foreign Trade, and Operations," *Kexuexue yu kexue jishu guanli* 8:37–39 (August 1986).
48. *Guoji maoyi,* January 1985.

49. Xu Xinli, "Management Should Be Strengthened in Foreign Trade Under the Open Door Policy," *Caijing yanjiu,* 18 December 1985, pp. 12–16, translated in *JPRS-CEA*-86-035, 1 April 1986, pp. 105–106.
50. Field research in Beijing, January 1986.
51. Jon Woronoff, *Asia's Miracle Economies* (Tokyo: Lotus Press, 1986). See also Lawrence Lau, ed., *Models of Development: A Comparative Study of Economic Growth in South Korea and Taiwan* (San Francisco: ICS Press, 1986).
52. See Denis Fred Simon, "Taiwan's Political Economy and the Evolving Links Between the PRC, Hong Kong, and Taiwan," *AEI Foreign Policy and Defense Review* 6.3:42–51 (1986).
53. "Liaowang Article on PRC-USSR Commission Work," *FBIS-PRC,* 8 April 1986, pp. C1–3.
54. "Superpower Triangle," *Far Eastern Economic Review,* 4 April 1985, pp. 17–18.
55. During the joint commission meeting, both sides agreed to cooperate in the refurbishing of 17 industrial plants in China formerly built by the USSR along with the construction of 7 new plants.
56. "Industry to be Self-Reliant," *China Daily,* 31 March 1987, p. 1.
57. "Ministry Adds 15 Items to Import Licensing List," *China Daily,* 5 March 1986, p. 3.
58. "Great Wall Towers Over Domestic Sales," *China Daily,* Business Weekly, 2 April 1986, p. 2.
59. "International Symposium on Technology Transfer," *China Daily,* 28 October 1986, p. 1.
60. "Fujian and Hitachi Work to Save Joint Venture," *South China Morning Post,* Business Section, 26 August 1986, pp. 1–2.
61. "Why Are We Here?," *Far Eastern Economic Review* 14:103 (1986). For a discussion of the problems experienced by American Motors Corporation of the United States, see "AMC's Troubles in China," *New York Times,* 11 April 1986, p. D1.
62. Dick Nanto and Hong Nack Kim, "Sino-Japanese Economic Relations," in US Congress, Joint Economic Committee, *China's Economy Looks Toward the Year 2000* (Washington: GPO, 21 May 1986), pp. 453–471.
64. *New York Times,* 15 March 1987, p. 1. See also *Washington Post,* 14 April 1987, p. 11.
65. "Robots Set To Receive Boost in Development," *China Daily,* 1 February 1986, p. 1.
66. "Peking Is Gearing Up to Launch Its Commercial Satellite Program," *Asian Wall Street Journal Weekly,* 17 March 1986, p. 3.
67. Yung Whee Rhee et al., *Korea's Competitive Edge* (Baltimore: Johns Hopkins University Press, 1985).

13. *Acquiring Foreign Technology: What Makes the Transfer Process Work? by Roy F. Grow*

1. In Shenyang, Song had been part of a group that had discussed *In Search of Excellence.* The group had been intrigued by the description of 3M in the book, and this was one of the reasons Song was visiting the corporation. See Thomas J. Peters and Robert H. Waterman, Jr., *In Search of Excellence: Lessons from America's Best-Run Corporations* (New York: Harper and Row, 1982), especially Chapter 7, "Autonomy and Entrepreneurship."
2. L. J. White, "The Evidence on Appropriate Factor Proportions in Manufacturing in Less Developed Countries: A Survey," *Economic Development and Cultural Change,* October 1978, pp. 25–59; B. S. Chung and C. H. Lee, "The Choice of Production Techniques by Foreign and Local Firms in Korea," *Economic Development and Cultural Change,* October 1980, pp. 135–140.
3. K. Kojima, "Transfer of Technology to Developing Countries–Japanese Type Versus American Type," *Hitotsubashi Journal of Economics,* February 1977, pp. 1–14. See also Robert T. Kudrle, "The Correlates of Direct Foreign Investment in Developing Countries: A Look at Recent Experience," in Robert Benjamin and Robert T. Kudrle, eds., *The Industrial Future of the Pacific Basin* (Boulder: Westview Press, 1984), Chapter 10; Louis T. Wells, Jr., "International Trade: The Product Life Cycle Approach," in Reed Moyer, ed., *International Business: Issues and Concepts* (New York: John Wiley and Sons, 1984), pp. 5–23; William H. Gruber and Raymond Vernon, "The Technology Factor in a World Trade Matrix," in Raymond Vernon, *The Technology Factor in International Trade* (New York, National Bureau of Economic Research: Columbia University Press, 1970), pp. 232–272.
4. The classic study is Wassily Leontief, "Domestic Production and Foreign Trade: The American Capital Position Re-examined," *Proceedings of the American Philosophical Society* 97:332–349 (September 1953). Leontief's work inspired a series of subsequent studies. See Masahiro Tatemoto and Chinichi Ichimura, "Factor Proportions and Foreign Trade: The Case of Japan," *Review of Economics and Statistics* 41:442–446 (November 1959); and R. Bharadwaj, "Factor Proportions and the Structure of Indo-U.S. Trade," *Indian Economic Journal* 10:105–106 (October 1962). See also Louis T. Wells, Jr., *Third World Multinationals: The Rise of Foreign Investment from Developing Countries,* (Cambridge: The MIT Press, 1983).
5. The best examples are James R. Kurth, "The Political Consequences of the Product Cycle: Industrial History and Political Outcomes," *International Organization* 33:1–34 (Winter 1979); Bruce Cumings, "The Origins and Development of the Northeast Asian Political Economy: Industrial Sectors, Product Cycles, and Political Consequences," *International Organization* 38.1:1–40 (Winter 1984).
6. Some of the most interesting work in this area is being done by William Fischer. See his "The Transfer of Western Managerial Know-how to China," Report prepared for the Office of Technology Assessment, May 1986.

7. This research is the basis of a forthcoming book: Roy F. Grow, *Competing in China: Japanese and American Businessmen in a New Market* (New York: Harper and Row/Ballinger, forthcoming).
8. There are, of course, two sides to every technology transfer story—a demand side and a supply side. On the demand side are the decisions by Chinese to acquire new technologies; on the supply side are decisions by foreign forms to offer them. Both sets of decisions are extremely complex and fascinating in their own way. Franklin B. Evans, "Selling as a Dyadic Relationship," *American Behavioral Scientist,* 6 May 1963, pp;. 76–79. Since 1980, the bulk of the technology transfers from abroad—commodities, turnkey projects, scientific processes, training of Chinese technicians—has come from corporations, not governments. See the discussion of firms in the transfer of technology in Robert Stobaugh and Louis T. Wells, Jr., *Technology Crossing Borders: The Choice, Transfer, and Management of International Technology Flows* (Boston, Harvard Business School Press, 1984), pp. 1–3; and Raymond Vernon, *Sovereignty at Bay* (New York: Basic Books, 1971).
9. I have been fascinated to discover the wide variety of factors that shape a firm's decision to transfer a given sort of technology, even within the same industry. Two computer firms, for example, may decide to transfer much different technologies: Honeywell deciding on one set of technologies and NEC and, say, Control Data another; Cargill different from Del Monte, Suntory, or Ajinomoto. My research suggests that a great deal of the process by which new technology flows to China is shaped by the *supplier* of this technology—primarily by decisions made within the foreign corporations themselves. I have written elsewhere about how Japanese and American decisions to sell to China are shaped: by the nature of the firm itself, its financial and technological underpinnings, the nature of its organizational and decision-making process, and the pattern of its planning and implementation procedures. See Roy F. Grow, "Japanese and American Firms in China: Lessons of a New Market," *Columbia Journal of World Business* 21.1 (Spring 1986).
10. The details on these economic bureaucracies were reported by Roy Grow and Thomas Gottschang in "Liaoning's Economic Bureaucracies: Flexibility and Rigidity in Response to Economic Change" for the conference on "Economic Bureaucracies in Seven Chinese Regions" at the East-West Center, Honolulu, 16–20 July 1985.
11. Their American hosts reported that the group was more energetic than knowledgeable. Still, they received high marks in many American cities for their enthusiasm, curiosity, and insistence that businessmen should come to Dalian to "see for themselves."
12. Names have been changed throughout.
13. Rosabeth Moss Kantor, *The Change Masters: Innovation for Productivity in the American Corporation* (New York: Simon and Schuster, 1983), p. 35.
14. Peter F. Drucker, *Innovation and Entrepreneurship* (New York: Harper and Row, 1985), pp. 21–30.

14. DOS ex Machina: *The Microelectronic Ghost in China's Modernization Machine, by Richard Baum*

Research for this project was initiated under a grant from the Social Science Research Council and the American Council of Learned Societies. The author is indebted to George Bing, Christina Holmes, and Stacie Martin-Giles for their expert research assistance; and to Detlef Rehn for his helpful comments on an earlier draft.

1. On the development of the Chinese computer industry, see P. D. Reichers, "Electronics Industry in China," in U.S. Congress, Joint Economic Committee, *People's Republic of China: An Economic Assessment* (Washington: US GPO, 1972), pp. 86–111; Bohdan O. Szuprowicz, "Electronics," in Leo A. Orleans, ed., *Science in Contemporary China* (Stanford: Stanford University Press, 1980), pp. 437–443; and Detlef Rehn, "China's Computer Industry at the Turning Point," in U.S. Congress, Joint Economic Committee, *China's Economy Looks Toward the Year 2000,* Vol. II (Washington: US GPO, 1986), pp. 216–23.
2. There is an important distinction between the number of computers actually installed—i.e., operational—and the total national inventory of machines. In recent years, more than 40% of China's computers have reportedly remained in warehouses or otherwise non-operational. See, e.g., *China Daily,* 11 January 1986.
3. Thus, for example, the Chinese DJS 220 mainframe computer was a rather straightforward copy of the old IBM system 360, while the DJS 180 was a near-duplicate of an early model Digital PDP-11. Similarly, the DJS 30 was reported to be a replica of the Nova 1200. See "A Tour of Computing Facilities in China," *Computer* 18.1 (January 1985); also "Chinese Foment Another Revolution," *Electronics,* 13 January 1983.
4. The following discussion draws on several main sources, including Rehn; Takashi Uehara, "Computers in China," *China Newsletter,* Nos. 55–56 (1985); Gong Bingzheng, "Woguo jisuanji yingyong xianzhuang, jiaoguo ji jinyibu puji tuiguang de jianyi" (The status, results and proposals for further popularization of computer applications in our country), *Jisuanji yingyong yu zhuanjian* (Computer applications and software) 1.5 (1984); James Stepanek, "Microcomputers in China," *The China Business Review,* May-June 1984, pp. 26–38; "Shanghai That Software," *Datamation,* 11 February 1985, pp. 40–42; "A Tour of Computing Facilities"; *The New York Times,* 6 January 1985; *Businessweek,* 11 June 1984; *China Daily,* 11 January 1986; and *Mini-Micro Systems,* February 1985.
5. The single most popular foreign microcomputer introduced into China in the early 1980s was the American-made Apple II, which quickly displaced the older Cromenco as China's foreign microcomputer of choice. (Several hundred Cromenco 8-bit micros had been imported in the late 1970s during China's initial, somewhat frenzied foray into the global high-technology market.) By 1982, Chinese engineers had successfully copied the Apple II's 8-

bit central processing architecture; and soon afterward the PRC began to manufacture and market a domestic clone known as the DJS 033.

6. *China Daily*, 11 January 1986.
7. *Businessweek*, 11 June 1984. China's production of ICs reportedly totaled 23 million chips in 1984. However, due to primitive manufacturing techniques, poor quality-control standards, and inadequate environmental controls (affecting temperature, humidity, and airborne particulate matter in IC production facilities), IC yields remained quite low, often in the 5–10% range.
8. Designed and developed for military use, the "Galaxy" had a reported speed of 100 million instructions per second (MIPS), while its civilian-engineered counterpart, the bimodal "757," was reportedly capable of operating at a speed of 2 MIPS in scalar mode and 10 MIPS in dual-vector mode. See Rehn; also *Jisuanji yanjiu yu fazhan* (Computer research and development) 21.2 (February 1984).
9. See, e.g., "Prognosis for China's Computer Fever," *BR* 46:17–21 (18 November 1986).
10. The following discussion draws from Loren R. Graham, "Science and Computers in Soviet Society," in Erik P. Hoffmann, ed., *The Soviet Union in the 1980s* (New York: Academy of Political Science, 1984), pp. 124–134; Erik P. Hoffmann, "Technology, Values, and Political Power in the Soviet Union: Do Computers Matter?" in Frederic J. Fleron, Jr., ed., *Technology and Communist Culture* (New York: Praeger Publishers, 1977), pp. 397–436; Frederick Starr, "Technology and Freedom in the Soviet Union," *Technology Review*, May-June 1984, pp. 38–47; Wilson Dizard, "Mikhail Gorbachev's Computer Challenge," *The Washington Quarterly* 9.2:157–163 (Spring 1986); John E. Austin, "Computer-Aided Planning and Decision-Making in the USSR," *Datamation*, September 1977, pp. 71–72; Vladimir Myasnikov, "Soviet Computers," *Computer* 15.10:101–102 (October 1982); "Soviets Launch Computer Literacy Drive," *Science* 231:109–110 (10 January 1986); *The Chicago Sun-Times*, 4 December 1983; *The Washington Post*, 14 March 1984; *The Christian Science Monitor*, 13 July 1984; and *The Los Angeles Times*, 30 March 1986.
11. Myasnikov, "Soviet Computers." These figures exclude military-operated computer facilities.
12. By comparison, there are more than 2 million PCs in US schools today.
13. The few Soviet-made PCs that have recently been made available for public sale have generally not been equipped with disk drives, printers, or modems. Loren Graham suggests that the Soviet practice of limiting public access to computers will, in the long run, prove counterproductive because it "severely limits the rate of growth of the computer culture . . . (and thus forfeits) the advantages of economies of scale that the mass production of computers is bringing" (Graham, "Science and Computers"). Whether computerization ultimately contributes to an erosion of the power monopoly traditionally enjoyed by party elites in Communist systems has been widely debated in the literature. See Hoffmann, "Technology, Values, and Political Power"; also Christopher Evans, *The Micro Millennium* (New York: Viking Press, 1979), pp. 208–209; Starr, "Technology and Freedom"; and Dizard, "Mikhail Gorbachev's Computer Challenge."

14. For an authoritative statement of China's current national priorities in computer-related R&D, see Zhang Xiaobin, "Microelectronics Policy in China," *ATAS Bulletin* 2:127–129 (November 1985).
15. See, e.g., "Taiwan Micros Flooding PRC," *Asian Computer Monthly* 83:48 (September 1984); and "High Tech Smugglers Aiming at China," *Asian Computer Monthly* 93:75 (July 1985).
16. These regulations were frequently sidestepped by the Chinese, who were able freely to purchase many of the restricted items on the Hong Kong gray market.
17. On COCOM and U.S. computer export controls, see, *inter alia, The Los Angeles Times,* 17 December 1985; and *The New York Times,* 1 January 1985.
18. See, e.g., "IBM's Peking Connection," *The New York Times,* 6 January 1985.
19. Similar forecasts of geometric growth in the United States home computer market a few years ago proved vastly overinflated, leading to a severe shakeout in the microcomputer industry.
20. Major computer grants to Chinese institutions have been made, *inter alia,* by IBM, Digital, Hewlett-Packard, Honeywell, Wang, Apple, Hitachi, Fujitsu, and Computerland. The largest single transfer of computer hardware into Chinese colleges and universities was funded from a $200-million World Bank loan.
21. For an analysis of recent international business transactions involving the manufacture and marketing of computer hardware and software in China, see *China Trade Report* 23.4:8–11 (April 1985).
22. See *Businessweek,* 8 July 1985, p. 42.
23. Zhang Xiaobin.
24. See *Los Angeles Times,* 29 March 1985; also "A Tour of Computing Facilities." At the low end of the consumer microelectronics market, a number of relatively unsophisticated single-board processors—used, e.g., in simple video games and mathematical calculations, or for basic computer literacy training—were available on the open market at prices ranging from US $50–$100.
25. As noted earlier, reliability of Chinese-produced ICs has been rather low, owing to primitive manufacturing conditions and poor quality controls.
26. China's 1982 census was conducted using an IBM 4341 mainframe, 20 IBM 4331s, and 8 Wang 2200 VSs. A number of other major Chinese computer applications (e.g., in such areas as seismology, petroleum exploration, shipbuilding, power generation, meteorology, and, most recently, airline reservations) have also been performed primarily on foreign-supplied mainframes. For a partial listing of Chinese end users of imported mainframe computers, see *China Trade Report,* April 1985, p. 9.
27. When the survey was published in 1985, the "product-quality" ratings were inexplicably deleted.
28. "Computer Association: A Bridge Between Producers and Users," *China Market* 9:59–61 (1985).
29. This finding of extreme end-user frustration over the quality of available hardware documentation reinforces the general impression that large numbers of Chinese computers lay idle due to lack of properly written or translated

manuals and other essential technical documents. See, e.g., Cary Lu, "China's Emerging Micro Industry," *High Technology*, March 1985, p. 70.

30. It is interesting to note that several foreign-made computers (including the Wang VS Minicomputer used in China's 1982 national census, as well as the Hitachi Series M, the Digital PDP-11 and VAX-11, the Honeywell DPS-6, and the CDC Cyber) were rated less than satisfactory in 4 categories. Somewhat surprisingly, China's much-heralded "Great Wall" 520—an IBM PC-compatible machine singled out for a great deal of domestic and foreign sales promotion—rated poorly in almost all categories.
31. See, e.g., "China's High-Tech Troubles," *The New York Times*, 5 May 1985.
32. Cary Lu, pp. 69–70.
33. Note, for example, the complaint about "too many 'grannies'" voiced by the Vice-Mayor of Shanghai in the lead quotation heading the present chapter. One authoritative Chinese source recently blamed many of China's endemic computer ills on "a lack of coordination and consultation at the national level" combined with "the lack of management capacity at the lower level." (See Zhang Xiaobin, p. 129.) For analysis of bureaucratic impediments to—and recent reforms designed to improve—effective organization and management of technological R&D in China, see the chapters by Tony Saich, Detlef Rehn, and Richard P. Suttmeier in the present volume.
34. See "China's Computer Cupid Finds Dates or Mates," *The Christian Science Monitor*, 14 December 1984.
35. See "Doc Computer: Ailing Chinese Consult TRS-80," *The Washington Post*, 20 August 1984.
36. Gong Bingzheng.
37. Ibid.; cf. also Rehn.
38. In this connection, it is also interesting to note that only a few dozen computers were installed in Chinese state planning and statistical organs in 1982—in marked contrast to the widespread use of large- and medium-sized computers in the vast network of 61 functional systems that comprise Gosplan's Automated System of Planning Estimation (ASPER) in the Soviet Union. On the use of computers in Soviet planning, See Austin.
39. See *China Daily*, 11 and 18 January 1986; Gong Bingzheng; and Madelyn Ross, "Shanghai's Push into High Technology," *The China Business Review*, March-April 1985, pp. 36–39. On the role of SEZs in the field of microelectronics, see "Bullish on China," *Datamation* 30 (1 September 1984).
40. Gong Bingzheng.
41. One Chinese analyst recently acknowledged that, with respect to computer networking capabilities, China in 1984 was "as efficient as foreign countries in the late 1960s. No network has been formed . . . [and] most computers are operated independently." (Gong Bingzheng.) Cf. also Edward A. Parrish, "The First International Conference on Computers and Applications: Reflections," *Computer*, September 1984; and *Xiaoxing weixing jisuanji xitong* (Minimicro systems), February 1984.
42. While it may be somewhat hyperbolic to suggest that the bicycle is the most common "technology of choice" for up/downloading data from one computer

to another in China today, it is nonetheless true that floppy disks, on which programs and data are stored, must frequently be transported manually from one machine to another, from one location to another, and from one operating system to another—often across great distances—due to a near-total lack of distributed, time-shared, or modem-linked computing facilities. See "A Tour of Computing Facilities."

43. See "Counting to A Billion," *Datamation* 29.3:183–185 (March 1983).

44. See *MIS Week* 6.22:32 (29 May 1985).

45. The system currently in use at the Peking Hotel links 25 IBM PCs, each equipped with 512K RAM, to a Digital Microsystems DMS-816 minicomputer using HiNet LAN software. Despite the sophistication of the system's hardware, there have been numerous reported bugs and breakdowns in the network (and in similar systems installed at other hotels in the nation's capital—none of which are networked with each other).

46. For discussion of various approaches to the problem of making computers speak Chinese, see, *inter alia*, "Software Readied in Chinese Language," *China Daily*, 11 January 1986; "Making Computers that Speak Chinese," *The New York Times*, 6 January 1985; and "Pinyin for the PC," *PC World*, June 1984, pp. 285–289. The April 1984 issue of the Chinese journal, *Xiaoxing weixing jisuanji xitong* (Mini-micro systems) is devoted entirely to Chinese-language software applications.

47. In 1984, IBM offered a customized version of its 5550 PC for sale in China, complete with a Chinese version of MS/DOS and a Chinese language keyboard set up to display (and combine) 287 different simplified character components, for a price of US $15,000. (A year later the price was cut by 20–30%.) Wang and Intech, among others, have also developed Chinese-language input/output systems for commercial distribution.

48. One highly promising recent development is the *Tianma* (Flying Horse) character-processing system, introduced in mid-1986. In this system, the operator enters romanized pinyin text directly onto a standard IBM-style keyboard. The program then *automatically* scans the text, analyzes the location and grammatical context of each word, selects the most appropriate character from its memory, and displays it instantly on the CRT—thereby eliminating several steps from the input process. Because installation of the system involves an elaborate hardware modification, however, it cannot be used in conjunction with existing, off-the-shelf software, but requires customized applications programs.

49. *Nanfang ribao*, 14 January 1984, cited in Uehara. This figure compares with approximately 3,000–5,000 Soviet work units having their own on-site computers in 1982. A "work unit" is here defined to include Party and governmental organs and branches at all levels; factories and commercial establishments (e.g., bank branches); schools and universities; hospitals; research institutes; military units; and other non-farming organizations.

50. For a discussion of computer applications in enterprise administration and management, see Hong Xingbao, "Lun weixingji zai kexue guanli zhong de yingyong" (On the application of microcomputers in scientific management),

Caizheng kexue shuangyuekan (Financial administration sciences bimonthly) 5:18–22. A few Chinese work units may already have committed high-tech overkill in their rush to jump on the computer bandwagon. In one rather curious case, the Capital Steel and Nail Company of Beijing reportedly purchased 126 foreign-made minis and 450 micros—roughly 1 computer for every 4 employees. Forty-one of the minis were earmarked for use in production control and business management; 14 for staff training; and the remaining 71 for process control. See "Personal Computer Fever Spreads to China," *Mini-Micro Systems* 18.2:79–82 (February 1985).

51. China's 7th Five-Year Plan (1986–1990), published in April 1986, calls—somewhat optimistically—for domestically produced ICs, computers, telecommunications equipment, and software to "catch up to advanced world standards" by 1990. The electronics industry is targeted to increase its total output value by 50% over the life of the Plan, with special emphasis given to "production of microcomputers and their peripheral equipment" and to "working out policies for developing application software." See *BR* 29:viii–ix (28 April 1986).
52. It is this aspect of China's current high-tech development boom—the willingness of Chinese leaders to experiment with limited forms of market-driven production, research, and development—that distinguishes it most clearly from its Soviet counterpart, which continues to be almost entirely plan-driven.
53. Current plans call for the establishment of several hundred independent software development centers in China by 1990.
54. On the bargaining that occurs in exchange transactions among organizational units in China, see Kenneth Lieberthal and Michel Oksenberg, *Bureaucratic Politics and Chinese Energy Development* (Washington: US Department of Commerce, Internal Trade Administration, 1986).
55. In this connection, Lucian Pye has observed that *guanxi* networks, with their exclusive, esoteric style of communication, lie at the heart of the phenomenon of factionalism in Chinese political culture: "Gross inequities in access to information tend to accentuate divisions between insiders and outsiders. . . . The policy of making information relatively scarce forces people to seek informal channels of communication, which in turn contributes to the building of networks that are at the core of factions. . . . The linkage between channels of communication and the networks of personal association behind factions is apparently well recognized, and higher cadres quite explicitly exploit their access to inside information to extend and reinforce their claims on dependent subordinates." (Lucian Pye, *The Dynamics of Chinese Politics.* [Cambridge, Mass.: Oelgeschlager, Gunn and Hain, 1981], pp. 218–219.)
56. On the Japanese experience, see Erik Baark, *Towards an Advanced Information Society in Japan* (Lund, Sweden: Research Policy Institute, Technology and Culture Occasional Series Report No. 13, 1985), esp. pp. 11–24.
57. A heightened sensitivity to the desire for computer security is believed to be at least partly responsible for the extraordinary upsurge of interest recently shown by Chinese officials in acquiring state-of-the-art LAN technology from abroad. Paradoxically, however, the introduction of hard-wired local networks

in individual *danwei,* by reinforcing existing communication barriers between individuals and work units, could further exacerbate existing organizational tendencies toward cellular insularity and impacted communications. A relevant example is provided by the major tourist hotels in Beijing, each of which has its own distinctive in-house LAN—no two of which are able to communicate with each other or share data. (See note 45, above.)

58. For a recent analysis of China's perennial struggle against the evils of bureaucratism, see Martin K. Whyte, "Who Hates Bureaucracy? A Chinese Puzzle" (University of Michigan, CRSO Working Paper No. 337, November 1986).
59. Evans, pp. 208–209. A similar point is made by Walter R. Roberts and Harold E. Engle, "The Global Information Revolution and the Communist World," *The Washington Quarterly* 9.2:141–155 (Spring 1986).
60. Hoffmann, "Technology, Values, and Political Power," pp. 401–411.
61. See Dizard; also Roy Malik, "Can the Soviet Union Survive Information Technology?" *Intermedia* 12.3:10–23 (May 1984).
62. Similar developmental disparities also exist in the areas of newspaper circulation and radio receivers. See *The World Almanac, 1987.*
63. Tom Stonier, *The Wealth of Information: A Profile of the Post-Industrial Economy* (London: Methuen, 1983); see also Evans.
64. Chinese TV fare in recent years has also included, *inter alia,* a number of American-made sit-coms and adventure series, along with a generous sampling of Hong Kong kung-fu movies and foreign-made commercial advertising. On the social effects of the spread of electronic mass media in Communist countries, see Roberts and Engle, "The Global Information Revolution."

Conclusion: Science, Technology, and China's Political Future—a Framework for Analysis, by Richard P. Suttmeier

1. Harvey Brooks, "Technology Evolution and Purpose," in Thomas J. Kuehn and Alan Porter, eds. *Science, Technology and National Policy* (Ithaca: Cornell University Press, 1981), pp. 35–56.
2. See, Richard P. Suttmeier, *Research and Revolution* (Lexington, MA.: Lexington Books, 1974).
3. E. E. Bauer, *China Takes Off: Technology Transfer and Modernization* (Seattle: University of Washington Press, 1986).
4. Personal communication.
5. Cf. Michel Oksenberg and Kenneth Lieberthal, *The Bureaucratic Politics of China's Energy Development* (Washington: US GPO, 1986).
6. One reason that Chinese problems with centralization and decentralization are as serious as they are is because they involve complex overlapping authority, a phenomenon the Chinese refer to as *kuai kuai, tiao tiao* (lit. "lumps and branches"). For a recent discussion of this system, see Oksenberg and Liebenthal.
7. See Wendy Frieman's chapter in this volume.
8. James Seymour, "China's Satellite Parties Today," *Asian Survey* 26.9 (September 1986).

9. This discussion is drawn from Brooks.
10. Ibid.
11. See Baum's chapter in this volume.
12. See Richard Ihde, *Technics and Praxis* (Boston: D. Reidel, 1979).
13. See Baum's chapter in this volume.
14. Brooks.
15. Ibid.
16. See Halpern's chapter in this volume.
17. Sanford A. Lakoff, "Scientists, Technologists and Political Power," in Ina Speigel-Rosing and Derek de Solla Price, eds., *Science, Technology and Society* (London and Beverly Hills: SAGE Publications, 1977), pp. 355–392.
18. Manfred Stanley, *The Technological Conscience* (New York: The Free Press, 1978).
19. Danny Kwok, *Scientism and Modern Chinese Thought* (New Haven: Yale University Press, 1965).

INDEX